MARCH TO ARMAGEDDON

MARCH
TO
ARMAGEDDON

The United States
and the Nuclear Arms Race,
1939 to the Present

RONALD E. POWASKI

New York   Oxford
OXFORD UNIVERSITY PRESS
1987

Oxford University Press

Oxford   New York   Toronto
Delhi   Bombay   Calcutta   Madras   Karachi
Petaling Jaya   Singapore   Hong Kong   Tokyo
Nairobi   Dar es Salaam   Cape Town
Melbourne   Auckland

and associated companies in
Beirut   Berlin   Ibadan   Nicosia

Library of Congress Cataloging-in-Publication Data
Powaski, Ronald E.
March to Armageddon.
Bibliography: p.
Includes index.
1. Nuclear weapons. 2. United States—Defenses.
3. Arms race—History—20th century. I. Title.
U264.P69   1987   355.8'25119'0973   86-28435
ISBN 0-19-503878-9

1 3 5 7 9 10 8 6 4 2

Printed in the United States of America
on acid-free paper

For Jo Ann

PREFACE

Almost 50,000 nuclear weapons fill the arsenals of the United States and the Soviet Union. In spite of numerous, well-publicized studies describing the apocalyptic possibilities of continuing the nuclear arms race and the increased public concern about the potential effects of a nuclear war, both superpowers are deploying, and planning to deploy, thousands of additional nuclear warheads. The American nuclear arsenal alone is expected to grow from 25,000 warheads to over 28,000 by the end of the 1980s. The size of this stockpile will increase further in the next decade if significant arms control or reduction agreements are not concluded in the near future.[1]

While both superpowers share responsibility for the origin and continuation of the nuclear arms race, this study will emphasize the role of the United States. Each president since Truman has promised to seek agreements that would restrain the growth of superpower nuclear arsenals as well as halt the spread of nuclear weapons to the world's nonweapon states. Paradoxically, Truman, and each of his successors, has augmented the destructive power of the American nuclear arsenal and has permitted nonnuclear weapon states to acquire the means by which to produce nuclear weapons. What accounts for this discrepancy between presidential words and actions?

This study is a synthesis of the existing literature on the nuclear arms race supported by my own research. In the chapter notes, I have attempted to acknowledge the work of those who have contributed to my understanding of this complex subject. I am also indebted to Yale University Library for permission to use the Henry L. Stimson Diary. I owe a special thanks to my research assistant, Kim Sirk Szendrey, as well as to George Kosman and George Barnum of Case Western Reserve University's Freiberger Library, for the help they provided in locating and facilitating the use of research materials. I am grateful to Rosalie Fette for helping with the mechanical aspects of preparing the manuscript, and to Frank Hoffert, James Kelley, and Kenneth Lowe for reading the manuscript and making many helpful suggestions. My appreciation also is extended to my editors, Nancy Lane and Joan Bossert, for their enthusiasm, wise counsel, and exceptional editorial work. I cannot adequately express gratitude to my wife, Jo Ann, for her assistance in preparing the manuscript, the encouragement she provided, and the patience she displayed during the writing of this book.

Euclid, Ohio   R. E. P.
January 1987

CONTENTS

1. Roosevelt and the Manhattan Project, 1939–1945　　　3

2. Truman, Hiroshima, and Nagasaki, 1945　　　12

3. Truman and International Control of the Atom, 1945–1947　　　29

4. Truman, the Cold War, and the Hydrogen Bomb, 1947–1952　　　46

5. Eisenhower and Massive Retaliation, 1953–1961　　　60

6. Eisenhower and Nuclear Arms Control, 1953–1961　　　74

7. Kennedy, Nuclear Weapons, and the Limited Test Ban Treaty, 1961–1963　　　93

8. Johnson, Nuclear Weapons, and the Pursuit of SALT, 1963–1969　　　113

9. Nixon and SALT I, 1969–1972　　　127

10. Nixon, Ford, and the Decline of Détente, 1972–1977　　　146

11. Carter and SALT II, 1977–1981　　　162

12. Reagan and the "Rearmament" of America, 1981–1983　　　184

13. Reagan and Nuclear Arms Talks, 1981 to the Present　　　197

Conclusion　　　222

Glossary of Acronyms and Technical Terms　　　233

Notes　　　237

Suggested Readings　　　275

Index　　　283

And he gathered them together into a place called in the Hebrew tongue Armageddon. . . . And there came a great voice out of the temple of heaven, from the throne, saying, It is done. And there were voices, and thunders, and lightnings; and there was a great earthquake, such as was not since men were upon the earth, so mighty an earthquake, and so great. . . . And the cities of the nations fell. . . . And every island fled away, and the mountains were not found. . . . And there fell upon men a great hail out of heaven, . . . and men blasphemed God because of the plague of the hail; for the plague thereof was exceedingly great.

Revelations 16: 16–21

If we do not change our direction, we are likely to end up where we are headed.

Ancient Chinese proverb

MARCH TO ARMAGEDDON

1

Roosevelt and the
Manhattan Project, 1939–1945

Nuclear Fission

In December 1938 two German physicists, Otto Hahn and Fritz Strassmann, performed a revolutionary experiment. Working at the Kaiser Wilhelm Institute in Berlin, Hahn and Strassmann bombarded uranium with neutrons and, in the process, split the uranium atom into two substances nearly equal in atomic weight, one of which they originally believed was radium. At first the two scientists could not explain what had happened to the uranium; it was inconceivable to them that an atom could be divided. Seeking an explanation of the experiment, Hahn in early January 1939 wrote a letter to his former colleague, the well-known physicist Lise Meitner, who, because she was Jewish, fled to Sweden when the Nazis occupied her native Austria.[1]

Meitner soon realized that what the German scientists had produced was barium, not radium, since only barium could have been produced by splitting the uranium atom, a process called "nuclear fission." She also realized that a small fraction of the original mass of uranium in the Hahn-Strassmann experiment had entirely vanished. As stipulated by the law of the conservation of matter and energy, the missing mass must have been transmitted into pure energy. Hahn and Strassmann had proven the validity of the theoretical relationship between mass and energy first stated by Albert Einstein in 1905. Using Einstein's equation, $E = mc^2$, Meitner calculated that a pound of uranium, if completely fissioned, would yield as much energy as burning several million pounds of coal. She also realized an ominous possibility: if the proper conditions were created, the fission of uranium could yield a weapon of unprecedented explosive power.[2]

Alarmed by the prospect of German atomic bombs, Meitner informed her nephew, the physicist Otto R. Frisch, about the Hahn-Strassmann experiment. Frisch, in turn, rushed off to Copenhagen to alert the Danish physicist and Nobel laureate Niels Bohr, who was about to set sail for America to attend a conference of theoretical physicists in Washington, D.C., on January 26, 1939. The foreboding news of the Hahn-Strassmann discovery that Bohr brought with him spread quickly among the scientists attending the Washington conference. Enrico Fermi, a brilliant physicist and refugee from Mussolini's Italy, was the first to realize that uranium, if present in sufficient quantity (called its "critical mass"), could generate enough neutrons to produce a chain reaction, without which nuclear fission could not be sustained. In an experiment performed on March 3, 1939, Leo Szilard, the Hungarian expatriate, demonstrated that Fermi's assessment was correct. He bombarded uranium with neutrons and produced more neutrons. He concluded correctly that, if sufficient uranium could be accumulated, an atomic bomb theoretically might be produced.[3]

Szilard, Fermi, and other like-minded refugees from fascist Europe were frightened by the possibility that the Hahn-Strassmann breakthrough could lead to the creation of a German atomic bomb. In fact, Fermi and Szilard feared that German physicists already enjoyed a substantial lead in nuclear research. This view was reinforced by news that Hitler had banned the export of uranium from Czechoslovakia after that country was occupied by the Nazis in March 1939. In the same month, Fermi and Szilard launched an effort to alert the American government to the potential danger of a German atomic bomb.[4]

The initial results of their effort, however, were not auspicious. In March 1939 they were able to persuade George B. Pegram, the dean of the Graduate Faculties at Columbia University, to arrange a meeting with Rear Admiral S. C. Hooper, the director of Technical Operations for the Navy. The Navy, however, was not impressed. Fermi and Szilard received a Naval Research Laboratory grant of only $1,500 for fission research.[5]

The two physicists decided to try another tack. They persuaded Albert Einstein, America's best-known scientist, to send a letter, drafted by Szilard, to President Franklin Roosevelt warning him of the potential danger of a German atomic bomb. The letter, which Einstein signed on August 2, stated in part "that a single bomb of this type carried by boat and exploded in port might very well destroy the whole port together with some surrounding territory." Perhaps the most alarming and persuasive part of the letter warned: "I understand Germany has actually stopped the sale of uranium from Czechoslovakian mines which she has taken over."[6]

To make sure the letter was not lost in bureaucratic channels, Szilard arranged to have it delivered personally to Roosevelt by Dr. Alexander Sachs, a director of the Lehman Brothers Investment Corporation, who Szilard thought was sufficiently close to the President to gain an adequate hearing. Because of the pressure of other presidential business, especially the outbreak of war in September, Roosevelt was unable to see Sachs until October 11. After Sachs read Einstein's letter to him that day, and as a result of a subsequent conversation with Sachs the following day, Roosevelt was impressed by its implications. "Alex," he responded, "what you are after is to see that the Nazis don't blow us up." After Sachs nodded his head affirmatively, Roosevelt answered: "This requires action." And with these words, the United States had entered the nuclear arms race.[7]

Initial Steps

In October 1939, shortly after Roosevelt's conversations with Sachs, the President took the first step in the American effort to build an atomic bomb. He approved the establishment of an ad hoc Uranium Committee, headed by Lyman C. Briggs, the director of the National Bureau of Standards, to advise him on atomic energy matters and particularly to determine whether an atomic bomb was feasible. Despite the foreseen difficulties, as well as the uncertainty of the outcome, the Uranium Committee concluded in a report the following month that the effort to build an atomic bomb should be made. It recommended as a first step that Fermi's plan to create a sustained chain reaction be supported by supplying him with four metric tons of pure-grade graphite and fifty tons of uranium oxide, which Fermi required for his nuclear reactor or "pile." The committee's recommendation was approved by Roosevelt.[8]

On June 27, 1940, after British intelligence determined that the Germans were accelerating their nuclear research program, Roosevelt intensified the American effort. The National Defense Research Committee (NDRC), under the chairmanship of Vannevar Bush, president of the Carnegie Institution of Washington, was established to coordinate and supervise all scientific research related to warfare. The Uranium Committee, still headed by Briggs, was made a subcommittee of the NDRC. By the fall of 1940, research grants and contracts were concluded with a number of American companies and universities, including the Standard Oil Company, Columbia University, Princeton University, the University of Chicago, and the University of California.[9]

The effort gained impetus by the arrival, early in the summer of 1941, of a report by Britain's Maud Committee, the agency that spearheaded British atomic research. The report stated that, in addition to uranium-235, plutonium, one of the by-products of a chain reaction, could also be used for bombs. The report outlined the manner by which fissionable uranium-235 could be separated from the more abundant, but less fissionable, uranium-238; it described how a chain reaction could be generated; and it explained how critical mass could be determined. It concluded that if an all-out effort were begun, a bomb could be produced in as little as two years.[10]

Spurred by the Maud Committee report, the American nuclear program was again intensified, and again reorganized. In the fall of 1941 the NDRC became a part of the newly created Office of Scientific Research and Development (OSRD), with Bush as the chairman, while the Uranium Committee became the Uranium Section, or Section One, of the NDRC. Thereafter the effort to build an atomic bomb was often referred to simply as "S-1." While S-1 concentrated on building the bomb, overall responsibility for the effort was assigned to the so-called Top Policy Group, which consisted of Roosevelt, Vice President Henry Wallace, Secretary of War Henry L. Stimson, Army Chief of Staff General George C. Marshall, Bush, the new chief of the OSRD, and James B. Conant of Harvard, who succeeded Bush as the head of the NDRC. On December 6, 1941, the day before the Japanese attack on Pearl Harbor, the President ordered S-1 to do everything possible to build an atomic bomb.

The primary motive for doing so was fear that Germany would develop nuclear weapons first.[11]

Six months later, in June 1942, Roosevelt and British Prime Minister Winston Churchill met at the President's home, in Hyde Park, New York, to discuss progress on "Tube Alloys," the British code-name for the atomic bomb project. By then it was clear to Churchill that Britain could not hope to match the scale of the American effort. It was therefore with great relief that the prime minister accepted Roosevelt's offer that the United States assume the major responsibility for the project while allowing British scientists to share the benefits that would be derived. Thus was born the Anglo-American atomic partnership.[12]

The Manhattan Project

By the end of 1942, the construction of an atomic bomb was beginning to look inevitable. That impression was strengthened on December 2, when Fermi and his co-workers at the University of Chicago engineered the first nuclear chain reaction. The experiment was conducted with considerable risk. The official history of the Atomic Energy Commission called the decision to conduct it a "gamble with a possibly catastrophic experiment in one of the most densely populated areas of the nation." Fortunately, Fermi's calculations were correct, and his "pile" did not malfunction.[13]

With the project beginning to outstrip the managerial and governmental resources of the scientists, Roosevelt, earlier in the summer of 1942, had transferred its supervision to the Army. By presidential order, the Manhattan Engineer District—the Manhattan Project, as it came to be known—was established on August 13. Command of the project was given to Brigadier General Leslie Groves, a brusque but efficient career officer and engineer by training who had supervised construction of the Pentagon. Groves almost immediately began to assemble the facilities needed to construct the bomb. At the Clinton Engineer Works, in Oak Ridge, Tennessee, uranium-235 would be separated from the less fissionable isotope, uranium-238. The Hanford Engineering Works near Pasco, Washington, was designed to be the production center for plutonium, the material used in the bomb that was dropped on Nagasaki. The design and construction of the bomb were made the responsibility of the Los Alamos, New Mexico, installation. Dr. J. Robert Oppenheimer was appointed its director despite his links, through a former fiancée, his brother, and others, to the Communist Party. Groves wanted Oppenheimer on the project, not only because he considered him loyal to the United States (Oppenheimer never joined the Communist Party), but, more important, because he considered the scientist's qualifications vital to the success of the project.[14]

Under presidential orders to ensure the utmost secrecy, Groves developed a security system called "compartmentalization." The system dictated that the flow of information among the scientists would be regulated on a need-to-know basis—only information crucial to their work would be disseminated. The scientists who worked on the Manhattan Project resented the system. Denied the opportunities to discuss the project freely among themselves, they found it difficult to understand the relationship of their specific tasks to that of the project as a whole. They believed their specific tasks, as well as the entire project, suffered as a result.[15]

But the scientists were not the only ones kept in the dark. When investigators from Senator Harry S. Truman's special committee on war expenditures arrived at the Hanford plant to determine why its construction was absorbing enormous sums of money, they were turned away at the gate. Later, Truman received a call from Secretary of War Stimson asking him to recall the investigators. Stimson told him that maintenance of the plant's clandestine nature was vital to the success of the war effort. Truman complied. He did not, however, learn the purpose of the plant until after he had become President.[16]

Because the Manhattan Project was top secret, not only Truman, but almost the entire Congress, was unaware of its existence. Until 1944, funds for the project came either from the military departments, which concealed their purpose, or from a special contingency fund appropriated for the President which was shielded from congressional scrutiny. In 1944, however, the cost of the project began to exceed by far the amount these sources could provide. As a result, the congressional leadership was informed and persuaded to steer the appropriations through the committees without the true purpose becoming general knowledge. The Republican minority floor leader, Joseph W. Martin, feared, as did other informed congressmen, the consequences of the project's failure. "If the bomb fizzled into a huge, grim joke," he wrote, "I would have been answerable to my Republican colleagues for having secretly put through a vast expenditure with absolutely nothing to show for it in return."[17]

While General Groves was able to keep his secret from most of Congress and the American people, as well as the Germans, the security system he established had a detrimental impact on future efforts to control the nuclear arms race. In effect, the system of secrecy he imposed on the Manhattan Project not only prevented public debate on the question of building and using atomic weapons, it also hindered discussion within the government. Since there was virtually no debate on the question, those few who were informed came to consider the construction and use of atomic weapons foregone conclusions.[18]

Niels Bohr and International Control of Atomic Energy

One of the first persons to give serious attention to the postwar implications of atomic energy was Niels Bohr, the Danish physicist who brought the news of the Hahn-Strassmann fission experiment to America in 1939. Four years later, Bohr fled to England from Nazi-occupied Denmark to work with the British scientists who were then at Los Alamos. Nevertheless, he also believed that development of the atomic bomb could lead to a postwar atomic arms race and, perhaps inevitably, to the destruction of the planet. Bohr emphasized that the only way to prevent a nuclear Armageddon was through international control of atomic energy, which alone could provide the degree of security that would preclude postwar reliance on atomic weapons. Because international control could only become effective when each nation was confident that no nation possessed atomic weapons, Bohr insisted that all military, industrial, and scientific installations be open to international inspection. But time was a critical factor. An agreement for international control of the atom, he maintained, could only be achieved by inviting the participation of the Soviet Union before the war was over, and even before the bomb was developed, to assure the Soviets that the United States and Britain were not attempting to monopolize the bomb.[19]

Early in 1944, U.S. Supreme Court Justice Felix Frankfurter, an old friend of Bohr's, had the opportunity to relay his views on atomic energy to Roosevelt. Although the President seemed impressed by Bohr's ideas, he was shocked to learn that Frankfurter was privy to the activities of a project as secret as the atomic bomb. Without letting the justice know how upset he was about this lapse of security, the President suggested that Bohr should discuss his ideas with "friends" in London. As a result, in April 1944, the scientist traveled to England to meet with Churchill.[20]

The prime minister was, to say the least, irritated by Bohr's presentation. While he could agree with him that the atomic bomb would be a major factor in the postwar world, he could not accept Bohr's view that the bomb was necessarily a menace. In fact, Churchill saw an Anglo-American atomic monopoly as the primary means of preserving Britain's status as a great power and the only effective way of checking the expansionist tendencies of the Soviet Union. When Roosevelt had temporarily restricted the exchange of American atomic information with the British in January 1943, primarily because prominent Americans (including Bush and Conant) had feared postwar, commercial nuclear competition between Britain and the United States, Churchill reacted with horror. "What are we going to have between the white snows of Russia and the white cliffs of Dover?" he retorted. Without the bomb, he feared that Britain would have precious little to counter the threat of Soviet hordes pouring into Western Europe.[21]

Under intense pressure from Churchill, Roosevelt resumed the unrestricted exchange of atomic information on July 20, 1943. The decision was formalized at the Quebec Conference in August 1943. The two leaders also agreed never to use atomic weapons against each other nor "against third parties without each other's consent."[22]

Roosevelt, the Soviet Union, and the Atomic Bomb

Roosevelt's decisions at Quebec were made in spite of the efforts of Bush and Conant to convince him that an Anglo-American atomic energy partnership during the war inevitably implied an Anglo-American military alliance afterward. Although the British did not expect to receive atomic weapons from the United States, the fact that British scientists had full access to American atomic research ensured that Britain would be able to construct atomic weapons independently after the war. A postwar atomic alliance with Britain, Bush tried to convince Roosevelt, should be the result of a decision considered on its merits, not the result of a wartime exchange of atomic information. He warned that an exclusive partnership after the war "might well lead to extraordinary efforts on the part of Russia to establish its own position in the field secretly, and might lead to a clash, say 20 years from now."[23]

Bush and Conant attempted to gain Secretary of War Stimson's support for their effort to change the President's mind on the policy of exclusive atomic cooperation with the British. In a memorandum to Stimson on September 30, 1944, they pointed out that it was highly unlikely that the Anglo-American atomic monopoly could be maintained for more than three or four years, a remarkably prescient forecast. For the United States to stay ahead of the Soviet Union, Bush and Conant predicted, new weapons would have to be produced, including a weapon a thousand times more destructive than the atom bomb: the hydrogen bomb. Like Bohr, Bush and Conant believed that the only way to prevent a nuclear arms race after the war was to search for a way to bring about international control of the atom before the war ended. Unlike Bohr, however, Bush and Conant did not advocate immediate disclosure of the Manhattan Project secret; they believed the Soviets should be informed only after the bomb was successfully tested. Then, after the war, the Soviets could be offered atomic information in exchange for their participation in a program of international inspection that would include all atomic facilities and military installations.[24]

However, by late 1944, when Bush and Conant received the opportunity to present their ideas on atomic energy to Stimson, the prospects for postwar cooperation between the United States and the Soviet Union on any issue, let alone one as sensitive as atomic energy, had deteriorated appreciably. The Soviets were making it quite evident that they intended to establish a buffer zone of pro-communist states in Eastern Europe. While Roosevelt was reluctantly prepared to accept the creation of a Soviet sphere of influence in that part of Europe liberated by the Red Army, he realized that, short of war with the Soviet Union, there was not much he could do about it; he nevertheless expected Stalin to abide by his pledge to allow democratically elected governments to be established in the areas under his influence. On December 30, 1944, Roosevelt told Stimson that he had written Stalin a note the day before vigorously protesting the Soviet decision to recognize the pro-communist Lublin Committee as the sole party in the provisional Polish government. Stalin thereby ruled out the possibility of democratic elections in Poland as well as participation of the Polish exile-government in London.[25]

With Soviet-American relations as strained as they were over Eastern Europe, Stimson was not inclined to go along with the suggestion of Bush and Conant to approach the Soviets on postwar international control of atomic energy. Although he thought they were thinking along the "right lines," he also thought it would be inadvisable to put their suggestion into full force until, as he put it, "we had gotten all we could in Russia in the way of liberalization [of Eastern Europe] in exchange for S-1 [the atomic bomb]." Stimson clearly intended to exploit the diplomatic possibilities of the bomb before he would agree to turn over American know-how to an international control agency.[26]

Faced with Stimson's foot-dragging, Bush attempted to put his ideas directly before the President. In a letter to Roosevelt, he recommended that the United Nations include a section devoted to international scientific research. Without the creation of an atmosphere of international trust, he warned, the establishment of an international atomic energy agency would be impossible and a nuclear arms race inevitable.[27]

Roosevelt, no doubt, understood what Bush was saying. One can only conclude, as historian Martin Sherwin has argued, that in resuming an unrestricted nuclear exchange with the British, the President had agreed with Churchill's assessment that a postwar Anglo-American atomic monopoly, rather than being a threat to world peace, would actually be the only viable means for preserving peace. At the Atlantic Conference in August 1941, Roosevelt had rejected the idea that an international peace organization alone could prevent another war. An Anglo-American combination, he told Churchill, would be far more effective. Like Churchill, Roosevelt came to view peace primarily as a product of balancing power. And he viewed the atomic bomb in the same way—as a powerful counterweight to the enormous ground forces of the Soviet Union. England, the President believed, must be strong if she were to serve as an effective barrier to Soviet postwar expansion.[28]

Without the knowledge of Bush, Conant, or even Stimson, according to Sherwin, Roosevelt had formally endorsed Churchill's view on the role of atomic power in the postwar world. In an *aide-mémoire* signed by Roosevelt and Churchill at Hyde Park on September 18, 1944, the two leaders agreed that the atomic bomb project "should continue to be regarded as of the utmost secrecy." In effect, the Soviets would not be informed about the purpose of the Manhattan Project. This was decided even though Roosevelt knew, at least since September 1943, that the Soviets were getting vital information from a Soviet spy working in the Radiation Laboratory of the University of California. Roosevelt evidently preferred to arouse Soviet distrust rather than surrender the means by which he hoped to counter Soviet ambitions after the war. Considering this, Sherwin concludes, it is not surprising that Roosevelt did not take action toward establishing an international atomic energy agency.[29]

Another historian, Robert Dallek, gives an interpretation somewhat at variance with Sherwin's conclusions. According to Dallek, Roosevelt was playing a double game with both the British and the Soviets. While he was ready to use Anglo-American atomic energy cooperation as a basis for maintaining British power and checking Soviet expansion after the war, he was also trying to promote Soviet-American friendship, primarily by trying to recognize Soviet security interests in Eastern Europe, but also by holding out the prospect of atomic cooperation with the Soviet Union after the war. In Dallek's eyes, Roosevelt's desire to avoid alienating the Soviets explains his expressed agreement with Bohr, in a conversation with the scientist on August 26, 1944, that contact with the Soviets should be made concerning international control of atomic energy. Roosevelt promised Bohr that he would pursue the matter with Churchill at their next summit, in Quebec, the following month. Dallek admits, however, that "there is no direct evidence that this [a desire for postwar Soviet-American cooperation] motivated his response to Bohr."[30]

Whatever Roosevelt's motives in dealing with Bohr, he certainly did not implement the scientist's suggestions on atomic energy. Quite the contrary. In the Hyde Park *aide-mémoire*, less than a month after the President had met with Bohr, he not only ratified the indefinite continuation of an Anglo-American atomic monopoly, he also accepted Churchill's assessment that Bohr could not be trusted and that he should be placed under close surveillance "to ensure that he is responsible for no leakage of information, particularly to the Russians."[31]

One Last Try

In the meantime, Bush had not abandoned his effort to win Stimson's support for approaching the Soviets. On March 3, 1945 he impressed upon Harvey Bundy, Stimson's aide, that unless plans for the international control of atomic energy were drafted as soon as possible, chaos would result when the secret of the atom bomb was revealed. After these ideas were conveyed to Stimson by Bundy on March 5, the secretary of war responded: "We are up against some very big decisions. The time is approaching when we can no longer avoid them and when events may force us into the public on the subject." From this point on, the atomic bomb became Stimson's most important responsibility.[32]

On March 15, Stimson had what turned out to be his last opportunity to discuss the bomb with Roosevelt. In essence, the secretary of war presented the ideas that were outlined to him by Bush, Conant, and Bundy. The decision to inform the Soviets about the bomb, he told the President, had to be made soon, certainly before the bomb was used. It was a decision that would require greater consultation with Congress. The President responded by making no objection to anything Stimson said, causing the secretary to conclude that "on the whole the talk . . . was successful." Yet as on several other occasions, Roosevelt disregarded Stimson's political advice. Nor did he receive another opportunity to turn the President around—on April 12, Roosevelt was dead.[33]

2

Truman, Hiroshima,
and Nagasaki, 1945

False Alarm

When Harry S. Truman assumed the presidency, the ultimate defeat of Germany was considered a certainty, and the creation of a German atomic bomb no longer a possibility. In fact, it was evident to American officials as early as November 1944 that Germany had never really been a nuclear threat. In that month, elements of a special unit created by General Groves, code-named ALSOS, arrived in France to ascertain the extent of German atomic research and development. From the evidence that ALSOS gleaned from a German research facility in Strasbourg, Groves was able to determine that Germany's effort to build atomic weapons had never passed the experimental stage.[1]

Although Germany posed no atomic threat, the work of the Manhattan Project was intensified, particularly after the Germans surrendered on May 8, 1945. As Los Alamos director J. Robert Oppenheimer recalled, "I don't think there was any time whe[n] we worked harder at the speedup than in the period after the German surrender." With Germany defeated, the only remaining enemy was Japan. Yet no American ever seriously considered the Japanese an atomic threat; they had even less of the requisite resources for building atomic weapons than the Germans.[2]

The Atomic Bomb and Japan

That the atomic bomb, if successfully developed, would be used was always an inherent assumption of the Manhattan Project. General Groves recalled that when he assumed command of the project he had been ordered to build a bomb that could be used. President Roosevelt never altered those orders. Secretary of War Stimson emphasized that it was his and the President's "common objective, through the war, to be the first to produce an atomic weapon and use

12

it." He added that "on no other ground could the wartime expenditure of so much time and money have been justified."[3]

Without the knowledge of Stimson or any other officials working on the Manhattan Project, Roosevelt had agreed with Churchill, in the Hyde Park *aide-mémoire* of September 19, 1944, that when a bomb "is finally available, it might, perhaps, after mature consideration, be used against the Japanese." According to Alexander Sachs, who related a conversation he had with Roosevelt in December 1944, the President planned to use the bomb against both Germany and Japan if it proved necessary, following a rehearsed demonstration of the weapon and a warning that it would be used against either nation if an ultimatum for immediate surrender was refused.[4]

There is other evidence to indicate that Roosevelt anticipated use of the bomb. He is reported to have said to his secretary, Grace Tully: "I can't tell you what this [the atomic project] is, Grace, but if it works, and pray God that it does, it will save many American lives." What Roosevelt was referring to was the Army's plan for an invasion of Japan, which, if carried out, would cost an estimated one million American casualties.[5]

Apparently with this in mind, Roosevelt approved the military preparations Groves had made for using the bomb. They included the creation of a special Air Force unit, the 509th Composite Group, which would deliver the two types of bombs that had been designed at the Los Alamos Laboratory. One, nicknamed "Thin Man" or "Little Boy," utilized uranium, while the other, "Fat Man," employed plutonium.[6]

It is quite certain that Truman knew nothing about Roosevelt's thoughts on the atomic bomb before he assumed the presidency. In the eighty-two days that he served as vice-president, he conferred with Roosevelt only twice. At no time did they discuss the Manhattan Project. In fact, Truman only learned about its purpose when Stimson informed him on the day Roosevelt died. In other words, the ultimate decision regarding the use of the atomic bomb fell on a man who by his own admission was almost totally unprepared for the presidency. And, since none of the top men associated with the direction of the Manhattan Project—Stimson, Bush, Conant, or Groves—questioned the assumption that the bomb would be used if built, neither did Truman. When Stimson and Groves gave the new President his first detailed briefing on the Manhattan Project on April 25, 1945, Truman responded by expressing confidence in their ability to complete it successfully.[7]

Truman and the Soviet Union: First Steps

To be sure, in the first days of his presidency, Truman had too little time to ruminate on the military, the diplomatic, let alone the moral, implications of using the atomic bomb. The one problem that dominated the early days of his administration was the Soviet Union. It was soon apparent to Truman, as it was to Roosevelt in the last months of his life, that the Soviets were not going to observe the Western interpretation of the Yalta agreements relating to Eastern Europe. With respect to Poland, the Soviets were not abiding by their pledge to Roosevelt that they would allow the Polish government-in-exile, the "London government," to participate fully in Poland's provisional government. Truman feared, quite justifiably, that the Soviets were taking steps to ensure that postwar Poland would be run by the pro-communist Lublin group.[8]

In the past, Roosevelt had tried to downplay Western difficulties with Stalin over Poland, as well as other issues where the United States possessed little leverage over the Soviets. Truman, on the other hand, both inexperienced and uninformed, was inclined to accept the hardline advice toward the Soviets that he received from Churchill and from key advisers within his own administration. W. Averell Harriman, U.S. ambassador to Moscow, saw the advance of the Soviet army as another "barbarian invasion of Europe," while Secretary of the Navy James V. Forrestal told Truman that it was time for a showdown with the Soviets over Eastern Europe. Prompted by the hardliners, Truman on April 23 berated Soviet Foreign Minister Vyacheslav Molotov for the Soviet Union's refusal to carry out its Yalta promises concerning Poland. "I have never been talked to like that in my life," Molotov protested. "Carry out your agreements and you won't get talked to like that," Truman retorted.[9]

In spite of Truman's inclination to tackle the Russian bear, he could also appreciate the argument of less hawkish advisers, like Stimson and Joseph E. Davies, former ambassador to Moscow, that it was important to avoid an outright break with the Soviet Union. Continued Soviet-American cooperation was vital, they argued, to the success of a postwar diplomatic settlement in Europe, the establishment of the United Nations, and the termination of the war with Japan. At Yalta, Stalin had promised Roosevelt that the Soviet Union would enter the war against Japan within three months after Germany's defeat. In exchange, Roosevelt had promised the Soviet leader that the United States would recognize Soviet hegemony over Mongolia and the restoration of Russian interests in the Far East that were lost after the Russo-Japanese War of 1904–1905. Truman was reluctant to deviate from this agreement, not only because it was Roosevelt's agreement, but also because his military advisers, particularly Army Chief of Staff General George C. Marshall, believed a Soviet attack on Japanese forces in Manchuria important to the success of the planned invasion of Japan.[10]

In Stimson's discussion with Truman on April 25, he stressed the need for caution in dealing with the Soviet Union. Because of the overwhelming destructive potential of the atomic bomb, he emphasized that it was imperative for the United States to do nothing that would preclude the possibility of international cooperation on atomic energy after the war. There was another, more important reason why Stimson wanted to avoid a showdown with the Soviets at this time. He wanted to make sure that the leverage he believed the atomic bomb would offer would be available when it came time to negotiate the issues that divided the Soviet Union and the United States. On May 14 he told Assistant Secretary of War John J. McCloy "that the time now and the method now to deal with Russia was to keep our mouths shut and let our actions speak for words. . . . They can't get along without our help and industries and we have coming into action a weapon which will be unique." The atomic bomb, in other words, was Stimson's ace in the hole; it was the one weapon that could not only end the war before an invasion of Japan became necessary, but would also be a powerful instrument for securing American interests after the war.[11]

Truman, it appears, was at least superficially aware of the value of the atomic bomb in dealing with the Soviets. In his memoirs, he recounted that Stimson "was at least as much concerned with the role of the atomic bomb in the shaping of history as in its capacity to shorten the war." In the spring of 1945, however, the atomic bomb was still only a hypothetical weapon. Until it became a reality, Truman accepted the necessity of doing everything possible to secure Soviet participation in the war against Japan. Consequently, when Churchill on May 6 suggested a summit meeting with Stalin to iron out Allied differences, Truman responded favorably and sent Harry Hopkins to Moscow to work out the arrangements. He was pleased when Hopkins reported back on May 28 that not only did Stalin agree to a summit at Potsdam, near Berlin, but he would also allow the London Poles to participate more extensively in Poland's provisional government. More important, the Soviet leader announced that the Soviet Army would be prepared to attack the Japanese by August 8.[12]

While Stimson apparently welcomed the chance to discuss America's difficulties with Stalin, he nevertheless felt that timing would be crucial. "Over any such tangled web of problems" with the Soviets, he wrote in his diary, "The S-1 secret would be dominant and yet we will not know . . . until after that meeting, whether this is a weapon in our hands or not. We think it will be shortly afterwards, but it seems a terrible thing to gamble with such big stakes in diplomacy without having your master card in your hand." Understandably, Stimson was relieved when Truman on June 6 informed him that he had postponed the summit meeting with Stalin until July 15. The postponement, Stimson recorded in his diary, would "give us more time"—undoubtedly, to test the atomic bomb.[13]

The Interim Committee

The members of the Interim Committee were no more inclined to question the use of the atomic bomb than Stimson or Groves. The committee was set up by Stimson, with Truman's approval, not only to deal with wartime use of the bomb, but also to make recommendations on postwar atomic research and development. Serving on the committee were Stimson, as chairman; George L. Harrison, Stimson's alternate; Conant and Bush; Karl T. Compton, president of M.I.T.; Assistant Secretary of State William L. Clayton; Under Secretary of the Navy Ralph A. Bard; and James Byrnes, who served as the President's personal representative. Even the men who served on the committee's Scientific Panel—Arthur H. Compton, Ernest O. Lawrence, Oppenheimer, and Fermi—believed that the successful completion of the Manhattan Project required the use of the atomic bomb. In fact, some of the scientists working on the project believed that use of the bomb would ultimately further the cause of world peace after Japan surrendered. In June, Arthur Compton told Stimson: "If the bomb were not used in the present war, the world would have no adequate warning as to what was to be expected if war should break out again."[14]

Accordingly, in its first formal meeting on May 31, the Interim Committee not only recommended use of the bomb against Japan, it also suggested that, to achieve the maximum psychological impact, the most desirable target would have to be a vital war plant employing a large number of workers with its surrounding workers' houses. The possibility of a nonmilitary demonstration of the bomb prior to its combat use was quickly rejected because the committee feared that the Japanese might interfere with the demonstration and because, with only two bombs available by August 1945, the United States could not afford a dud.[15]

Even thought the committee's scientists warned that nuclear weapons with an explosive yield of one hundred million tons of TNT would eventually be possible, they believed that nuclear energy, both for military and for industrial purposes, was vital to America's security. The committee responded by recommending an expansion of civilian nuclear research as well as the continued expansion of America's atomic weapons arsenal after the first bombs were produced.[16]

When it came to a discussion of sharing atomic information with the Soviets, Oppenheimer and Karl T. Compton suggested that there should be an exchange of information between the two countries, without giving the Soviets the specific details they would need to produce their own nuclear weapons. However, James Byrnes, who Groves believed was the key person in the committee's deliberations, feared that any disclosure of atomic information by the United States would only prompt Stalin to demand full Soviet participation in the Manhattan Project. Byrnes felt that, while the United States should do everything to improve political relations with the Soviet Union, nothing should be done to jeopardize America's lead in the development of atomic energy. In the end, he was able to persuade the committee to accept his viewpoint.[17]

On June 1 Truman accepted the recommendations of the Interim Committee. The atomic bomb would be used against Japan as soon as it was feasible; it would be dropped, Truman recalled, "without specific warning and against a target that would clearly show its devastating strength." On June 6 Stimson recorded in his diary that he told Truman that he "was a little fearful that before we could get ready, the Air Force might have Japan so thoroughly bombed that the new weapon would not have a fair background [against which] to show its strength." The President, Stimson remembered, "laughed and said he understood."[18]

In addition, both Stimson and Truman agreed that no information pertaining to the bomb would be made available "to Russia or anyone else" until the first bomb had been successfully "laid on Japan." If the Soviets raised the question of their participation at the Potsdam Conference, Stimson suggested, Truman should reply in a "simple statement that as yet we were not quite ready to do it." In exchange for eventual partnership with the Soviets on atomic energy, Stimson recommended that the United States demand a *quid pro quo*. Truman, Stimson recalled, "said he had been thinking of that and mentioned the same thing that [Stimson] was thinking of, namely the settlement of the Polish, Rumanian, Yugoslavian, and Manchurian problems."[19]

Scientific Opposition to Use of the Atomic Bomb

Not all the scientists associated with the Manhattan Project accepted the recommendations of the Interim Committee. Among the leading opponents was Leo Szilard, one of the first scientists to back the atomic bomb project. Although Szilard opposed use of the bomb against Japan without forewarning, he was more concerned about the impact of atomic weapons on postwar international relations. In a memorandum to Roosevelt in March 1945, he predicted an early end to America's atomic monopoly and pointed out that the cities of the United States would become vulnerable to atomic attack if nuclear weapons were not banned. He stressed that if a nuclear arms race were to be avoided, steps would have to be devised to ensure that atomic energy developed for peaceful purposes could not have military applications. If this proved to be unfeasible, Szilard suggested that it might then be necessary to forgo the peaceful benefits of atomic energy altogether.[20]

Because Roosevelt probably did not read Szilard's memorandum before his death, the scientist tried to obtain an appointment with Truman. Instead of the President, however, Szilard met with James Byrnes on May 28. The meeting did not go well. Byrnes considered unthinkable Szilard's desire to "participate in policy making" and particularly his suggestion that scientists should discuss the use of the bomb with the President's cabinet. Szilard, for his part, was dumbfounded by the lack of knowledge Byrnes displayed on the subject of atomic energy. Rather than being concerned with the problem of developing a long-term strategy for precluding a nuclear arms race, Byrnes seemed to be more impressed by the possibility of using the bomb as a means of intimidating the Soviets. Needless to say, Szilard found himself unable to discuss with Byrnes his idea "that the interest of peace might best be served and an arms race avoided by not using the bomb against Japan, keeping it a secret, and letting the Russians think that . . . work on it had not succeeded." Prompted by the frustration produced by the meeting with Byrnes, Szilard launched a drive to enlist the aid of scientists at the University of Chicago's Metallurgical Laboratory (Met Lab) to prevent use of the bomb against Japan.[21]

Szilard's effort was complemented by James Franck, a brilliant physicist and Nobel laureate, who had been driven from his native Germany by the Nazis. Franck had joined the Met Lab on the condition—which was accepted by its director, Arthur Compton—that he be allowed at a later date to present his views on the use of atomic weapons to the President. In the spring of 1945, Franck exercised that right. He organized a committee of scientists at the Met Lab to draft a report on atomic energy. The Franck Report, which was completed on June 11, emphasized the need to consider nuclear weapons as a com-

ponent of long-range national policy, rather than simply as a wartime military
expedient. It stated that use of the bomb against Japan without adequate warning might "easily destroy all our chances of success" in establishing international control of atomic energy. Use of the bomb would shock world opinion,
would reduce trust in America's sincerity to seek international agreement on
atomic energy, and would set a precedent for the future use of atomic weapons.
The result would be a nuclear arms race in which the United States could not
expect to maintain a significant lead for more than ten years.[22]

The Franck Report was delivered to the office of George Harrison, Stimson's assistant, on June 12. But instead of giving it to his boss, who was out
of town at the time, Harrison sent the report to the Scientific Panel of the
Interim Committee. The panel, which met on June 16, rejected the report's
recommendation that the bomb not be dropped on Japan. Yet it did accept the
report's suggestion to inform the Soviet Union about the atomic bomb before
it was used against Japan. The committee also agreed that the Soviets should
be told that the United States was willing to cooperate with them to bring about
international control of atomic energy after the war. On June 21 the Interim
Committee accepted these new recommendations. Its action in doing so would
have a direct bearing on Truman's decision to tell Stalin about the bomb at
Potsdam. There is, however, no evidence to indicate that Truman knew about
the contents of the Franck Report, let alone had the opportunity to read it.[23]

Although the Franck Report failed to sway the Interim Committee, Franck,
Szilard, and other like-minded scientists at the Met Lab continued their effort
to stop the use of the bomb against Japan. On July 17 Szilard sent a petition
to Truman containing the signatures of sixty-seven scientists at the University
of Chicago. Its essential statement held that "the United States shall not resort
to use of atomic bombs in this war unless the terms which will be imposed
upon Japan have been made public in detail and Japan, knowing these terms,
refused to surrender."[24]

Szilard also attempted to circulate his petition at the Clinton Laboratories
in Oak Ridge and at Los Alamos. Only two people signed it, at least one of
whom later made a formal retraction. An alternative petition to Szilard's, calling for the presentation of a clear statement of intent to the Japanese government before the use of the bomb, received eighty-eight signatures. But its cir-
culation was stopped by military authorities because it allegedly revealed the
state of progress on the bomb. At Los Alamos, Oppenheimer blocked circu-
lation of Szilard's petition because he doubted the right of scientists to influence political decisions. As with the Franck Report, there is no evidence to
indicate that Szilard's petition ever reached Truman.[25]

Strategy: Japan and the Soviet Union

The decision of the Scientific Panel to reject the Franck Report's recommendation to refrain from using the bomb against Japan without warning was made
without much, if any, regard for the administration's diplomatic or military

strategy. The panel's members obviously believed that there was no alternative, short of the bomb, to a costly invasion of Japan. At the same time, the nation's military planners were largely unaware of the divergent views of the scientists on the use of the bomb. In fact, few military leaders even knew the bomb was being built. General Douglas MacArthur, the Supreme Commander of Allied Forces in the Pacific, was not told that the bomb existed until five days before it was dropped on Hiroshima. The inability of scientists and military leaders to consult with each other on the wisdom of using the bomb was a testament to the efficiency of Groves' "compartmentalization" system.[26]

It is not surprising, therefore, considering the complete divorce that existed in the government's deliberations between the political and military ramifications of atomic energy development, that the atomic bomb was not even mentioned in a strategy meeting Truman held with his military advisers on June 18. And yet, while the use of the bomb was not discussed at this meeting, it remained an important factor in the minds of the men who considered the planned invasion of Japan.[27]

As the fourth summer of the war against the Japanese approached, an invasion of the home islands looked more necessary than ever to Stimson and General Marshall, particularly if the bomb did not become a viable option. Both men shared the opinion that, though Japan had been weakened by aerial bombardment and naval blockade, the Japanese still had sufficient power to resist, as Stimson put it, "to the bitter end." The Japanese, he argued, still had over four million troops in the field. The suicidal *kamikaze* attacks on American naval forces and the stiff resistance the Japanese had put up on Iwo Jima and Okinawa convinced Stimson and Marshall that, if the atomic bomb were not available for use against the Japanese, an invasion of the home islands would be necessary to bring the war to a conclusion. Considering the casualties this would entail, one can understand Stimson's eagerness to get the bomb tested quickly, even to the point of asking Marshall "whether or not we couldn't hold matters off from very heavy involvement in casualties until after we had tried out S-1."[28]

Others, primarily in the Navy and the Air Force, believed that an invasion of Japan was unnecessary regardless of whether the bomb would be available. Wrote Admiral William D. Leahy, Roosevelt's and Truman's chief of staff: "My conclusion, with which the naval representatives agreed, was that America's least expensive course of action was to continue to intensify the air and sea blockade and at the same time occupy the Philippines. I believed that a completely blockaded Japan would then fall by its own weight." After the war, the Air Force published two surveys which concluded that "certainly prior to 1 November 1945, Japan would have surrendered, even if the atomic bombs had not been dropped, even if Russia had not entered the war, and even if no invasion had been planned or contemplated." In spite of their misgivings, however, the Navy and Air Force chiefs went along with the Army's plan for the invasion of Japan.[29]

As approved in the meeting of June 18, the plan called for an invasion of the southern home island of Kyushu in November 1945 followed by an attack on the Tokyo Plain on March 1, 1946. It was estimated that the invasion would require 2.7 million men and would be completed by the fall of 1946. The success of the invasion, General Marshall emphasized, required Soviet participation: "The impact of Russian entry on the already helpless Japanese may well be the decisive action levering them into capitulation at that time or shortly thereafter if we land in Japan." However, Admiral Ernest J. King, Chief of Naval Operations, disagreed: "While the cost of defeating Japan would be greater without Russian aid," he stated, "there was no question . . . but that we could handle it alone."[30]

Others, like Under Secretary of State and at times Acting Secretary of State Joseph Grew, opposed any Soviet participation in the war against Japan. Grew feared that "once Russia is in the war against Japan, then Mongolia, Manchuria and Korea will gradually slip into Russia's orbit, to be followed in due course by China and eventually Japan." In May, Grew had gone so far as to suggest to Stimson that the United States hedge on its Yalta agreements with respect to Soviet desires in the Far East. But Stimson rejected that suggestion. Considering the overwhelming military power the Soviets were rapidly moving into eastern Siberia in the wake of the German surrender, Stimson believed that, with the possible exception of the Kurile Islands, the areas conceded at Yalta by the United States were the Soviet Union's for the taking.[31]

In deciding how far the United States should go in procuring Soviet assistance against Japan, Truman sided with Marshall. To the President, who still did not know if the atomic bomb would be available before the invasion of Japan occurred, the Soviet Union's entry into the war could mean the saving of thousands of American lives. Accordingly, he promised that at the Potsdam Conference he would do all he could to bring the Soviets into the war.[32]

An Option Not Pursued: A Negotiated End
to the War with Japan

Besides use of the atomic bomb and an invasion of Japan, there was at least one other option available to the United States, one that was not discussed in the strategy meeting Truman and his military advisers conducted on June 18. Although Joseph Grew was not present at the meeting, he believed that an invasion of Japan—and, by inference, use of the atomic bomb—would be unnecessary if the United States modified its insistence on unconditional surrender. Grew stressed that the key to modification involved allowing the Japanese to retain their emperor. To the Japanese, the emperor was the human incarnation of God, the very symbol of Japan itself. To permit him to be overthrown and put on trial as a war criminal, Grew argued, was tantamount to liquidating Japan as a nation. On the other hand, he believed that retention of the emperor would not only hasten the end of the war—before an invasion, use of the atomic bomb, or extensive Soviet involvement were necessary—it would also facilitate the surrender of Japanese forces after the war and the maintenance of order within Japan after the surrender.[33]

There is much to substantiate Grew's assessment. By the spring of 1945, the position in which Japan found itself grew more precarious and prompted key Japanese leaders, including Emperor Hirohito and his chief political adviser, Marquis Kido Koichi, to actively seek peace. However, because the militarists, who dominated the government, opposed surrender, the emperor and Kido had to move secretly and cautiously. On April 7, with the support of Hirohito, Kido was successful in gaining the premiership for Admiral Suzuki Kantaro, a moderate who could be looked upon to seek peace. Two days later, another advocate of the peace policy, Togo Shigenori, was appointed foreign minister.[34]

In May, without the approval, or even the knowledge, of the Supreme Council for the Direction of the War, Togo initiated an effort to secure Soviet mediation to bring the war to an honorable end. In a July 17 cable to the Japanese ambassador in Moscow, Sato Naotake, Togo gave a hint of what conditions he thought would bring about an honorable termination of the conflict: "If only the United States and Great Britain would recognize Japan's honor and existence, we would terminate the war. . . . But if the enemy insists on unconditional surrender to the very end, then our country and His Majesty would unanimously resolve to fight a war of resistance to the bitter end."[35]

The evidence indicates quite conclusively that the U.S. government was aware of the Japanese effort to secure Soviet mediation. From Moscow, Harry Hopkins cabled on May 30 that "certain elements in Japan are putting out peace feelers." More important, President Truman had access to the cables Togo was sending to Sato in Moscow because American intelligence had broken the Japanese secret code. The cables were delivered to Truman at Potsdam by Forrestal on July 28. Ten days earlier, Stalin told Truman that the Japanese were trying to secure Soviet mediation, but that he had given them an evasive reply. The Soviet leader was not about to be denied his share of the spoils by facilitating a Japanese surrender before the Soviet Union could enter the war.[36]

Yet despite an American awareness of the Japanese effort to seek a negotiated end to the war, little was done by the United States to test the sincerity of the Japanese "peace feelers." The one major exception was an effort conducted by the Navy. With Forrestal's approval, Captain Ellis M. Zacharias, of naval intelligence, made a series of fourteen broadcasts to Japan, between May 8 and August 4, seeking to convince the Japanese government that unconditional surrender did not necessarily entail the elimination of the throne. However, Zacharias' broadcasts were ignored by the Japanese government, perhaps, among other reasons, because the broadcasts failed to state specifically that the emperor's position would be preserved after the war.[37]

In a meeting with Grew on May 29, Stimson, Forrestal, and Marshall persuaded him that a public statement on the status of the emperor at that time would be premature. Stimson believed, and Marshall evidently concurred, that the question of what to say to the Japanese, and when to say it, should be governed by whether and when the United States had the atomic bomb. Stimson did not believe that Japan would heed a warning until "after she had been sufficiently pounded, possibly with S-1." Disheartened, Grew informed the

President that the consensus of opinion was that a statement on the emperor should be postponed. Truman agreed. In a special message to Congress on June 1, the President made no modification of the unconditional surrender formula and implied that, if necessary, Japan would be invaded.[38]

On June 18, however, Grew again tried to persuade the President to modify the unconditional surrender formula. By then Okinawa was almost securely in American hands, and Grew thought that the Japanese would be more amenable to a negotiated surrender. But Truman decided to postpone the matter until he could discuss it with Churchill, Stalin, and the Combined Chiefs of Staff at the Potsdam Conference, which was scheduled to begin on July 16. Deferring the decision, he wrote later, served a number of purposes: demonstrating to the Japanese the unity of the Allies, providing time to ascertain the intentions of the Soviets, and gaining time to learn whether the atomic bomb would be a reality.[39]

TRINITY

The sense of urgency that pervaded the highest levels of the Truman administration was also experienced by the men working in the desert of New Mexico preparing the test of the first atomic bomb, code-named TRINITY. "It was certainly very true," Oppenheimer remembered, "that we were very earnest about doing it [the bomb test] in time for the Potsdam Conference. . . . And we did it under weather conditions which were not ideal because we saw there was danger of postponement."[40]

At long last, in a remote corner of the Alamagordo Air Force Base, at 5:29 in the morning of July 16, 1945, the first atomic explosive was detonated. Groves reported its effects: an enormous fire ball as bright as several midday suns, a mushroom cloud that extended 41,000 feet into the substratosphere, a tremendous crash that broke a window 125 miles away, a crater 1,200 feet in diameter, a crumpled forty-ton tower one-half mile from the explosion. The terror of the event reminded Oppenheimer of the words from the Hindu scripture, the *Bhagavad-Gita*: "Now I am become death, destroyer of worlds; waiting the hour that ripens to their doom." George Kistiakowsky, a scientist who worked on the Manhattan Project, remembered the experience as "the nearest thing to doomsday that one could possibly imagine. I am sure that at the end of the world—in the last millisecond of the earth's existence—the last man will see what we have just seen." Groves was more laconic: "The war's over. One or two of these things and Japan will be finished."[41]

The first news of the successful test reached Stimson at Potsdam on the evening of July 16. On July 21, Stimson received Groves' report and personally read it to the President. Truman, Stimson recalled, "was tremendously pepped up by it." Without knowing what had occurred, Churchill remembered that at the next meeting of the Big Three Truman had become a "changed man. He told the Russians just where they got off and generally bossed the whole meeting."[42]

On July 22, when Churchill finally did receive a full description of the successful test, he was effervescent: "Stimson, what was gun powder? Trivial. What was electricity? Meaningless. The atomic bomb is the Second Coming in wrath." The prime minister saw in the bomb "a speedy end to the Second World War, and perhaps to much else besides"—probably a reference to Soviet ambitions in the Far East. Stimson clearly shared Churchill's sentiments. By July 23 he had come to the conclusion, along with General Marshall, that "with our new weapon we would not need the assistance of the Russians to conquer Japan." The next day he so informed Truman, and added his hope that the war could be ended before the Soviet army moved into Manchuria.[43]

But the problem of informing the Soviets about the bomb still remained. Shocked by a firsthand glimpse of Soviet totalitarianism in occupied Berlin, Stimson had lost interest in the prospect of sharing atomic information with the Soviets and in the possibility of achieving international control of the atom. On July 19 he sent a long memorandum to Truman in which he recommended that "before we share our new discovery with Russia we should consider carefully whether we can do so safely under any system of control until Russia puts into effective action [a democratic constitution]."[44]

In spite of Stimson's reservations, which were fully shared by Truman, the Soviets had to be told something about the bomb, if only to preclude giving them an excuse for ending the wartime alliance. Accordingly, it was decided that Stalin would be informed that the United States had the atomic bomb, but that no details concerning it would be provided. Truman recalled that after he had informed Stalin on July 24 that the United States "had a new weapon of unusual destructive force," the Soviet dictator "showed no special interest. All he said was that he was glad to hear it and hoped that we would make 'good use of it against the Japanese.'"[45]

Stalin obviously was not as surprised as Truman had anticipated he would be. The evidence indicates that, by the time of the Potsdam Conference, Stalin had been informed about the American effort to build an atomic bomb by spies who had infiltrated the Manhattan Project. Klaus Fuchs, a German-born British scientist who worked at Los Alamos, not only relayed to the Soviets the scale and timing of the American project, he also provided details of the plutonium bomb, including its design and method of construction, as well as a plethora of other information relating to the bomb.[46]

Stalin had been doing all he could for some time, at least since 1940, to spur construction of a Soviet atomic bomb. In fact, before the German invasion of the Soviet Union in June 1941, the Soviets may have been ahead of the United States in the development of key aspects of atomic energy, particularly the separation of uranium isotopes. As the Germans began to retreat, the Soviet atomic effort was able to accelerate. By the end of the war, Soviet scientists were only four years from successfully testing an atomic explosive. After Truman's casual announcement of the American atomic bomb at the Potsdam Conference table, Stalin called aside Marshall Georgi Zhukov and Foreign Minister Molotov and said, "We've got to work on [Igor] Kurchatov [director of the Soviet atomic weapons program] and hurry things up."[47]

The Potsdam Declaration

The success of the Alamagordo test also had a direct bearing on the American approach to the Japanese. The atom bomb could ensure that Japan would surrender without an American invasion and, with a little bit of luck, before the Soviets invaded Manchuria. Recalled James Byrnes, "in view of what we knew of Soviet actions in eastern Germany and the violations of the Yalta agreements in Poland, Rumania and Bulgaria, I would have been satisfied had they determined not to enter the war. Notwithstanding Japan's persistent refusal to surrender unconditionally, I believed the atomic bomb would be successful and would force the Japanese to accept surrender on our terms. I feared what would happen when the Red Army entered Manchuria."[48]

To get Japan to surrender before the Soviets could get deeply involved in the war, speed was essential. Since Truman and Byrnes wanted to coordinate the issuance of the Potsdam Declaration with the delivery of the bombs, both were anxious to learn when the weapons would be ready. With regard to the declaration, Stimson recalled that Truman "proposed to shoot it out as soon as he heard the definite day of the operation." On the evening of July 23, Stimson learned that "Little Boy," the uranium bomb, might be ready as early as August 1 and almost certainly before August 10. "Fat Man" would probably be ready by August 6. The President was delighted and told Stimson that the news "gave him the cue for his warning" to Japan.[49]

The need for speed also ruled out any inclination Truman may have had to engage the Japanese in lengthy negotiations concerning the postwar status of the emperor. In a discussion with the President on July 2, Stimson, with the support of Grew and Forrestal, had made yet another attempt to moderate the unconditional surrender policy. Stimson wrote a draft of a warning to the Japanese that contained a clause which would permit "a constitutional monarchy under the present dynasty." But after July 3, when Byrnes became secretary of state, Stimson's influence with the President declined precipitously as that of Byrnes increased. With respect to the unconditional surrender policy, Byrnes was greatly influenced by the view of former Secretary of State Cordell Hull, who advised him to delete from Stimson's draft warning any reference to the Japanese emperor. Hull believed it "sounded too much like appeasement of Japan, especially after the resolute stand we had maintained on unconditional surrender." Moreover, Hull added, Stimson's draft" seemed to guarantee continuance not only of the Emperor, but also of the feudal privileges of a ruling caste under the Emperor."[50]

Truman was sensitive to the points raised by Hull. He was extremely reluctant to revise a policy so obviously associated with Roosevelt. He also felt that the American people would oppose any leniency toward the hated enemy, and particularly toward Hirohito. Truman, to some extent, shared their hostility. When Churchill suggested some change in the unconditional surrender policy that would allow the Japanese to end the war with honor, Truman bluntly responded that he did not think that the Japanese retained any honor after their dastardly attack on Pearl Harbor. As a result, the Potsdam Declaration, which was released on July 26, made no change in the demand for unconditional surrender and made no mention of the future status of the emperor. The only reference the declaration made to the atomic bomb was the nebulous threat that Japan's continuation of the war would lead to her "prompt and utter destruction."[51]

Final Preparations

While the Potsdam Declaration was being sent out to Japan, the final preparations for use of the atomic bombs were hastily being completed under the direction of General Groves. The first units of the 509th Composite Group, which would deliver the nuclear weapons, had been on the island of Tinian in the Marianas since mid-May. The components of the bombs were shipped from Los Alamos, arriving on Tinian on July 26. (The ship that carried the components across the Pacific, the cruiser *Indianapolis,* was sunk by a Japanese submarine on its return trip). One day earlier, on July 25, the order was made to drop the bombs on Japan. It called for the 509th to bomb one of four target cities—Hiroshima, Kokura, Niigata, or Nagasaki—weather permitting, as soon as August 4. The plan envisioned that additional bombs would be delivered on these targets as soon as they were available: one as early as August 6, another about August 24, possibly three in September, the number increasing each month thereafter to as many as seven in December. The hope of the administration was that the rapid use of a few bombs would create the impression in the minds of the Japanese government that the United States had many atomic weapons.[52]

While Truman recalled that, as he put it, "the final decision of where and when to use the atomic bomb was up to me," the evidence indicates that, at least by July 22, the operation did not require additional presidential action. Explained Groves: "As far as I was concerned, his [Truman's] decision was one of noninterference—basically a decision not to upset the existing plans." Indeed, the final order to drop the bombs was issued under Groves' authority; it contained no mention of the Commander-in-Chief of the United States.[53]

Publicly, Truman expressed no doubt about the decision to drop the bombs on Japan. "Let there be no mistake about it," he recalled, "I regarded the bomb as a military weapon and never had any doubt that it would be used." Although Truman stated that there was a consensus among his advisers that the bomb should be used, the question was never discussed in meetings of either the U.S. Joint Chiefs of Staff or with the British Chiefs of Staff. The advice Truman received came primarily from Byrnes, Stimson, Marshall, and Groves, all of whom had acted as though the decision had been made long ago, as indeed it had.[54]

One of the few to protest the use of the bomb was General Dwight D. Eisenhower, the Supreme Commander of Allied Forces in Western Europe. Although Eisenhower played no direct role in the deliberations at Potsdam, he was able to visit Stimson during the conference. Regarding use of the bomb against the Japanese, Eisenhower remembered telling Stimson, "It wasn't necessary to hit them with that awful thing." He later wrote: "It was my belief that Japan was, at that very moment, seeking some way to surrender with a minimum loss of 'face.'"[55]

Hiroshima

On July 27, when the Japanese Supreme War Council met to consider Japan's response to the Potsdam Declaration, Foreign Minister Togo was still attempting to find some way to end the war with honor. He stated his opinion that the declaration did not demand Japan's unconditional surrender explicitly and therefore should be regarded as a moderation of Allied terms. The Japanese military leaders, however, were not yet prepared to consider surrender; they demanded a strong statement from the government condemning the declaration. It is a testament to Togo's fortitude and skill that, despite the vehement opposition of the militarists, he was able to persuade the council and the cabinet that Japan should await further developments before responding to the declaration, while trying to determine the intentions of the Soviet Union. Unfortunately, the next day, the Japanese press reported the government's response in a way that would challenge the most creative imagination. *Domei,* the semi-official Japanese news agency, interpreted Togo's statement that the government had not yet made a decision on the Potsdam Declaration to mean that the government would ignore the declaration. When translated by American officials, the Japanese response was treated not only as a rejection of the Potsdam Declaration but as a contemptuous one at that.[56]

While the Japanese government gave further consideration to the Potsdam Declaration, the U.S. Air Force went into action. At 2:45 in the morning of Sunday, August 6, the *Enola Gay,* the B-29 carrying "Little Boy," lifted off the Tinian runway. Five and one-half hours later, at 8:15 Hiroshima time, the bomb was dropped from an altitude of 31,000 feet. Forty-three seconds later it detonated 1,800 feet above the city. Aboard the *Enola Gay,* tailgunner George Caron described the scene from a distance of eleven miles and 29,000 feet of altitude as a "peep into hell."[57]

News of the attack on Hiroshima reached General Groves late Sunday evening. He reported to Stimson and Marshall: "Entire city except outermost ends of dock area was covered with a gray dust layer which joined the [mushroom] cloud column. It was extremely turbulent with flashes of fire visible in the dust. Estimated diameter of this dust layer, at least three miles." When Truman received the news onboard the cruiser *Augusta,* which was carrying him home from Europe, he told sailors: "This is the greatest thing in history."[58]

The Japanese government did not receive a complete description of the attack on Hiroshima until August 8. Lieutenant General Seijo Arisue, who headed the investigating team that flew to the city, reported: "When the plane flew over Hiroshima there was but one black dead tree, as if a crow was perched over it. There was nothing but that tree. . . . The city itself was completely wiped out." The Japanese soon realized that only an atomic bomb could produce such destruction.[59]

Years later, it was possible to make a more complete assessment of what had happened at Hiroshima. At the hypocenter of the explosion, with a TNT equivalent of twenty kilotons, the temperature reached several million degrees centigrade. The shock wave created by the explosion was strong enough to break windows fifteen kilometers from the hypocenter. Four kilometers away, buildings were charred. Three kilometers away, about ninety percent of the buildings had experienced fire and blast damage. Within a two-kilometer circle, only ashes, fist-size pieces of rubble, and a few shells of reinforced concrete buildings remained. Thirteen square kilometers of the city were razed, including forty-two of the city's forty-five hospitals. Of Hiroshimas's 340,000 population, 130,000 were dead by November 1945 and an additional 70,000 had died by 1950. Included among these casualties were twenty American airmen who were being held in Hiroshima as prisoners of war. Both Groves and Stimson were aware that American POWs were being held in the city, but they refused to remove Hiroshima from the target list. Those who survived the blast had to deal with nausea, vomiting, diarrhea, weakness, and blood disorder— all symptoms of acute radiation sickness. They also had to endure the profound psychological effects of their ordeal, which included guilt that they had survived while others had not, and anxiety about their future health and the impact of radiation on their descendants. The Research Institute for Nuclear Medicine and Biology at Hiroshima University estimated that, by the mid-1950s, the risk of cancer increased 30 or 40 times the norm. An increase in stillbirths, birth defects, and infant mortality was also clear in the 70,000 pregnancies examined in the study. There is, of course, no way to measure the entire magnitude of the human suffering caused by the atomic conflagration of Hiroshima.[60]

Nagasaki and the Japanese Surrender

While the Japanese government was assessing the effects of the attack on Hiroshima, it received another shock, one that added immeasurably to the pressure toward surrender. At one hour after midnight, Tokyo time, on August 9, Soviet forces attacked the Japanese army in Manchuria. Although the date of the Soviet Union's entry into the war against Japan coincided with Stalin's pledge to Hopkins of May 28, the Soviet leader realized that the use of the atomic bomb on Hiroshima, coupled with what he considered the lenient terms offered in the Potsdam Declaration, made Japanese surrender only a matter of days. Any Soviet delay could bring about a Japanese capitulation before the Soviet Union received the opportunity to claim its full share of the spoils.[61]

Ten hours after the Soviet attack, the Japanese Supreme War Council met in Tokyo. Foreign Minister Togo insisted that it was absolutely essential for Japan to comply with the Potsdam Declaration before another Japanese city was destroyed by an atomic bomb. The militarist faction, however, continued to demur. In addition to demanding the retention of the emperor after the war, the militarists insisted on two other conditions before they would accept surrender: (1) Japan would not be occupied after the war, and if that could not be obtained, Japan must be occupied only minimally; and (2) Japanese war criminals would be tried by Japanese courts. Because Togo considered the militarists' demands excessive, the meeting of the council adjourned without reaching a decision.[62]

By this time, however, it was too late to avoid the loss of a second Japanese city. At 10:58 A.M. local time, on August 9, "Fat Man," the plutonium bomb, was dropped on Nagasaki. Ironically and tragically for that city, Nagasaki was selected as the target only after the B-29 carrying the bomb could not see Kokura, the primary target, because of dense cloud cover over that city. The second atomic attack on Japan occurred two days earlier than originally planned because poor weather was anticipated on August 11, and because the Truman administration wanted to create the impression, by a rapid followup to the Hiroshima attack, that the United States possessed more atomic bombs than it actually did.[63]

The destruction of Nagasaki sealed the fate of Japan. Over the continued opposition of the militarists, who still wanted to prolong the war, the emperor decisively intervened to bring the conflict to an end. The Japanese government, on the morning of August 10, agreed to accept the terms of the Potsdam Declaration. Only one condition was attached: "that the said declaration does not comprise any demand which prejudices the prerogatives of His Majesty as Sovereign Ruler."[64]

When Truman assembled his advisers on the morning of August 10 to weigh the Japanese message, the question that dominated the meeting was the one raised by Grew in May: Should the United States redefine unconditional surrender to permit retention of the emperor? After heated debate among the participants, Truman accepted a compromise devised by Forrestal: the United States would agree to retention of the emperor provided that "from the moment of surrender the authority of the Emperor and the Japanese Government to rule the state shall be subject to the Supreme Commander of the Allied powers, who will take such steps as he deems proper to effectuate the surrender terms." In addition, the American reply stated: "The ultimate form of government of Japan shall, in accordance with the Potsdam Declaration, be established by the freely expressed will of the Japanese people."[65]

On August 11, following acceptance by Britain, China and, with reluctance, the Soviet Union, the American reply was transmitted to the Japanese. Even at this late date, however, neither the cabinet nor the Supreme War Council could reach a decision on the American terms. Again the intervention of the emperor was necessary. In a short speech to his ministers on August 12, Hirohito said the American terms should be accepted. "Unless the war be brought to an end at this moment," he said, "I fear that the national polity will be destroyed, and the nation annihilated." Hirohito's ministers accepted the imperial will. During the night of August 14, Japan transmitted her decision to surrender to the United States. Truman announced the Japanese capitulation at 7:00 P.M. Washington time, the same day. On September 2, 1945 the war with Japan officially ended when the instruments of surrender were signed aboard the U.S.S. *Missouri* in Tokyo Bay.[66]

3

Truman and International Control of the Atom, 1945–1947

Stimson Reconsiders the Bomb

After the Potsdam Conference, and shortly before his September retirement from government service, Secretary of War Henry L. Stimson had a chance to rethink what he considered an excessively negative attitude that the Truman administration had adopted toward the Soviets. He was particularly alarmed by the apparent eagerness of Secretary of State Byrnes to engage the Soviets in atomic diplomacy. Stimson recorded in his diary that Byrnes was preoccupied with the forthcoming London Conference of Foreign Ministers, and "looks to having the presence of the bomb in his pocket . . . as a great weapon to get through the thing."[1]

On September 5, 1945 Stimson met with Truman to discuss the future role of the atomic bomb in Soviet-American relations. In this discussion, and in a later memorandum, Stimson told Truman that he had revised his earlier opinions about using the bomb "as a direct lever" to produce "a change in Russia's attitude towards individual liberty." He now believed that the only alternative to an agreement with the Soviets on the bomb would be a "secret armament race of a rather desperate character." To avoid that, he proposed that the United States should inform the Soviets of its desire to enter an international agreement to control the use of atomic energy for peaceful purposes, stop the manufacture of atomic weapons, and surrender those which already existed.[2]

Stimson stressed that on the issue of atomic energy it was vital to approach the Soviets directly, quickly, and privately. He felt that "loose debates" in the United Nations "would provoke scant favor from the Soviets." He warned the President, in an allusion to the Byrnes approach, that if the United States failed to approach the Soviets immediately on the atomic bomb issue and instead "merely continued to negotiate with them, having the weapon rather ostentatiously on our hip," then "it would be much less likely that we will ever get the kind of covenant we may desperately need in the future."[3]

After listening to Stimson's presentation, Truman stated that his own ideas were "in full accord" with those of the secretary of war. In all probability, they were not. Although Truman greatly admired Stimson, he felt uncomfortable in his presence and was disinclined to disagree with him directly, certainly not then, right after Stimson had announced his decision to retire from government service. Truman, in effect, simply nodded his head to Stimson's proposals, saying that he would present them to the cabinet for its reaction.[4]

The Stimson Plan

On September 21, Stimson's seventy-eighth birthday and last day in office, he presented to the cabinet his proposal for an immediate and direct approach to the Soviet Union on the issue of international control of atomic energy.[5]

Stimson's proposal divided the cabinet. Its opponents for the most part disliked the idea of sharing atomic information with the Soviets. Secretary of the Navy James Forrestal flatly stated that the bomb and the knowledge that produced it were "the property of the American people." The United States, he asserted, "should not try to buy" the "understanding" of the Soviets by sharing atomic information with them. Appeasement, he argued, did not work against Hitler; it would not work against Stalin. Instead of surrendering atomic weapons to an international authority, Forrestal recommended that the United States exercise "a trusteeship over the bomb on behalf of the United Nations."[6]

Henry Wallace, who had been one of the five men originally appointed by Roosevelt in 1941 to oversee the development of the bomb, vehemently rejected Forrestal's line of reasoning. A failure to share scientific information with the Soviets, he predicted, would make them "a sour, embittered people." Forrestal was unmoved. Later he recorded that Wallace was "completely, everlastingly and wholeheartedly in favor of giving it [the atomic bomb] to the Russians."[7]

Those who favored the sharing of atomic know-how with the Soviets relied on the analysis of Vannevar Bush and other scientists that their acquisition of atomic energy was inevitable anyway, and not in the too distant future. Bush pointed out that the real secret of the bomb involved the details of bomb design and the manufacturing process. He predicted that the Soviets could make an atomic bomb in five years "provided they devote a very large part of their scientific and industrial effort to it."[8]

On the other hand, those who opposed the sharing of atomic technology based their opposition on the assumption that the American atomic monopoly would be of rather lengthy duration, and that the United States should use that interval to mold the kind of world in which American interests could remain secure. The man most responsible for this assumption was General Leslie Groves, director of the Manhattan Project. The Soviets, he estimated, would need at least twenty years to build an atomic bomb because they lacked the resources and know-how to do so sooner. Unfortunately for Groves, and for those who believed him, his analysis was wide of the mark. The FBI reported that the Soviets had more than adequate supplies of uranium, and that they had been engaged in atomic research since at least 1940.[9]

One can only conclude that Groves did not want to believe the Soviets could produce an atomic bomb so soon. To have believed otherwise would have strengthened the case for international control of atomic energy, which Groves vehemently opposed. Exclusive American possession of the atom bomb, he said, meant "complete victory in our hands until the time another nation has it." To those who opposed the sharing of atomic information, Groves' views were comforting. They were reinforced by the reputation for success the general had acquired from his work on the atomic bomb. On the other hand, most of the scientists who challenged Groves' prediction of an American monopoly were regarded by those in Groves' camp as, in the words of historian Daniel Yergin, "unstable, naive, untrustworthy, and ill-equipped to determine high policy."[10]

Byrnes and Atomic Diplomacy

Despite the advantages Groves believed atomic diplomacy would offer, Secretary of State Byrnes learned otherwise at the September 1945 London Conference of Foreign Ministers. Even though the United States enjoyed exclusive possession of the atomic bomb, Byrnes could not get Soviet Foreign Minister Molotov to modify Soviet policy in Eastern Europe. Specifically, the Soviets refused to accept Byrnes' demand to reorganize the governments of Bulgaria and Rumania along democratic lines. As a result, the conference ended in stalemate on October 2. The Soviets, Byrnes learned, were "stubborn, obstinate, and they don't scare." And neither Byrnes nor Truman were about to try to frighten them by using the atomic bomb just to change the governments of Bulgaria and Rumania. Budget director Harold Smith tried to cheer up the President following the failure of the London Conference by saying, "Mr. President, you have an atomic bomb up your sleeve." But Truman could only respond, "Yes, but I am not sure it can ever be used." Apparently, Truman shared Byrnes' newfound realization concerning the futility of atomic diplomacy.[11]

Yet in spite of the failure in London, Byrnes was not ready to recommend international control of the atom. He had little faith in any system of international control without an effective inspection system, and he had little reason to believe the Soviets would allow inspection of their own atomic installations when they had allowed little Western access to Rumania and Hungary. As a result, when Oppenheimer warned the secretary of state in August that unless international control were established over atomic energy bombs many times more powerful than atomic weapons would be produced, Byrnes responded: "For the time being . . . an international agreement [is] not practical and . . . he [Oppenheimer] and the rest of the gang should pursue their work full force."[12]

Truman and International Control of the Atom

Although Truman was fully aware, as he told former Secretary of State Edward Stettinius in October, that there was "no precious secret" for the Soviets to steal, he was under intense pressure to reject any sharing of atomic information with the Soviet Union. According to public opinion polls taken at the end of September, some seventy percent of the American people and over ninety percent of the congressmen questioned the policy of "sharing the atomic secret."[13]

At the same time, Truman was exposed to pressure from the atomic scientists, who had finally persuaded the administration to lift the wartime restrictions on their right of public expression. At the University of Chicago, James Franck initiated another petition, which was signed by sixty-four faculty members, urging President Truman to share atomic energy with other nations in order to avoid a nuclear arms race. In November the newly created Atomic Scientists of Chicago joined with similar associations at other Manhattan Project installations to form the Federation of Atomic Scientists, which was destined to become a major force in the effort to secure international control of the atom."[14]

The Truman administration was pressured from still another source, the British. Since the bombing of Hiroshima, Prime Minister Clement Attlee, who defeated Churchill in the July 1945 parliamentary election, had been urging Truman to give some indication of America's plans for the future prospects of the atomic bomb. "Am I to plan for a peaceful or a warlike world?" he queried the President on September 25. Attlee was particularly eager to know if the United States intended to continue the Anglo-American atomic alliance agreed to in the Hyde Park memorandum.[15]

On October 3 Truman finally gave some indication of the atomic policy his administration intended to follow. In so doing, he attempted to please both sides of the debate on the issue of sharing atomic information. On the one hand, he stated that the atomic "secret" would not remain one for long, and therefore "the hope of civilization lies in international arrangements looking, if possible, to the renunciation of the use and development of the atomic bomb." On the other hand, he rejected the direct and immediate approach to the Soviets on atomic energy proposed by Stimson. Instead, Truman said that after discussions with Britain and Canada, the United States would approach other nations, without specifically naming the Soviet Union, in order to reach agreement on the exchange of scientific information. He also promised that there would be no "disclosures relating to the manufacturing processes leading to the production of the atomic bomb itself."[16]

In an October 8 news conference, Truman was more explicit. He stated that the Soviets would have to build an atomic bomb "on their own hook, just as we did." When a reporter asked him if this meant "that the armaments race is on," Truman answered affirmatively, but added that the United States "would stay ahead."[17]

Rather than the prospect of a world with international control of atomic energy, Truman clearly had more faith in a Pax Americana based on the atomic bomb. In an address on October 27, he stated that the American atomic bomb was "no threat to any nation," but "a sacred trust" which, he implied, the United States would retain indefinitely.[18]

The Bomb, the British, and the Soviets

In spite of Truman's October pronouncements, the administration, at the end of that month, only days before the President's scheduled November 10 meeting in Washington with Attlee and Canadian Prime Minister Mackenzie King, was still without a formulated policy on the international control of atomic energy. In early November, Vannevar Bush attempted to present a draft of a policy that he described as "go[ing] beyond great generalities."[19]

The Bush plan consisted of two parts. The first part concerned the Anglo-American atomic partnership. It called for termination of the Quebec Agreement, which Bush asserted "was intended for the war period only." A new document, he stated, should be negotiated, "providing merely for sharing of materials, leaving political clauses and the dissemination of information to be worked out on a more general international basis." The second part of the Bush plan focused on the American approach to the Soviet Union. To prevent a nuclear arms race, Bush proposed a three-stage movement toward international control of the atom, one that would "open up Russia" while doing nothing to endanger American security. In the first stage, Bush proposed "the full dissemination of fundamental information of science in all fields including atomic fission." It would include free access to all laboratories doing basic research. This step, Bush commented, "would probably cost us nothing," but would test the Soviet Union's willingness to cooperate. If there were progress in this stage, the second stage could begin. It called for the gradual implementation of a system of international inspection. Once the United States was assured that an inspection system would work, the third and final stage could start, during which atomic weapons, facilities, and materials would be converted to civilian application.[20]

Although Bush considered his plan generous, in reality, it offered the Soviets the prospect of little immediate gain in return for accepting major security risks. In exchange for an American pledge that the United States would dismantle its atomic arsenal in the indefinite future, the Soviets would have to open to international inspection all of their nuclear facilities. While the Bush plan would enable the United States to preserve its atomic monopoly for some time, the United States would also learn, through the international inspection system of which Americans would be a part, the nature, state of progress, and location of the Soviet Union's nuclear installations. The Bush plan, in effect, would increase the short-term atomic vulnerability of the Soviet Union for a long-term promise that Soviet security would be enhanced by the abolition of nuclear armaments. Because the Soviet Union would be forced to accept the lion's share of the risks, the Bush plan appealed to Truman and Byrnes. On

November 7, they accepted it as the basis of America's atomic energy policy. However, instead of the direct, private, and immediate approach to the Soviets that Stimson had recommended, Truman decided to work through the United Nations.[21]

In the Three Nation Declaration of November 15, the British and Canadians accepted the policy adopted by the United States. The British really had no choice but to accept the termination of the Quebec Agreement, considering Britain's economic and military dependence on the United States. At the very time Attlee was discussing atomic policy with Truman, another British delegation was in the United States negotiating the terms of a six billion dollar American loan. Yet the new agreement did not completely terminate Anglo-American atomic cooperation. In exchange for a vague American promise of continued nuclear assistance, the British agreed to continue to provide the United States with materials needed for the construction of atomic weapons. Nevertheless, despite appearances to the contrary, the Anglo-American atomic partnership created by Roosevelt and Churchill ceased to exist.[22]

While the British accepted, albeit with much reluctance, their new, and inferior, atomic relationship with the United States, the Soviets were outraged by their exclusion from the atomic deliberations of their wartime allies. On November 7 Molotov reacted by stating that "it is not possible at the present time for a technical secret of any great size to remain the exclusive possession of some country or some narrow circle of countries . . . we will have atomic energy and many other things too." With no indication that the West intended to make a sincere effort to share the potential of atomic energy, Molotov in effect said that the Soviet Union would develop its own atomic weapons. The effort to establish international control of atomic energy, in other words, was doomed before it ever began.[23]

Byrnes Reconsiders Atomic Diplomacy

In the meantime, Byrnes, upset by his failure to play the role of peacemaker at the London Conference in September, decided to make another attempt at personal diplomacy in the style of his late mentor, Franklin Roosevelt. On November 23 he cabled Ambassador Harriman in Moscow to suggest to the Soviets another meeting of the Council of Foreign Ministers. The Soviets accepted, and the conference was scheduled for the middle of the next month in the Soviet capital.[24]

At the urging of the State Department, Byrnes agreed to raise the matter of the atomic bomb at the Moscow Conference. But he intended to follow an approach quite different from the one he had pursued at the London meeting. Rather than carrying the bomb "on his hip," he intended to lay it on the table. Rather than publicizing the negotiations, he planned to keep them close to his vest, away from the State Department, from the British, and even, as much as possible, from the President. The Moscow Conference was going to be Byrnes' own big show.[25]

On December 10, a State Department committee, headed by Benjamin V. Cohen and Leo Pasvolsky, completed work on a draft approach Byrnes would present to the Soviets. It closely parallelled the Three Nation Declaration of the previous month, but called for four stages instead of three. The first stage would involve an exchange of scientists and scientific information, techniques and materials. The second called for the development and exchange of knowledge concerning natural resources. The third stage envisioned an exchange of technological and engineering information. The fourth would implement safeguards against employing methods of mass destruction.[26]

There were other key differences between the Three Nation Declaration and the approach favored by the Cohen-Pasvolsky committee. Unlike the declaration, the Cohen-Pasvolsky approach did not require the Soviets to prove themselves stage by stage before the United States would offer anything of substance in return. The committee believed that agreement at any stage should be pursued whenever it appeared achievable. In addition, unlike the Three Nation Declaration, the new approach offered the Soviets a reason for believing that they would gain quick access to more than just basic scientific information. The new plan was also different in suggesting that the United States negotiate with the Soviets before the United Nations began its work, and in this way more closely followed the direct approach recommended by Stimson. On the other hand, like the Bush plan on which the Three Nation Declaration was based, the Cohen-Pasvolsky plan would not give the Soviets anything that would endanger American security. The committee realized that even the hint of disclosing "secrets" would arouse a storm of opposition within the United States and doom any possibility of atomic cooperation with the Soviets.[27]

In spite of the caution with which the State Department approached the forthcoming Moscow negotiations, opposition arose anyway from hardliners within the government. As soon as Forrestal learned the contents of the Cohen-Pasvolsky draft, he phoned Byrnes to tell him "most strongly" that there should be no discussion with the Soviets of specific information concerning atomic energy until the United States could be guaranteed "genuine reciprocity in exchange." Senator Arthur Vandenberg, a leading spokesman for the Republican Party on American foreign policy, considered an "'exchange' of scientists and scientific information as sheer appeasement because Russia had nothing to 'exchange.'" Senator Tom Connally, chairman of the Foreign Relations Committee, wanted to know why adequate safeguards were listed last. "Don't you have your four points in reverse order?" Connally asked Byrnes. "Number Four [international controls] should be Number One."[28]

After Connally made no headway with Byrnes, he bluntly told Truman, "we must have an inspection system *before* we exchange information about the atomic bomb and atomic energy." Truman agreed and promised that he would so instruct Byrnes. When Byrnes received the President's instructions after arriving in Moscow, he assured Truman that scientific exchange would not begin before "effective enforceable safeguards" had been devised, a major reversal of the Cohen-Pasvolsky approach. But Byrnes' departure from the State Department draft was, at first, more apparent than real. Without informing Truman, he decided to omit from the American proposal the passages on "separate stages" and "enforceable safeguards." By so doing, Byrnes attempted to leave open the possibility of an informal understanding with the Soviets on international control of atomic energy. It was only after Senator Vandenberg went public with the congressional demand for specific stages on December 20 that Byrnes felt compelled to reinsert the controversial passages into the American proposal, explaining to Molotov that they "had been omitted by mistake" from the initial version.[29]

Much to Byrnes' surprise, Molotov insisted that atomic energy should be moved from the top of the conference agenda to the bottom. The move was probably motivated by the Soviet desire to make atomic diplomacy appear fruitless by minimizing the importance of the atomic bomb. When the issue was finally discussed, Molotov agreed that the Soviet Union would participate in the United Nations atomic energy commission. However, he only accepted Byrnes' proposal to proceed in stages toward international control after the secretary of state agreed to his demand that the commission be accountable to the Security Council, where Soviet interests could be protected by the veto.[30]

The Reaction

Byrnes was pleased with the results of the Moscow Conference. In addition to the agreement on a U.N. atomic energy commission, the United States and the Soviet Union had formulated a procedure for a conference that would draft the peace treaties with the former Axis nations. He also considered it no minor achievement that he had persuaded the Soviets to add non-communists to the Bulgarian cabinet. But he was unprepared for the reaction his efforts caused at home. A portion of the American press as well as influential members of the government considered his Balkan settlement a sellout. Reflecting this view, Admiral Leahy described the Moscow accord as "an appeasement document which gives to the Soviets everything they want and preserves to America nothing."[31]

Byrnes was shocked most by Truman's reaction. On January 5, 1946, after the secretary of state had returned from Moscow, Truman personally told him that his accomplishments at the conference were "unreal." He considered the Balkan agreement nothing more than "a general promise" from the Soviets. He bluntly accused Byrnes of taking "it upon himself to move the foreign policy of the United States in a direction to which he could not and would not agree"—that is, toward recognition of Soviet predominance in Eastern Europe. He also criticized Byrnes for failing to keep him fully informed on the proceedings of the conference, particularly those dealing with atomic energy. With respect to that issue, Truman told Byrnes that he thought he had settled American policy in October. He was against scientific exchange with the Soviets until all other issues had been resolved, and quite naturally considered Byrnes' attempt to insert the Cohen-Pasvolsky report into the administration's atomic policy as tantamount to insubordination.[32]

Truman was also upset by Byrnes' inability to stop Soviet encroachments in the Near East. The Soviets were demanding from Turkey strips of Turkish territory on the Black Sea and in Armenia, an expanded right to use the Straits of the Bosphorus and the Dardenelles, and permission to build a naval base near the Straits. Truman considered the Soviet demands on Turkey a threat to the independence of that country. In Iran, the Soviet army was displaying a marked reluctance to end its wartime occupation of the northern portion of that country after the United States and Britain had indicated that they would withdraw their troops from the southern part. Truman flatly refused to consider the Soviet demand that, before they withdrew their forces, they must be awarded oil concessions in Iran equal to those that were granted to the United States and Britain. Rather than accommodate the Soviets on the oil concession issue, Truman denounced the continued Soviet military presence in Iran as a blatant act of aggression. He demanded that Byrnes protest to the Soviets "with all the vigor of which we are capable," and to emphasize his order, told Byrnes, "I'm tired of babying the Soviets."[33]

Distressed by Truman's reaction to his Moscow trip, Byrnes could only conclude that the hardliners in the administration and in Congress had taken advantage of his absence from Washington and had finally succeeded in winning Truman over to their side. In so doing, he felt that the President had reneged on the understanding they had worked out when Truman had asked him to be secretary of state—that is, Truman would exercise overall supervision of the nation's foreign policy, while he would have a free hand to formulate and to conduct it. Byrnes suspected that Truman now felt sufficiently secure to formulate foreign policy himself. Perhaps at the bottom of the dispute was the realization, which neither man could forget, that, had it not been for the opposition of organized labor to Byrnes' vice-presidential candidacy in 1944, Byrnes, not Truman, would be sitting in the presidential chair.[34]

Byrnes realized that his influence with the President had been destroyed by his brief flirtation with Roosevelt's accommodationist approach at the Moscow Conference. He told Truman that, for reasons of health, he intended to resign as soon as he concluded the negotiations on the peace treaties. Truman, without reluctance, accepted his decision. With Byrnes' departure in 1947, the last influential advocate within the administration of Stimson's direct approach would be gone.[35]

Containment

Truman's confrontation with Byrnes on January 5, 1946 was a turning point in the administration's handling of foreign policy. Truman, in effect, had broken the last link with Roosevelt's accommodationist policy toward the Soviet Union. Now he would rely more and more on the views of those who insisted that the main force behind Soviet foreign policy was the expansion of world communism, and that any American concessions to the Soviets would ultimately support that end.[36]

The hardline view was most effectively presented by George Kennan, U.S. chargé d'affaires in Moscow. In his famous "Long Telegram" of February 22, 1946, Kennan warned that, for the Soviet Union, "it is desirable and necessary that the internal harmony of our society be disrupted, our traditional way of life destroyed, the international authority of our state be broken, if Soviet power is to be secure." In Moscow's view, Kennan concluded, coexistence with the West was only a temporary expediency. The United States, he asserted, was at war and should accept a wartime frame of mind.[37]

The closest thing to the declaration of war alluded to by Kennan was delivered by Winston Churchill in a speech at Fulton, Missouri, on March 5, 1946. With Truman's tacit approval, which was indicated by the President's presence on the speaker's dais, Churchill declared that an "Iron Curtain" had descended from "Stettin in the Baltic to Trieste in the Adriatic." Considering the West to be militarily weak following its postwar demobilization, Churchill regarded the atomic bomb essential to its defense. He branded it "wrong and impudent to entrust the secret knowledge or experience of the atomic bomb" to an international authority. Coming closely on the heels of Kennan's Long Telegram, Churchill's speech did much to mold American public opinion against not only a diplomatic solution of the problems the West was experiencing with the Soviet Union but also international control of the atom.[38]

Churchill's tone also did much to harden the Soviet attitude toward the West. Stalin considered the speech an ultimatum: "Accept our rule voluntarily, and then all will be well; otherwise war is inevitable." Within a three-week period after the Churchill speech, Stalin terminated his effort to secure a one billion dollar American loan, rejected membership in the World Bank and the International Monetary Fund, increased, then decreased, Soviet pressure on Iran, timed the withdrawal of Soviet troops from Manchuria to support the infiltration of the Chinese Communist forces of Mao Zedong, and launched an ideological purge of the Kremlin leadership designed to remove pro-Westerners from positions of influence. The first months of 1946, in short, saw the eruption of what has come to be called the Cold War.[39]

Domestic Control of Atomic Energy

The Cold War undoubtedly clouded the atmosphere in which negotiations on international control of atomic energy would be conducted. But this effort was also greatly hampered by a debate within the United States that developed over the issue of domestic control of atomic energy. As early as May 1945, the need for a government agency that would safeguard the atomic bomb and develop nonmilitary applications of atomic energy was recognized. The debate centered on the question of who would run the proposed atomic energy agency—civilians or the military.[40]

In October 1945, a bill giving the military a veto over the proposed agency's decisions was introduced in Congress by Representative Andrew May, the chairman of the House Military Affairs Committee, and by Senator Edwin C.

Johnson, chairman of the Senate Military Affairs Committee. The May-Johnson bill was subjected to immediate and vociferous opposition. Assistant Secretary of State Dean Acheson opposed it because he feared that it would prevent the exchange of atomic information with other nations and would thereby jeopardize the prospects for international control of the atom. The Federation of Atomic Scientists feared that the bill would produce new restrictions on scientific exchange similar to those that had existed under the military-administered Manhattan Project.[41]

Prompted by these and other considerations, a rival bill was introduced by Senator Brien McMahon, a freshman Democrat from Connecticut, who was also the chairman of the newly created Senate Committee on Atomic Energy. The McMahon bill provided for strong civilian supervision of atomic energy and, for that reason primarily, drew the fire of those favoring military supervision of domestic atomic energy. In fact, as historical research would later reveal, the opposition was prepared to go to almost any length to kill the McMahon bill.[42]

On February 3, 1946 Americans were shocked to hear columnist Drew Pearson's report that a Soviet spy ring, led by British physicist Alan Nunn May, had been discovered in Canada. Another reporter, Frank McNaughton, disclosed that a "confidential source" had revealed to him that the real target of the Canadian spies was the American atomic energy program and that, in fact, another Soviet spy ring was already operating within the United States. According to McNaughton's informant, the FBI would have moved more quickly against the spies had it not been for the reluctance of the State Department to upset Soviet-American relations.[43]

Although it was later revealed that the Canadian spies had learned little of value about America's atomic projects, the spy scare did much to curb public support for civilian control of atomic energy. Under pressure from a frightened public and Congress, the McMahon committee, with only McMahon dissenting, decided to pass a Vandenberg-sponsored amendment establishing a military liaison board that would review decisions of the civilian atomic energy commissioners and appeal them as far as the President if necessary. McMahon complained in vain to Truman that the revised bill gave the military the sole veto power in the commission. Truman, who alone could have stopped the spy scare which was destroying any chance of civilian control of atomic energy, remained aloof from the struggle, and the revised bill was enacted as the Atomic Energy Act of 1946.[44]

Rather than blaming Truman for the defeat of the original McMahon bill, however, Henry Wallace placed the responsibility squarely on General Groves. Said Wallace: "Groves could use the plea of protecting against Russian spies to do anything he wants to . . . to almost any excess." Years later, when McNaughton's private papers were opened to researchers, they revealed that the "confidential source" he cited in his spy reports was none other than Groves. The blossoming of the spy scare, which Groves shrewdly manipulated to ensure a prominent military voice in the nation's domestic atomic energy program, when combined with the marked deterioration of Soviet-American relations during the winter of 1945–1946, did much to discourage public support for international control of atomic energy. By midsummer of 1946 the attitude of the American people, one congressman reflected, was "to lock the bomb in a burglar-proof safe and not discuss it."[45]

The Acheson-Lilienthal Plan

On January 24, 1946 the General Assembly of the United Nations meeting in London passed a resolution identical to the one accepted by the two super-powers at Moscow in December. It created a U.N. Atomic Energy Commission, consisting of the member nations of the Security Council plus Canada. It was authorized to make proposals for exchanging basic scientific information, confining atomic energy to peaceful purposes, eliminating atomic and other weapons of mass destruction from national armaments, and effectively safeguarding the complying states. Like the Moscow proposal and the Three Nation Declaration, the U.N. resolution was specific in requiring a gradual step-by-step process.[46]

Before boarding a plane bound for the London U.N. session, Secretary of State Byrnes announced that he had appointed a special five-member committee "to study the subject of control and safeguards necessary to protect the government." The members were Acheson (the committee chairman), Vannevar Bush, James Conant, General Groves, and John McCloy, the former assistant secretary of war. Byrnes expected the committee to produce an American plan on the control of atomic energy that he could eventually present to the U.N. Atomic Energy Commission.[47]

Since Acheson knew little about the technical aspects of atomic energy, he appointed a Board of Consultants to handle those matters. The board was chaired by David Lilienthal, the former head of the Tennessee Valley Authority and the first director of the newly created Atomic Energy Commission. Other members were J. Robert Oppenheimer; Charles Barnard, the head of New Jersey Bell Telephone; Harry A. Winne, vice-president in charge of engineering at General Electric; and Dr. Charles Thomas, vice-president of Monsanto Chemical Company. Like the Interim Committee, the Acheson and Lilienthal groups drew their membership primarily from America's business establishment. And like their predecessor, their deliberations were shielded from the scrutiny or influence of congressional or public opinion.[48]

In its deliberations, the Lilienthal group rejected a number of proposals for controlling atomic energy. One of the first to be discarded was an international agreement outlawing atomic weapons. Explained Lilienthal, "there was no security whatever for people everywhere, no prospect of a moment's freedom from fear of an atomic armament race if this is all we had to offer." Also rejected was Lilienthal's suggestion to prohibit all atomic development, civilian as well as military. The industrialists on the board were simply too excited by the commercial and humanitarian prospects of atomic energy to permit its total abolition. Nor did Lilienthal's group have much faith in a system of international inspection as the sole solution to the problem. From Oppenheimer they learned that it was possible to divert peaceful applications of atomic energy to military purposes, and that it would be possible, no matter how stringent an international inspection system, to conceal such diversion.[49]

It was Oppenheimer who proposed the solution that the Lilienthal board finally accepted. He envisioned an international agency, an Atomic Development Authority (ADA), that would emphasize the development of atomic energy rather than its control. It was based on the assumption, as Oppenheimer saw it, that atomic processes must be separated into safe and dangerous activities. The dangerous activities included the acquisition of raw materials, production of fissionable materials, and the manufacture of the mechanical components of atomic weapons. These would be owned and operated by the ADA. The so-called safe aspects of atomic energy—commercial power plants, research and medical facilities—could remain under national control. While the problem of diverting these peaceful aspects of atomic energy for military activities would still remain, Oppenheimer thought it could be ameliorated in at least two ways. First, fissionable materials could be "denatured"; that is, by adding a less fissionable isotope of uranium to highly fissionable uranium-235 and plutonium, the latter two elements could be rendered useless for bomb production. Second, states could be dissuaded from developing their own atomic energy installations by providing them with virtually unlimited access to safe technology and materials owned and operated by the ADA.[50]

While the Lilienthal board had deemphasized the importance of international inspection during its deliberations, the plan they accepted did require access by the ADA to all plants and materials it owned or operated. In addition, while the Lilienthal plan made no specific provisions for punishing violators of the ADA's regulations, the agency would nevertheless possess the means to retaliate against offenders by having the power to withhold raw materials from any country participating in the plan. At the insistence of the Acheson committee, the Lilienthal plan incorporated the concept of stages first proposed by Bush in November. The first stage called for an international survey of raw materials, with some disclosure of atomic information. Only in the last stage would America surrender her atomic monopoly.[51]

Although the Acheson and Lilienthal groups portrayed their plan as a generous offer, in actuality, like the Bush plan of November, it was very detrimental to the Soviet Union. By requiring the early surrender of Soviet atomic raw materials and facilities to ADA ownership, the Acheson-Lilienthal Plan, in effect, would have deprived the Soviets of the capability of producing atomic weapons long before the United States would be required to dismantle its own atomic arsenal. In addition, an ADA staffed primarily with Americans, who alone were considered by the Acheson group to have the proven atomic know-how, would expose the Soviet economy to the possibility of capitalist interference. By attempting to minimize the possibilities of a diversion of civilian atomic energy to military purposes, the ADA could effectively retard or cripple the development of all atomic energy in the Soviet Union. Moreover, opening the Soviet Union to American inspectors would cause Soviet leaders to fear that Soviet military capabilities—and vulnerabilities—would be revealed to American military planners. On the other hand, Soviet inspectors could not hope to gain much more in the way of American military information than they were already obtaining through the American news media.[52]

While the Acheson-Lilienthal Plan required the Soviets to rely on America's good faith that the United States would indeed surrender its atomic arsenal to the ADA in the final stage of the plan, the very idea of stages implied that the Soviets could not be trusted. While the Acheson-Lilienthal groups were genuinely hopeful that the Soviets would accept their plan, they were also aware that Soviet acceptance would be tantamount to another Russian Revolution.[53]

The Baruch Plan

To present the Acheson-Lilienthal Plan to the United Nations, President Truman on March 18, 1946 appointed Bernard Baruch as American ambassador to the U.N. Atomic Energy Commission. The appointment of the 75-year-old businessman and close confidant to several presidents in no way enhanced the likelihood that the Soviets would accept the Acheson-Lilienthal Plan. Indeed, there was every indication that the plan, at least in its original version, would not be presented at all. Baruch accepted the U.N. position only after Truman had assured him that he would not be simply a "messenger boy" for the Acheson-Lilienthal group, and that he would have a major role in defining the final American proposal.[54]

Lilienthal was "quite sick" after he received the news of Baruch's appointment. "We need a man who is young, vigorous, not vain, and who the Russians would feel isn't out simply to put them in a hole, not really caring about international cooperation," Lilienthal confided to his journal. "Baruch has none of these qualifications." Lilienthal probably cringed when Baruch said, "I knew all I wanted to know [about the atomic bomb]. It went boom and it killed millions of people." Oppenheimer flatly refused to serve as Baruch's scientific adviser. He was offended by Baruch's lack of understanding, his eagerness to revise the Acheson-Lilienthal Plan, and his readiness to speak "about preparing the American people for a refusal by Russia."[55]

But Truman was not primarily concerned about Baruch's technical qualifications. What he wanted most was a man who would assure the Congress and the American people that the Truman administration was not about to give away the nation's atomic secrets. And Baruch was that man. Vandenberg was so pleased by Baruch's appointment that he saw no need to have him testify before the Senate Foreign Relations Committee before his confirmation. All Baruch had to promise, which he did, was that there would be no disclosure of atomic information before a dependable system of safeguards was established, and that all agreements reached with the Soviets would be submitted to the Senate.[56]

Six weeks later, Baruch completed his work on the Acheson-Lilienthal Plan.

Although there were not many revisions in his version of the plan, they were major. Baruch proposed to deny the ADA control of the mining and refining of fissionable materials, believing private industry could perform those functions more efficiently. Second, in a test of "Russia's good faith," he called for a survey of raw materials before the implementation of international controls. Third, he emphasized the punishment, including atomic attack, of violators as the "heart" of his plan. The people of the world, he said, want "an international law with teeth in it." Fourth, Baruch demanded that the Security Council veto could not apply to ADA matters. The Soviet veto, he argued, would render the ADA meaningless. Finally, Baruch expanded the scope of the negotiations to include not only atomic weapons but all weapons. "If we succeed in finding a suitable way to control atomic weapons," he declared, "it is reasonable to hope that we may also preclude the use of other weapons adaptable to mass destruction."[57]

Lilienthal's Board of Consultants opposed each of Baruch's major revisions. Harry A. Winne warned that the Soviets would interpret a raw materials survey as an attempt to ascertain whether the Soviet Union had enough resources to be dangerous before the United States agreed to international control of the atom. Acheson objected that Baruch's insistence on punishment needlessly emphasized the prospects of failure. It added nothing to the security of the arrangement, he believed, since any violation of the plan would cause nations to take their own action. Abolition of the veto was also deemed unnecessary for the same reason. No nation would permit a Soviet veto to hinder their reaction to a Soviet violation of ADA regulations. Raising the issue of total disarmament, Lilienthal believed, "would hopelessly confuse and mix issues and obscure the hope of working out something on the bomb."[58]

Nor did the Joint Chiefs of Staff have much enthusiasm for Baruch's proposals, although for fundamentally different reasons. In general, America's military chiefs were reluctant, because of the nation's postwar demobilization, to relinquish control of atomic weapons. Moreover, Admiral Chester Nimitz warned that the American people would not support automatic punishment "of nations for acts which do not directly concern the United States." He also pointed out "the incongruity [in the Baruch Plan] that the atom bomb is necessary to enforce an agreement to outlaw its use."[59]

Although Baruch was disappointed by the negative reaction of the Joint Chiefs, he found comfort in the attitude of General Groves. It was Groves who had recommended that the phrase in the Baruch Plan proposing "prompt and certain" punishment should be amended to "immediate and certain punishment." No doubt was to remain that use of atomic weapons was a legitimate form of punishment.[60]

The view that counted most, of course, was that of the President. Although Truman at first tried to moderate the Baruch proposals that would be blatantly unacceptable to the Soviets, he gave up the effort when Baruch threatened to resign if his plan were revised. On June 7 Truman assured Baruch that he would support the plan in its entirety.[61]

The Reaction to the Baruch Plan

On June 14, 1946 Baruch presented his plan to the U.N. Atomic Energy Commission, declaring "we are here to make a choice between the quick and the dead." Most American commentators considered the Baruch Plan an act of magnanimity. The United States was offering to surrender its atomic monopoly to achieve world peace. On the other hand, the Hearst papers, expressing the minority reaction, considered the plan a surrender "to foreign masters of the American secret of the atomic bomb." At the other extreme, *Pravda,* the organ of the Soviet Communist Party, regarded the Baruch Plan as a "product of atomic diplomacy." It considered the stages of transition called for in the plan so loosely defined that the United States, in effect, could determine the plan's schedule. "Why does the U.S. government want to continue production and storage of atomic weapons if this weapon is being forbidden?" *Pravda* asked. Why should the Soviet Union trust the United States when, by demanding inspection of Soviet facilities, the United States clearly did not trust the Soviet Union?[62]

In the eyes of the Soviets, and many others as well, the sincerity of the Baruch Plan was glaringly brought into question by the American atomic weapon tests that were conducted at the very time the United States was trying to negotiate their abolition. On July 1 the world's fourth atomic bomb (with a picture of actress Rita Hayworth pasted on its side) was detonated in the midst of a fleet of seventy-three retired warships off Bikini Atoll in the Marshall Islands. On July 25 another atomic bomb was exploded. Senators James Huffman and Scott Lucas had tried in vain to persuade the President to cancel the tests, fearing their negative impact on the U.N. proceedings. The best they could do was gain a six-week delay. The few other congressmen who questioned the wisdom of the tests were more concerned about the destruction of seaworthy vessels than they were about the political implications of exploding atomic weapons.[63]

The first formal Soviet response to the Baruch Plan came on June 19. Actually, it was a nonresponse. The Soviet delegate, Andrei Gromyko, totally ignored the Baruch proposals. Instead, he offered a plan that, in essence, consisted of two stages. The first called for an international convention to prohibit the production and use of atomic weapons. It would be followed within six months by an international agreement providing penalties for violations. In the second stage of the Gromyko Plan, two U.N. committees would be created. One would plan for the exchange of scientific information. The other would work on a system of safeguards against violations of the treaty. Gromyko also presented a draft of his proposed international agreement. It called for the destruction of all atomic weapons within three months of the treaty's ratification. The treaty would come into force after one-half of the nations who signed it also ratified it and after the treaty gained the approval of the U.N. Security Council. At the end of his speech to the U.N. Atomic Energy Commission, Gromyko stated flatly that the Soviet Union would accept no revision of the veto power as it applies to atomic energy issues.[64]

The Gromyko Plan was clearly unacceptable to America's leadership, as the Soviets, no doubt, anticipated it would be. By calling for nuclear disarmament and the sharing of atomic secrets before establishing international controls, the Soviet plan reversed the order of the Baruch Plan. In effect, the Soviets were asking the United States to surrender its nuclear advantage in exchange for a vague Soviet promise to participate in a system of international control. On July 5 Baruch formally rejected the Soviet plan and told the Soviet delegation that the United States would rather abandon the negotiations than submit an alternate plan. On July 24 Gromyko responded by rejecting the American plan and informing Baruch that "in time in America your plan will be seen to be unfair."[65]

By the end of July, then, it was obvious that the negotiations were going nowhere. For the Americans the only question that remained was how they were to be brought to an end. It had to be done, Baruch believed, in such a way that the onus of failure fell squarely on the Soviet Union. In November Baruch, with Truman's permission, decided to bring his plan to a vote in the United Nations before the end of the year. Soviet efforts to get Baruch to revise his plan were considered delaying tactics by the Americans and were rejected outright. On December 30 the Baruch Plan was finally brought to a vote. It passed, but with two abstentions—the Soviet Union and Poland. Five days later Baruch resigned, a hero in the eyes of an overwhelming majority of the American people.[66]

4

Truman, the Cold War, and the Hydrogen Bomb, 1947–1952

The Aftermath of the Baruch Plan

The failure of the Baruch Plan had important repercussions on the American attitude toward the Soviet Union. To most Americans the Baruch Plan, despite its stillbirth, was considered a major diplomatic victory for the United States. The Soviets, by rejecting the plan, strengthened the American impression that Russia was not only the primary obstacle to international control of atomic energy, but also to the peace of the world. Gradually, most Americans accepted the contention of the Truman administration that a diplomatic approach to the Soviet Union was bound to be futile; that the only thing the Soviets understood was force. The breakdown in Soviet-American relations that began here would have a major impact on the history of the nuclear arms race.

The Truman Doctrine

In 1947 Greece was added to the growing list of nations that witnessed a collision of American and Soviet interests after World War II. Occupied by the British after the war, Greece was overwhelmed by major problems of relief and reconstruction, paralyzed by an economy on the verge of collapse, threatened by hostile Balkan neighbors, and ruptured by a civil war between supporters of the right-wing government of Constantine Tsaldares, on one side, and a coalition of his opponents that included socialists, communists, and liberals on the other.[1]

In February 1947 Britain, hard-pressed by its own economic crisis, informed the United States that it could no longer bear the burden of trying to keep order in Greece. The Truman administration felt it had no choice but to assume the responsibility the British were about to surrender. Blaming all of Greece's problems on the Soviet Union rather than on the complicated internal factors that had produced them, State Department analyst Mark Ethridge warned that "if Greece falls to communism, the whole Near East and part of North Africa as well is certain to pass under Soviet influence." Another member of the department, George Kennan, insisted that the United States "confront the Russians with unalterable counterforce at every point where they show signs of encroaching upon the interests of a peaceful and stable world." He promised that if the United States pursued a policy of "long term, patient but firm and vigilant containment of Russia's expansive tendencies," the result would be "either the breakup or the gradual mellowing of Soviet power."[2]

The Truman Doctrine was the first application of Kennan's containment policy. In a joint session of Congress on March 12, 1947, Truman asserted that "it must be the policy of the United States to support free peoples who are resisting attempted subjugation by armed minorities or by outside pressure." The President requested congressional approval for $300 million in aid for Greece and $100 million for Turkey to help them meet the communist challenge. Before doing so, however, he accepted the advice of Senator Vandenberg that it first would be necessary to "scare the hell out of the American people" in order to get the funds from a cost-conscious Congress opposed to excessive foreign expenditures. Accordingly, Truman declared that aid for Greece and Turkey was only a part of a global struggle "between alternative ways of life," and that the "fall" of these nations to communism would produce similar results elsewhere.[3]

The Marshall Plan

In addition to the Truman Doctrine, the other component of the containment strategy that emerged in 1947 was the Marshall Plan. Named after George C. Marshall, who succeeded James Byrnes as secretary of state in January of that year, the plan was designed primarily as a massive economic aid program (over $12 billion by 1952) to rebuild war-torn Europe. But the Marshall Plan also had definite, even if tacit, anti-Soviet undertones. The economic recovery of Europe, its authors' realized, would help ensure that Western Europe remain politically stable, sufficiently conservative to protect America's European economic investments, and, as a result, less susceptible to Soviet pressure.[4]

While the Soviet Union and its Eastern European satellite states were not specifically excluded from participation in the Marshall Plan, it was soon apparent to the Soviets that their involvement would seriously compromise Soviet economic and political interests. In return for American economic assistance under the Marshall Plan, the Soviets would be expected to expose their economy and, by association, their political structure to American penetration and influence. As a result, the Soviets appeared to fear the Marshall Plan more than the American atomic monopoly. Accordingly, Stalin responded to the American economic assistance program in the only way he could, short of surrendering the Soviet stranglehold on Eastern Europe. On July 2, 1947 the Soviet Union rejected participation in the Marshall Plan and subsequently pressured its satellites to follow suit. On October 5 the Soviets announced their own alternative to the American program, the so-called Molotov Plan.[5]

The rival Marshall and Molotov plans did much to freeze the military division of Europe into contending political and economic spheres of influence. In the West, with economic assistance from the Marshall Plan, democratic governments at least tolerant of free enterprise principles were either revitalized or created for the first time. In the East, the Molotov Plan served as the basis for the creation of COMECON, the instrument by which the economies of Eastern Europe were welded to that of the Soviet Union. The economic regimentation of Eastern Europe by the Soviet Union was also accompanied by intensified political repression, now that the Soviets saw no further need for continuing any semblance of democracy in the occupied states. By the spring of 1948, with a communist coup that brought Czechoslovakia firmly into the Soviet camp, the last vestiges of democracy vanished in Eastern Europe.

The German Problem

While the Marshall and Molotov plans helped to widen the fractures in the wartime Grand Alliance, the issue that completed its breakup, the same issue that had brought the alliance into being in the first place, was Germany. At Potsdam, the Big Three had decided that Germany would remain one economic entity in spite of its division into military occupation zones. The decision was ignored by the French, who ruled their zone independently of the other allied zones, both to keep Germany weak as well as to facilitate payment of reparations from current German production. Like the French, the Soviets wanted to keep Germany weak; unlike the French, the Soviets believed that it would be easier to collect reparation payments (totalling $10 billion for the Soviet Union) from a unified Germany than one divided into separated economic entities.[6]

American policymakers, prompted largely by Stalin's refusal to permit the reunification of Germany along democratic lines, came to see his support for a unified Germany as a Soviet plot to dominate the entire country. Beginning in 1946 the Truman administration initiated a series of policies designed to prevent that. In May of that year, German reparations to the Soviet Union from the American and British occupation zones were halted because, as British Foreign Minister Ernest Bevin explained, the Soviets "would loot Germany at our expense" and use the proceeds to augment Soviet military power. On September 6, 1946 Secretary of State Byrnes promised the German people self-government, suggested that the cessation of German territory to Poland would not be permanent, and indicated that American troops would remain in Germany indefinitely. To establish a strong barrier to Soviet expansion, the United States and Britain took the first steps toward creating a West German state. On January 1, 1947 the American and British occupation zones were fused into one administrative entity. In February 1948 the Western powers prepared West

Germany for participation in the Marshall Plan by instituting a program of currency reform in the Western occupation zones. In addition, the Western powers, without the concurrence or even the participation of the Soviet Union, agreed to convene an assembly that would draft a constitution for a West German state.[7]

The Soviets quickly realized the significance of these Western measures. The creation of a West German state tied militarily, politically, and economically to the enormous power of the United States, they believed, would constitute a new German menace to the Soviet Union. To forestall that development, the Soviets applied pressure at the only place they could, the divided city of Berlin, 125 miles deep inside the Soviet zone. Taking advantage of the absence of a formal agreement concerning the land access rights of the Western powers to Berlin (only Western air and railroad access rights had been specifically spelled out and recognized by the Soviet government), the Soviets began to restrict ground travel to West Berlin in March 1948. On June 24 the so-called Berlin blockade began when the Soviets responded to the implementation of the West German currency reforms by halting all Western land traffic to the city.[8]

The Truman administration regarded the Soviet blockade of Berlin as a test of the West's determination to defend the freedom not only of West Berlin but of all Western Europe. As a result, the administration took vigorous countermeasures against the Soviet Union. Traffic into West Berlin from the Soviet zone was halted. Moreover, a monumental airlift of supplies into West Berlin was initiated, enabling the city to withstand the Soviet stranglehold. And, in an obvious demonstration of America's atomic power, sixty B-29s were dispatched to Britain by the President with the approval of the newly created National Security Council. Although the B-29s were called "atomic bombers," they carried no atomic weapons, nor were they equipped to do so. Nevertheless, the threat of nuclear action against the Soviet Union, previously tacit, was for the first time made explicit.[9]

It was not the threat of nuclear devastation, however, that moved the Soviets to see the futility of their blockade of Berlin. Rather than forestalling the formation of a unified West German state, the blockade probably hastened it. Rather than bringing West Berlin to its knees, the Soviet blockade produced an economically painful Western counterblockade of the Soviet zone. Moreover, the blockade was a monumental propaganda defeat for the Soviet Union, for it served to reinforce the hardline interpretation of Soviet intentions. It was also a major military blunder because it contributed directly to the creation the following year of an anti-Soviet military alliance in Western Europe, the North Atlantic Treaty Organization (NATO). Realizing by May 1949 that the blockade had lost all beneficial meaning, the Soviets brought it to an end after the West agreed to lift its own counterblockade.[10]

The Berlin blockade proved to be another watershed in the history of the Cold War. Its failure destroyed the last hope for rapid German reunification. The Soviet inability to forestall the creation of a West German state (the Ger-

man Federal Republic) in 1949 prompted the Soviets to establish a puppet state
in East Germany (the German Democratic Republic) the same year. The division of Germany, in turn, sealed the postwar division of Europe into rival
American and Soviet spheres of influence.

Deterrence

The Berlin blockade also completed the transformation of America's military
and diplomatic strategy toward the Soviet Union that had begun with Truman's
assumption of the presidency in 1945. At the time of the Yalta Conference in
February 1945, Soviet objectives were seen by Americans as essentially defensive. By 1948, however, a National Security Council study, NSC-20, saw
the Soviet goal as nothing less than domination of the entire world. America's
primary objective, the study concluded, must be one of reducing "the power
and influence of Moscow" by all means possible: including the "liberation" of
Eastern Europe, the dismantling of the Soviet military establishment, and the
dissolution of the Soviet Communist Party. While NSC-20 believed that these
goals could be achieved without force, it did not preclude the possibility of
war.[11]

The decision contained in NSC-20 to rely on the atomic bomb and air power
as the primary means of waging war with the Soviet Union was made in the
absence of any perceived alternatives. Neither the American people nor Congress were prepared to pay the price of maintaining strong, conventional deterrent forces. In 1947 the Congress rejected Truman's plan to establish Universal Military Training (a peacetime draft). The administration, determined to
slash military spending in order to curtail inflation and fund pet domestic programs, was also opposed to expanding America's conventional forces.[12]

One effect of the budgetary constraints on defense spending was increased
interservice rivalry. In winning the role of the nation's primary deterrent force,
the Air Force first had to overcome not only the cost-conscious mentality of
the President and Congress, as well as the general apathy of the American
public toward defense-related issues, but also the considerable opposition of
the Navy, which foresaw a diminished role for itself if the Air Force got its
way. While the Air Force attempted to establish the B-36 bomber as the mainstay of the nation's atomic deterrent force, the Navy sought to emphasize the
deterrent value of carrier-based aircraft equipped with nuclear weapons. Both
services not only lobbied intensively, but also resorted to scare tactics, including exaggerated estimates of Soviet intentions and capabilities, to win support
for their respective strategies. In the end, the Air Force prevailed over the
Navy. In the summer of 1947 the President's Air Policy Commission, chaired
by Thomas Finletter, a Philadelphia lawyer and a former assistant secretary of
state, issued a report calling for a nuclear deterrent strategy based primarily on
the atomic bomb delivered by the long-range bomber.[13]

The adoption of a nuclear deterrent strategy by the Truman administration revolutionized America's approach to war. In the past, Americans had generally prepared for war only after a war had begun. Now, the advocates of deterrence argued, the United States would have to prepare for war in advance of war, in order to prevent war. However, it was never quite clear how much force would be needed to deter the outbreak of war, an ambiguity that would ultimately do much to stimulate the production of nuclear weapons.

The Finletter Report was also the first study to specifically emphasize the deterrent value of atomic weapons targeted on enemy cities rather than military installations. "Fleetwood," a war plan drawn up in the fall of 1948, proposed to drop the entire American arsenal of atomic bombs on the cities of the Soviet Union in the first month of war. Moscow, itself, would be the target of no fewer than eight atomic bombs; Leningrad would be the target of at least seven. If the war continued for as long as two years, approximately two hundred bombs would be dropped on the Soviet Union, obliterating an estimated forty percent of Soviet industry and killing seven million people. The plan had no conception of limiting the war or of achieving a negotiated end to the hostilities, as NSC-20 had anticipated. It simply assumed that, if deterrence failed, not only diplomacy but even strategy, itself, would end.[14]

The United States also intended to rely on the atomic bomb and air power to deter and, if necessary, to defend against a Soviet attack on Western Europe. Despite the inconclusive effect of the "bombers to Britain" move during the Berlin crisis, another National Security Council study, NSC-30, concluded that only the atomic bomb, considering the nation's antipathy to conventional forces, could deter the Soviet Union from overrunning the continent. The Soviets, the study emphasized, "should . . . never be given the slightest reason to believe the United States would even consider not [using] atomic weapons against them if necessary." It was advice Truman felt he had no alternative but to accept.[15]

The Embryonic American Nuclear Arsenal

Even though the United States was now formally committed to the use of atomic weapons, Truman was shocked to discover, two years after Hiroshima, how small the American nuclear arsenal actually was. On April 3, 1947 David Lilienthal, chairman of the Atomic Energy Commission, informed him that the United States had the components of only seven nuclear weapons, only one of which was "probably operable"—too little, it was assumed, to either deter or defend against a Soviet attack on Western Europe.[16]

To a great extent, the small number of nuclear weapons resulted from the difficulty in acquiring weapons-grade uranium and plutonium. That problem was greatly alleviated in January 1948, when the British agreed to supplement American uranium stocks. It was also ameliorated by the construction of new reactors that would increase the supply of plutonium. As a result of the SAND-STONE atomic weapons test series in the spring of 1948, a much more efficient method of using fissionable materials in nuclear explosives was developed, making it possible to increase the total number of weapons in the American stockpile by sixty-three percent and their yield by seventy-five percent. Over the next two years, because of these and other measures, the size of the American nuclear arsenal increased dramatically. By June 1950 the United States possessed almost 300 atomic bombs.[17]

The Pentagon, however, wanted more nuclear weapons. As pointed out in a 1949 report of a committee headed by Air Force Lieutenant General Hubert R. Harmon, the Pentagon feared that the Air Force would be unable to deter Soviet aggression in the Middle East and in Europe with only a few hundred atomic bombs. The report prompted the Joint Chiefs of Staff to request a substantial increase in bomb production. Yet Lilienthal suspected that the request of the Joint Chiefs was based less on strategic need than on the military's desire to have as many atomic weapons as possible.[18]

Lilienthal was able to persuade Truman to delay approving the Joint Chiefs' request for increased atomic bomb production, since the President was already concerned that the cost of expanding atomic weapons production would wreck the budget. At the same time, Truman was becoming sensitive to the rising doubts about the heavy emphasis on the air-atomic strategy—the decision of the Air Force to put "all its eggs in one basket" with the long-range bomber. Realizing that the Air Force's strategy deprived other armed services of needed funds and also ruled out other, less drastic options in the event of war, Truman feared that an air-atomic strategy meant total war involving the indiscriminate use of atomic weapons.[19]

Others expressed the same concern. In a highly publicized speech in January 1949, Rear Admiral Daniel Gallery called nuclear deterrence a "strategy of desperation and weakness" and argued that "we should abandon the idea of destroying enemy cities one after another until [the enemy] gives up and find some better way of gaining our objective." George Kennan was also having second thoughts about the bomb, fearing the psychological impact both at home and abroad of accelerating the production of atomic weapons. He believed "that it might be to the nation's advantage if it were decided that the bomb would never be used"; that the only purpose of nuclear weapons was to deter their use by the other side.[20]

In 1949, however, there was no possibility that the United States would abandon a strategy based on nuclear weapons. On April 7 of that year, three days after the NATO pact was signed, a Pentagon committee ruled out any plan for a protracted land war with the Soviets in Europe. The plan the Pentagon approved called for "delaying action in which the possibilities offered by air forces and mass destruction weapons would be put to their full use." This strategy would not only enable the administration to reduce military expenditures, it would preclude the need for a peacetime draft. The atomic bomb, in effect, had replaced a conventional force deterrent. In a September 1949 National Security memorandum, NSC-57, the Pentagon's atomic strategy was ratified, sanctioning the first use of nuclear weapons by the United States in the event of Soviet aggression against Western Europe.[21]

The Soviet Atomic Bomb

The planning of America's military strategy during the first half of the Truman presidency took place with little regard for the prospect that the Soviets would be able to develop their own nuclear weapons in the near future. The Finletter Commission considered January 1, 1953 the earliest practical date the United States could fear a Soviet nuclear attack. Needless to say, Truman and his advisers were shocked when a B-29 flying over the South Pacific on September 3, 1949 reported that it had detected a higher than normal radioactivity count. It indicated that, at some time in late August (the 29th, it was later determined), the Soviets had exploded an atomic device.[22]

On September 23, after a special committee verified the evidence of the Soviet explosion, Truman broke the news to the American people that the Soviet Union probably had an atomic bomb. He assured the nation, however, that the Soviet accomplishment had been anticipated by the administration and that adequate countermeasures had been prepared. Lilienthal labeled all of this "bunk," realizing that instead of a rational approach to the Soviet atomic bomb, either before or after it had become an established fact, the administration would simply decide to approve the Joint Chiefs' request for an expanded atomic weapons program. Lilienthal was upset by the reversal. "We keep saying," he recorded in his journal, "'We have no other course.' What we should say is, 'We are not bright enough to see any other course.'" The only consolation Lilienthal could glean from this decision was his expectation that at least the increased production of atomic weapons would take the wind out of the effort to develop a far more destructive weapon, the hydrogen bomb. It was an expectation that was not to be fulfilled, for there were forces other than the military that were pushing for the development of a hydrogen bomb.[23]

Teller's Campaign for the Hydrogen Bomb

The main impetus behind the H-bomb came from Edward Teller, a Hungarian refugee and a brilliant physicist who had worked on the Manhattan Project. According to Teller, in early 1942 Enrico Fermi asked him "whether the high temperatures that were expected to occur in an atomic bomb could be utilized to start reactions similar to those proceeding in the Sun?" What Fermi was speculating about was the possibility that, at temperatures on the order of one hundred million degrees, the nuclei of hydrogen atoms could be fused together to form helium, a process that would release enormous amounts of energy, nearly three times the energy per unit weight produced by uranium fission. After concluding that the production of a hydrogen bomb was possible, if sufficient emphasis were given to the project, Teller made its development a personal obsession. Because the development of the atomic bomb demanded almost the full attention of the Los Alamos Laboratory, little could be done toward developing a hydrogen bomb during the war.[24]

Even after the war, research on the hydrogen bomb continued at only a snail's pace, primarily because the American atomic monopoly provided little incentive for developing thermonuclear weapons. Teller felt the delay was inexcuseable. "If the Los Alamos Laboratory had continued to function after Hiroshima with a full complement of such brilliant people as Oppenheimer, Fermi and [Hans] Bethe," he argued, "I am convinced that . . . we would have had the hydrogen bomb in 1947 instead of 1952." Norris Bradbury, who succeeded Oppenheimer as director of the laboratory, was not the only one of Teller's colleagues who disagreed with this assumption. Bradbury believed that a crash hydrogen bomb project in 1946 would have been a waste of time: "We would have spent time lashing about in a field in which we were not equipped to do adequate computational work. . . . We would have spent time exploring by inadequate methods a system which was far from certain to be successful." According to Hans Bethe, who was one of the first to conceive of the possibility of producing a thermonuclear reaction, Teller urged the crash project on the basis of calculations that proved to be incorrect. "Nobody will blame Teller because the calculations of 1946 were wrong," Bethe wrote. "But he was blamed at Los Alamos for leading the laboratory, and indeed the whole country, into an adventurous program on the basis of calculations which he himself must have known to have been very incomplete."[25]

Minds less scrupulous than Teller's would develop his argument that the H-bomb could have been built sooner than it was into a charge that the delay was premeditated. In this view, the man most responsible for the delay was Oppenheimer. On the day Truman announced that the Soviets had exploded an atomic weapon, Teller called Oppenheimer to ask what could be done to meet the Soviet challenge. Oppenheimer responded coolly, "Keep your shirt on." In fact, Oppenheimer was, by this time, dead-set against developing the hydrogen bomb. On October 21, 1949 he wrote James B. Conant that "the Super [the hydrogen bomb] is not very different from what it was when we first spoke of it more than seven years ago: a weapon of unknown design, cost, deliverability and military value."[26]

Undismayed by Oppenheimer's opposition to the project, Teller struck out on his own. He soon won the support of some very influential people. One was Senator Brien McMahon, the chairman of the Joint Committee on Atomic Energy. Prodded by William Borden, head of the Joint Committee's staff and the man who four years later would accuse Oppenheimer of being a Soviet agent, McMahon could become nearly fanatic about the subject of thermonuclear weapons. To the senator, the hydrogen bomb was a way of producing more destructive capability for less cost than fission weapons. Moreover, he viewed the hydrogen bomb as the only effective way of countering the Soviet Union's enormous ground forces.[27]

In weighing the potential value of the hydrogen bomb, the Joint Chiefs of Staff placed great emphasis on its psychological impact. "Possession of a thermonuclear weapon by the USSR without such possession by the United States would be intolerable," argued General Omar Bradley, chairman of the Joint Chiefs. On the other hand, an American hydrogen bomb monopoly, the Joint

Chiefs told Truman in January 1950, "might have a sobering effect in favor of peace" and "might be a decisive factor if properly used" in war. There is no evidence that the Joint Chiefs gave any consideration to the prospect that a weapon as indiscriminately destructive as the hydrogen bomb would make meaningless America's strategy of limited nuclear warfare, which presumed that a Soviet-American war would end with a political settlement. Nor were they bothered by the moral objections to the bomb. They considered it "folly to argue whether one weapon is more immoral than another," for "in the larger sense, it is war itself which is immoral, and the stigma of such immorality must rest upon the nation which initiates hostilities."[28]

The Hydrogen Bomb Decision

In the fall of 1949, Truman submitted the question of developing the hydrogen bomb to the General Advisory Committee (GAC) of the Atomic Energy Commission (AEC). The committee, chaired by Oppenheimer, had as its primary function providing the AEC with scientific and technical advice. In a report completed on October 30, the GAC unanimously opposed the development of thermonuclear weapons. The moral factor was the paramount consideration behind its decision. The committee believed that the immense destructive force of the hydrogen bomb made its effects necessarily indiscriminate and "necessarily an evil thing considered in any light."[29]

But the GAC opposed the H-bomb proposal for practical as well as moral reasons. The United States, the committee believed, already had more than enough atomic weapons to deter Soviet aggression even if the Soviets alone developed a hydrogen bomb. Oppenheimer also felt that it would unnecessarily drain funds from another project, the development of tactical nuclear weapons, which he considered a more usable alternative to thermonuclear weapons. Oppenheimer was among the first to appreciate the potential consequences of a thermonuclear exchange. He believed, however, that "it would be folly to oppose exploration of this weapon" (the hydrogen bomb) because of the way it had "caught the imagination, both of the congressional and military people."[30]

Nevertheless, in a minority addendum to the GAC report, two members of the committee, Enrico Fermi and physicist I. I. Rabi, attempted to do just that. They proposed an international agreement prohibiting the development of thermonuclear weapons. The United States, the scientists recommended, should sign the agreement "conditional on the response of the Soviet Government." However, neither Fermi nor Rabi pressed their views on the matter. Later, Rabi expressed regret that he had let the proponents of the hydrogen bomb do all the lobbying.[31]

Before making a decision on the GAC report's recommendations, Truman turned for advice to a special committee of the National Security Council specifically created to deal with the hydrogen bomb. The members of the committee were Lilienthal, Secretary of Defense Louis Johnson, and Dean Acheson, who succeeded Marshall as secretary of state in 1949. Lilienthal considered the hydrogen bomb, like the atom bomb, simply a "cheap, easy way out." Believing that the fate of civilization hung on the H-bomb decision, he urged restraint while another attempt to secure international control of nuclear weapons was made. Louis Johnson, on the other hand, backed the hydrogen bomb wholeheartedly. "We want," he said, "a military establishment sufficient to deter [an] aggressor and sufficient to kick the hell out of her if she doesn't stay deterred." What was surprising was the change in attitude toward nuclear weapons that Acheson displayed. Even though he had fought hard for international control of atomic energy in 1946, Acheson's Cold War experiences since then had left him deeply pessimistic about the possibilities of achieving any meaningful agreement with the Soviets, particularly one prohibiting the development of thermonuclear weapons.[32]

On January 31, 1950 the special committee presented its recommendations to the President. In essence, it recommended that the United States continue work on the hydrogen bomb to determine its feasibility. Stricken from a draft of the committee's report was a paragraph calling for the President to defer a decision to produce thermonuclear weapons beyond the number required to determine their feasibility. In other words, the committee made no recommendation to limit the deployment of thermonuclear weapons.[33]

Apparently believing that further opposition was futile, Lilienthal reluctantly went along with the committee's recommendations. Nevertheless, he requested, and received, an opportunity to express the "grave reservations" he had about the committee's recommendation to produce the hydrogen bomb. He told Truman that "this course was not the wisest one and that another one was open"—increasing the production of atomic weapons and developing a greater variety of them, particularly for use in tactical situations. At the same time, he reiterated his view that another intensive effort should be made to negotiate an agreement providing for international control of atomic energy.[34]

Lilienthal's presentation made little or no impact on Truman. The President had obviously made up his mind before the meeting; it lasted only seven minutes. He cut short Lilienthal's presentation and did not even bother to read the report's supporting analysis. He simply asked, "Can the Russians do it?" When all the heads nodded affirmatively, Truman responded, "In that case, we have no choice. We'll go ahead." Recalling the meeting later, Lilienthal wrote that his effort to block development of the hydrogen bomb was like saying "no to a steamroller." The Atomic Energy Commission, he felt, had become "nothing more than a major contractor to the Department of Defense." Six weeks later, on March 10, after Truman was warned by the Pentagon that the Soviets might have several atomic bombs as well as a hydrogen bomb, he ordered the AEC to prepare for the production of hydrogen bombs.[35]

Secrecy and the Hydrogen Bomb Decision

Truman's decision to produce the hydrogen bomb was not based simply on the assumed capability of the Soviets to build thermonuclear weapons. As with the decision to develop the atomic bomb, many factors were responsible for the H-bomb decision. Budgetary considerations were important. Truman admitted that much of the fiscal pain of his hydrogen bomb decision was alleviated by an existing appropriation of $300 million in the Atomic Energy Commission budget for increased atomic bomb production. He simply shifted the funds to the H-bomb project.[36]

Another factor was public opinion or, more accurately, the absence of public opinion. There was no public, or even congressional, debate on the decision to develop the hydrogen bomb. And this is exactly what Truman wanted. The deliberations on the H-bomb were top secret, and everyone who participated in them understood that. Oppenheimer was one of the few who questioned the need for the secrecy. He stated that "wisdom itself cannot flourish, nor even truth be determined, without the give and take of debate or criticism. The relevant facts could be of little help to an enemy; yet they are indispensable for an understanding of questions of policy."[37]

One reason for the secrecy surrounding the H-bomb deliberations was the belief of Truman and his advisers that the public and the Congress were not competent to deal with the complex international issues of the postwar era. Moreover, the administration as well as most congressmen believed that it was vital to preserve unity—"bipartisanship"—in the face of the communist challenge; that criticism of the administration's foreign policy would only weaken the "free world."[38]

The administration also insisted on secrecy because it was sensitive to the charge that its security procedures were, at best, sloppy. On January 21, 1950, ten days before Truman announced his H-bomb decision, former New Dealer Alger Hiss was convicted of perjury for denying that he had given government secrets to a Soviet agent, Whittaker Chambers, in 1938. On February 2 Klaus Fuchs, the brilliant German-born physicist who had worked with British scientists on the Manhattan Project, was arrested in England for spying for the Soviet Union. There is no firm evidence to indicate that Truman was aware of the Fuchs case before he made his decision on the hydrogen bomb. But coming on the heels of the Hiss case, Fuchs' confession may have contributed to the administration's perceived need to keep the H-bomb deliberations secret, if not to the decision to produce the H-bomb itself.[39]

In addition, despite the substantial lead the United States enjoyed over the Soviet Union in the development of atomic energy, Americans were shocked and frightened by the explosion of the Soviet atomic bomb. It had a particularly terrifying impact on those Americans who believed the administration's repeated assertions that the American atomic monopoly was the sole guarantee of the nation's security. Many, if not most, Americans came to see the Soviet atomic bomb as less an achievement of Soviet science and technology than a result of poor security measures by the Truman administration. More than a few Americans agreed with the charge of Senator Joseph McCarthy, first made on February 9, 1950, and many times thereafter, that the administration's security procedures were lax because its whole approach to the communist "threat" was lax. The same administration, McCarthy asserted, that had permitted the

Soviets to "steal" the atomic bomb had also "surrendered" Eastern Europe to
the Soviets, and had "lost" China to the communists in 1949.[40]

According to historian Athan Theoharis, Truman primarily had himself to
blame for the success McCarthy had in attacking the administration's foreign
and security policies. The irresponsible rhetoric that characterized McCarthy's
crusade was first used by the President and his spokesmen. Truman repeatedly
resorted to "red-baiting" in attacking his opponents. In the presidential campaign of 1948, for example, he referred to Henry Wallace, the candidate of
the Progressive Party, "and his communists" as advocates of "peace at any
price" and ultimately as "appeasers" for favoring a diplomatic approach to the
Soviet Union. For Truman, by 1948, ideological vituperation had replaced diplomacy as the favored method of dealing with the Soviets. By then, for the
President, the Soviet Union was "a modern tyranny led by a small group who
have abandoned their faith in God," rather than a nation with which the United
States could do business.[41]

Alarmist rhetoric like this helped to create the political environment of
heightened insecurity and tension that surrounded Truman's decision to develop
the hydrogen bomb. Prodded by administration scare tactics, the public came
to regard maximum military pressure as the only way to deal with "godless"
communism. When Truman said he had no alternative but to build the H-bomb,
what he really meant was that a militarized public opinion, which he helped
to mold, left him no choice but to continue to rely on military methods in
dealing with the Soviet Union. His refusal to permit public debate before the
H-bomb decision, and the haste with which he made that decision, was in no
small way due to his realization that any attempt to block development of the
hydrogen bomb would have been political suicide.

NSC-68 and the Korean War

The same day that Truman approved the development of the hydrogen bomb,
he also set in motion a complete review of his administration's military strategy
toward the Soviet Union. The study that resulted, NSC-68, called for an enormous increase in American defense spending in order to prevent Soviet domination of the world. Ultimately, NSC-68 provided the rationale not only for
the funding of the hydrogen bomb, but for America's conventional rearmament
as well. Ironically, the administration's decision to rebuild America's conventional forces was a tacit admission that even the hydrogen bomb could not
guarantee America's security.[42]

The Korean War reinforced the growing realization that nuclear weapons
possessed only limited military and diplomatic value. Even when Chinese forces
were overrunning the American lines in late November 1950, the Truman
administration refused to condone the use of atomic weapons in Korea. One
reason for the administration's atomic inaction in Korea was the obvious un-
suitability of that land for atomic warfare. Korea possessed no military targets
large enough to warrant the use of atomic weapons. This realization, however,

did not stop General Douglas MacArthur, the American field commander in Korea, from recommending that atomic weapons be dropped in the Yalu River in order to halt the Chinese onslaught. "I've never heard anything so preposterous in my life," was the way General Omar Bradley reacted to MacArthur's idea.[43]

Although the administration rejected MacArthur's recommendation outright, the President did muse, privately, about the possibility of using atomic weapons to end the Korean War. Frustrated by the military stalemate into which the war had developed, Truman confided to his journal in January 1952 that an ultimatum to the Soviet Union and China, backed by the threat of using atomic bombs against them, might be the way to end the war. What stopped the President's musing was his all too clear realization that the use of atomic weapons against the Soviets and Chinese could produce a total war. And all-out war, he wrote, "means that Moscow, St. Petersburg [Leningrad], Mukden, Vladivostok, Peking, Shanghai, Port Arthur, Dairen, Odessa, Stalingrad, and every manufacturing plant in China and the Soviet Union will be eliminated." Fortunately for these cities, and much else besides, Truman's reason prevailed over his emotions. He not only refused to use atomic weapons during the Korean War, he also rejected a request by the Joint Chiefs of Staff for automatic permission to transfer the nonnuclear components of atomic bombs to "areas of danger" in time of crisis.[44]

MIKE

By 1952, with the Korean War dragging on, Edward Teller, with the assistance of mathematician Stanislaw Ulam, had surmounted the most formidable obstacle in the effort to develop the hydrogen bomb—generating sufficient heat and pressure to initiate and sustain a fusion reaction. He solved the problem with a design based on the fission-fusion-implosion principle. Teller's device contained a core of liquid fuel—the hydrogen isotopes, deuterium and tritium—surrounded by a liquid-hydrogen cooling agent. The core, in turn, was surrounded by a shell of fissionable material encased in a conventional explosive. When detonated, the conventional explosive would drive the fissionable material inward, compressing it into a critical mass and creating an atomic explosion. Compressed and heated by the atomic explosion, the deuterium and tritium fuel would fuse the hydrogen into helium, liberating enormous quantities of energy in the process.[45]

The first test of the Teller device, called MIKE, took place in the western Pacific, on the island of Elugelab, in the Marshall chain, on November 1, 1952. MIKE was quite literally an earth-shattering success. The explosion yielded an unexpected energy equivalent of ten megatons (ten million tons of TNT), an amount roughly one thousand times greater than the energy released by the bomb dropped on Hiroshima (some thirteen kilotons). When MIKE's massive fireball cleared the ground, the spectators were shocked to discover that the entire island, one mile in diameter, had disappeared.

5

Eisenhower and Massive Retaliation, 1953–1961

Eisenhower and the Nuclear Arms Race

When Dwight D. Eisenhower assumed the presidency in January 1953, he was in at least two respects better prepared for the office than his predecessor: he possessed both extensive diplomatic and military experience. In addition to having served as the supreme commander of the victorious Allied Forces in Europe during World War II, he also had displayed considerable talent in handling such irascible personalities as Winston Churchill and Charles de Gaulle. Perhaps this experience explains why Eisenhower, unlike Truman, considered the Soviet Union a challenge that could be managed rather than a menace that required an all-consuming vigilance. Nevertheless, despite the greater degree of flexibility with which Eisenhower approached the Soviet Union, he was not able to wind down the nuclear arms race. By 1961, when he left office, the explosive power of the American nuclear arsenal approached 30,000 megatons, the equivalent of ten tons of TNT for every person on earth.[1]

The Temper of the Times

A number of factors were responsible for the enormous growth of the American nuclear arsenal during the Eisenhower years. For one, the temper of the times did not seem to permit a diplomatic solution of the problems that divided the Soviet Union and the United States. Eisenhower's first year in office coincided with the zenith of McCarthyism, which cast aspersions on the loyalty of Americans who advocated a moderate approach to the communist world. While the President was immune from McCarthy's onslaught, he nevertheless was reluctant to challenge him openly or to curb the witch-hunt the senator conducted in the federal bureaucracy.[2]

J. Robert Oppenheimer, the architect of the atomic bomb, was among the victims of McCarthy's red scare. In 1954, as a result of an Atomic Energy

Commission investigation, Oppenheimer's security clearance was removed, which implied that he had colluded with communists. Although Oppenheimer did have personal relationships with American communists, the AEC investigation failed to prove that he had ever been disloyal to the United States. (A testament to his innocence was the restoration of his security clearance by President John F. Kennedy in 1963.) Rather than disloyalty, the prime reason for Oppenheimer's disgrace seems to have been his unrelenting opposition to the development of thermonuclear weapons. By fighting the decision to develop the hydrogen bomb, he had made enemies of influential people, including the newly appointed AEC chairman, Lewis Strauss, and Edward Teller, the most prominent scientist to testify against him in the AEC investigation. Oppenheimer's dismissal not only removed a voice of moderation in the formulation of America's nuclear strategy, it also may have intimidated others who would have otherwise challenged the assumptions on which the strategic arms race was based.[3]

The temper of the times was also reflected in the rhetoric of Eisenhower's first secretary of state, John Foster Dulles. In his public utterances, Dulles appeared to be a man obsessed with the menace of communism, insensitive to the diversity which existed within the communist world, and distrustful of the Soviets in particular—someone who feared that any attempt to negotiate a resolution of the problems that perpetuated the Cold War could only be detrimental to American interests.

However, Dulles' rhetoric was largely for domestic consumption, an attempt, historian Michael A. Guhin has written, to "dissociate himself from the Democrats [he had served in the Truman State Department] and their policies in order to retain a position for his views in the Republican Party," particularly its avid, anti-communist, right wing. The external communist menace served another purpose, which is seen in Dulles' admission "that in promoting our programs in Congress we have to make evident the international communist menace. Otherwise, such programs would be decimated."[4]

Emphasizing the menace of communism, Dulles believed, would also help to keep America's alliance relationships strong. The moment of greatest danger to the West, Dulles predicted, would occur when the perceived threat of communism began to fade. "Fear," he concluded, "makes easy the tasks of diplomats, for then the fearful draw together and seek the protection of the collective strength."[5]

To be sure, Dulles was sufficiently pragmatic, his ideological rhetoric aside, to realize that negotiations with the Soviets could serve America's interests. However, his definition of American interest was so broad, his distrust of communists so deep, and his public attacks on the Soviet Union so vitriolic, that he may have effectively precluded any meaningful thaw in the Cold War.[6]

Massive Retaliation

In formulating a military strategy toward the Soviet Union, the Eisenhower administration was greatly influenced by its perception of the Korean War.

According to Eisenhower, shortly after taking office, he let it be known through diplomatic channels that, as he put it, "we intended to move decisively without inhibition in our use of weapons" if the Communist Chinese did not agree to a negotiated settlement of the conflict. He suggested in his memoirs that his threat was instrumental in bringing the war to an end, through a negotiated settlement that was not unfavorable to the United States, with Korea divided roughly along the same line that existed before the conflict began.[7]

Although there is little evidence to indicate that Eisenhower's atomic threat was actually made, the Korean War nevertheless reinforced his belief that nuclear weapons should be the foundation of America's military strategy. The use of nuclear weapons in a localized war, he believed, could enable the United States to avoid a repetition of the enormous casualties it suffered in Korea. Moreover, emphasizing the nuclear deterrent would preclude the necessity of maintaining costly conventional forces and would thereby enable Eisenhower to fulfill a campaign pledge to reduce taxes by cutting the defense budget. Excessive defense spending, the President stressed repeatedly, undermined the nation's economic strength. A bankrupt America, he said, was more the Soviet goal than an America defeated on the battlefield. Nuclear weapons would give the United States, Defense Secretary Charles Wilson stated, "a bigger bang for the buck."[8]

The administration's deterrent strategy was unveiled publicly in a speech delivered by Dulles on January 12, 1954. He announced that it would be the intention of the administration to react massively, with nuclear weapons, in the event of communist aggression. In late 1954, the President told congressional leaders that the general idea was "to blow [the] hell out of them [the communists] in a hurry if they start anything." And the administration clearly did not intend to limit the use of nuclear weapons to strategic warfare. "The United States," Eisenhower stated, "cannot afford to preclude itself from using nuclear weapons even in a local situation, if . . . such use will best advance U.S. security interests."[9]

Both Eisenhower and Dulles were aware of the risks of threatening the use of nuclear weapons in response to localized communist aggression. But, stated Dulles, "you have to take chances for peace just as you must take chances for war." He added that "the ability to get to the verge without getting into war is the necessary art. . . . If you are scared to go to the brink, you are lost." Eisenhower and Dulles both believed that their willingness to threaten nuclear war would make it unnecessary to wage any war, nuclear or conventional.[10]

Nor was the administration necessarily bluffing in threatening the use of nuclear weapons. Not only did it ponder the use of atomic weapons in Korea, but also in Indochina in 1954 to break the Communist Vietminh seige of the French garrison in Dien Bien Phu. Commented Air Force Chief of Staff General Nathan Twining: "You could take all day to drop a bomb, make sure you put it in the right place. . . . And clean those Commies out of there and . . . the French would come marching out of Dien Bien Phu in fine shape." The National Security Council, however, acknowledged that use of tactical nuclear weapons could make Indochina "a battlefield of destruction on the Korean scale" and would depress America's allies by removing "the last hope that these weapons would not be used again in war." Nevertheless, the NSC said it would consider their employment "to keep Southeast Asia from falling under communist control and to preserve the principle of collective security." In the end, Eisenhower did not authorize the use of nuclear weapons in Indochina, partly because the British and the Congress did not support American military intervention in the war, partly because the Army insisted that the use of nuclear weapons would not preclude the need for American ground forces, and partly because the French government preferred a negotiated settlement of the conflict.[11]

The administration also considered the use of nuclear weapons in 1955, and again in 1958, when the Communist Chinese threatened to invade the Nationalist-held islands of Matsu and Quemoy, off the China coast in the Taiwan Straits. On March 10, 1955 Dulles told the President: "If we defend Quemoy and Matsu, we'll have to use atomic weapons. They alone will be effective against the mainland airfields." Eisenhower agreed, and publicized the fact that the use of nuclear weapons was being considered. Fortunately, the Chinese Communists ceased hostilities on May 22, before the administration's nuclear resolve could be put to the test. When the crisis flared again in August 1958, the administration again resorted to the nuclear threat, this time by deploying nuclear-capable howitzers on Quemoy. The action apparently was instrumental in dissuading the Chinese from attacking the off-shore islands, but it also may have been a major factor in strengthening the Chinese determination to develop their own nuclear arsenal.[12]

The New Look

To back up the "massive retaliation" strategy, as it was called in the press, the administration embarked on a military program designed to give the nation's armed forces a "New Look." It called for a major cut in conventional forces and a massive buildup of nuclear forces. As a result, Army strength was reduced from 1.4 million men to an even million between 1953 and 1957, while the Navy dropped from 765,000 men to 650,000. The cuts permitted the administration to reduce taxes by $7.4 billion in fiscal 1955. The size of the Air Force, on the other hand, expanded appreciably during the "New Look" years, from 913,000 men to 975,000, and from 110 wings to 137. In addition, in June 1953 the Air Force began ordering the nation's first intercontinental jet bomber, the B-52, with the capability of delivering hydrogen bombs on Soviet targets. All of this was a reflection of the fact that air power was going to be the primary component of the administration's massive retaliation strategy.[13]

While relying on the long-range bomber as the nation's primary delivery vehicle, the Eisenhower administration placed major emphasis on the development of ballistic missiles. After lightweight nuclear warheads were first successfully tested in 1954, the administration the following year decided to approve development of the Atlas missile, America's first intercontinental ballistic

missile (ICBM). The Atlas was designed to deliver a one-megaton warhead
5,500 nautical miles. In the same year, the administration also approved the
development of the nation's first intermediate-range ballistic missile (IRBM),
the Thor. It could carry a nuclear warhead 1,500 miles. In 1957 administration
approval was given to yet another Air Force ICBM, a solid-fueled missile called
the Minuteman. When Minuteman became operational in 1962, it replaced the
manned bomber as the primary component of the nation's strategic forces.[14]

Not to be excluded from a nuclear role, both the Army and the Navy demanded, and received, permission to develop their own ballistic missiles. The
Army's project, Jupiter, was an almost identical version of the Air Force's
Thor. The Army attempted to justify the duplication by insisting it needed a
missile similar to the Thor, but with a reduced diameter (supposedly, to permit
its passage through Swiss railway tunnels!). The Thor-Jupiter controversy was
a classic example of how interservice rivalries can fuel the nuclear arms race.
While the Army attempted to duplicate, without duplicating, the Thor, the Navy
concentrated on the development of a submarine-launched ballistic missile
(SLBM), the Polaris. Each Polaris was a solid-fueled missile designed to deliver a one-megaton warhead more than a thousand miles. The first Polaris
SLBMs were deployed aboard the *George Washington* in November 1960.[15]

Both the Army and the Navy were also successful in getting the administration to approve the nuclearization of their nonstrategic arsenals. Ground and
naval forces were equipped with low-yield (less than twenty kilotons of equivalent TNT) tactical nuclear weapons. Such weapons, Dulles said, "can utterly
destroy military targets without endangering unrelated civilian targets." Moreover, tactical nuclear weapons were seen as a relatively inexpensive way to
offset the perceived Soviet superiority in conventional forces. For this reason,
NATO agreed in December 1954 to integrate tactical nuclear weapons, including atomic cannons, missiles, and even land mines, into its defense system.
NATO ground forces would remain an integral part of the alliance's defenses,
but they would really serve as the "trip wire" for triggering the use of nuclear
weapons rather than as the primary means of defending Western Europe.[16]

The Attack on the New Look

The Eisenhower "New Look" was bombarded with criticism almost from its
inception. More than a few defense analysts, like then Harvard professor Henry
Kissinger, believed that the explosion of a Soviet hydrogen bomb in August
1953 made the prospect of applying massive retaliation to resist local communist aggression a suicidal proposition. Since neither side, he pointed out,
had any agreed format for limiting a nuclear war, even the use of tactical nuclear weapons could quickly escalate to an all-out nuclear exchange. Threatening to use nuclear weapons one could not safely use, Kissinger argued, would
be interpreted by the Soviets as nothing more than a bluff. Moreover, as the
United States became more and more vulnerable to direct atomic attack by the
Soviet Union, the credibility of the American nuclear deterrent could disappear

altogether. Faced with the same prospect, some Europeans were coming to believe that the United States would not be inclined to risk the destruction of New York in order to save Paris. Two results were the British decision to develop a hydrogen bomb and the French decision to develop their own nuclear arsenal.[17]

Army generals, such as Matthew Ridgeway, Maxwell D. Taylor, and James M. Gavin, argued that the New Look dangerously sacrificed the Army's ability to deter local, conventional wars. The ineffectiveness of nuclear weapons as a deterrent against localized aggression, they believed, was clearly demonstrated by the administration's refusal to allow the use of nuclear weapons against the communists at Dien Bien Phu.[18]

Nor did the increased emphasis on nuclear weapons in any way assist the "liberation" of Eastern Europe, one of Dulles' goals, even when the East Europeans demonstrated their eagerness to throw off the Soviet yoke. The limitations of the American nuclear arsenal were graphically demonstrated by the administration's essentially passive reaction to the East Berlin uprising in 1953 and the Hungarian Revolution three years later. The fact is the United States was not prepared to risk war with the Soviet Union, especially after the Soviets had acquired an appreciable nuclear arsenal, in order to undermine Soviet control of Eastern Europe.[19]

Both the President and his secretary of state vigorously defended their massive retaliation strategy in the face of the criticism it met. The use of nuclear weapons, Dulles stated, would not always be an automatic response to communist aggression. Implying that some areas were not vital to the United States and therefore not worth defending, he admitted that there could be "setbacks to the cause of freedom." The idea, Dulles stated, was to "keep the Reds guessing" about what interests would be defended.[20]

However, as the fifties came to a close, the criticism massive retaliation received, and the defensive posture the administration was forced to adopt as a result, gradually lessened the President's enthusiasm for it. The threat to use nuclear weapons, Eisenhower admitted, had little impact on guerrilla activities in the Third World. Moreover, it tended to frighten the countries on whose territory nuclear weapons would be used. Furthermore, as the Soviet nuclear arsenal expanded, the threat to use any nuclear weapon, strategic or tactical, became riskier. "If the Kremlin and Washington ever lock up in a war," Eisenhower said in 1954, "the results are too horrible to contemplate. I can't even imagine them." In October 1960, however, the President was compelled to admit that "we were unfortunately so committed to nuclear weapons that the only practical move would [have been] to start using them from the beginning without any distinction betweem them and conventional weapons."[21]

The Bomber Gap

The Soviets were not enamored by the administration's increased emphasis on nuclear weapons. Nikita Khrushchev, who ascended to the pinnacle of Soviet power in 1955, considered Dulles' brinksmanship strategy nothing but "bare-faced atomic blackmail." Yet what the Soviet leader feared more than the diplomatic consequences of the American nuclear buildup was the capability it gave the United States to destroy the still embryonic Soviet nuclear force. As a result, the Soviets pressed ahead with their own nuclear weapons program. By 1955 they had amassed a nuclear arsenal of some 300 to 400 atomic and thermonuclear weapons.[22]

Americans, in general, overreacted to the growth of the Soviet nuclear arsenal. When the first Soviet intercontinental bombers appeared in the July 1955 Moscow air show, they flew in "waves" that prompted U.S. Air Force spokesmen to claim that the Soviets would soon have an impressive lead over the United States in bombers unless American air power was augmented rapidly. The result was the first of a series of "gap" scares that would sweep the United States during the nuclear arms race. In this case, the "bomber gap" scare of 1955 turned out to be nothing more than a hoax fabricated by the Soviets. Unknown to the Air Force and American intelligence at that time, the "waves" of bombers that flew in the Moscow air show were actually the same planes repeatedly flown over the spectator stands to create an illusion of massive Soviet strategic air power.[23]

In reality, the Soviet Union never achieved the number of long-range bombers predicted by the Air Force and American intelligence. At the time of the bomber gap scare, the Soviets had no more than 300, and probably less than 200, bombers capable of reaching American cities. By contrast, the United States had over 1,600 medium-range B-47s and over fifty long-range B-52s, all of which were capable of reaching the Soviet Union from bases around the Soviet periphery. While the Soviets could have initiated a destructive first-strike against the United States, the size of the American Air Force made it impossible for the Soviet Union to escape a far more devastating American retaliation.[24]

American intelligence concluded later that the Soviets had failed to develop a massive bomber force for primarily economic reasons. Like Eisenhower, Khrushchev apparently did not believe defense spending should take precedence over a healthy economy. Rather than building expensive long-range bombers in significant numbers, the Soviet leader decided to concentrate his country's economic resources on the development of ballistic missiles. Emphasis was placed initially on intermediate-range missiles—by which Khrushchev intended to hold Western Europe hostage and prevent an American first-strike against the Soviet homeland.[25]

Ironically, the zeal displayed during the bomber gap scare by the U.S. Air Force, as well as its allies in the defense industry and in the Congress, helped Khrushchev to reap substantial diplomatic advantages. In 1956, at a time when the Soviets were in no position to threaten either the United States or its allies, the Soviet leader, by raising the specter of a global nuclear war, was able to neutralize the possibility of American intervention in behalf of the Hungarian revolutionaries and was instrumental in forcing the British, French, and Israelis to withdraw from Egypt during the Suez War.[26]

However, while Eisenhower was not inclined to risk nuclear war with the Soviet Union over Hungary or Egypt, neither was he fooled by the bomber gap myth. In mid-1956 the United States acquired the ability to gauge the extent and nature of the Soviet aerial threat. At that time, U-2 photo-reconnaissance aircraft began to fly missions over Soviet territory. "The detail of the photos was so sharp," according to Raymond Bissel, Jr., the CIA official in charge of the U-2 program, "that one could almost read the tail-markings of bombers photographed from a height of fourteen miles." The U-2 photos revealed no widespread deployment of Soviet long-range bombers. On the basis of this evidence, which Eisenhower withheld from public knowledge in order to preserve the secrecy of the U-2 program, he was able to reassure the American people that the nation's strategic forces were more than adequate to deter a Soviet attack. But he also told the nation that "our former unique physical security has almost totally disappeared before the long-range bomber and the destructive power of a single bomb." The President's assurances, however, did not satisfy the Congress, and ultimately he was compelled to increase production of long-range bombers. Khrushchev paid for his unwarranted blustering by encouraging the escalation of a nuclear arms race he probably neither wanted nor could afford.[27]

Yet Eisenhower believed that there was nothing, short of a preemptive nuclear strike against the Soviet Union, that the United States could do to prevent the development of a substantial Soviet nuclear arsenal—he rejected the preventive, nuclear-strike option out of hand. Rather than incurring the enormous financial burden of trying to maintain overwhelming American nuclear superiority, Eisenhower simply decided to maintain enough nuclear weapons to deter a Soviet first-strike. The new strategy, called "sufficient deterrence," was unveiled in August 1956 as a part of the administration's "New, New Look" strategy. By relying on a variety of nuclear delivery systems—including carrier-based aircraft, long- and medium-range bombers, and ballistic missiles—the administration expected to retain a massive retaliatory capability even if the Soviets gained a numerical advantage in long-range bombers.[28]

Sputnik

Eisenhower's effort to keep a lid on military spending was severely tested by the impact of Sputnik, an earth-orbiting satellite launched by the Soviets on October 4, 1957. American scientists were not only impressed by the fact that the Soviets were the first to accomplish this scientific achievement, they were also amazed by the weight of the Soviet satellite, 184 pounds—six times heavier than *Vanguard,* the satellite the United States was at that time preparing to launch. Even more impressive was the November 3 launch of *Sputnik II,* a dog-carrying satellite weighing 1,120 pounds. America's scientists, as well as political and military leaders, realized that the launching of such a massive satellite validated Khrushchev's boast of the previous August that the Soviet Union had developed an intercontinental ballistic missile. Only a powerful mis-

sile, with an accurate guidance system, could place a payload as heavy as
Sputnik II into earth orbit. Obviously, a missile that could launch satellites into
orbit could also deliver nuclear warheads to American targets. When contrasted
with the failure of the U.S. Navy to launch *Vanguard* on December 6 and the
meager weight of America's first successful satellite, the Army's thirty-pound
Explorer, which was placed into orbit on February 1, 1958, the Soviet space
achievements were nothing short of shocking to most Americans. It was un-
believable that a people supposedly as backward as the Soviets could demonstrate such technological prowess.[29]

Some commentators were fascinated by the emotionalism of the American
reaction to the Sputniks. Television journalist Edward R. Murrow commented
that the "acute depression" produced by the Soviet satellites and the exhila-
ration resulting from the launch of *Explorer* were "symptoms not usually associated with normal health, in individuals or nations." It was, he stated, a
product of the great national superiority complex that Americans possessed and
that Sputnik severely shattered. Said Eisenhower: "I had no idea that the American people were so psychologically vulnerable."[30]

Closer analysis, however, reveals that the Soviet technological lead was
more apparent than real. The Soviets simply placed greater emphasis on their
satellite program than the United States. The Eisenhower administration, realizing that the United States enjoyed an enormous superiority in strategic
bombers, saw no need for a crash, missile-development program, and it was
not inclined to spend extravagant sums of money on military programs it considered unnecessary. The Soviets, on the other hand, believed they were vulnerable to an American first-strike and saw ballistic missiles rather than bombers as the most effective, and least expensive, way to deter it. In addition, the
Eisenhower administration was unwilling, at least initially, to use a military
missile like the Atlas for launching satellites for fear of revealing military secrets. The Soviets had no such inhibition. Considering their strategic inferiority
relative to the United States, and their vulnerability to an American first-strike,
they had more to gain by revealing some of their missile achievements than
by keeping them secret.[31]

"The Missile Gap"

As far as the strategic arms race is concerned, the most far-reaching consequence of the Sputniks was the creation of another weapons-gap myth—the
"missile gap." The missile gap myth was based on the premise that the Soviets
must have had many operational ICBMs to have launched Sputnik, while the
United States had none. In its extreme form, the myth presumed that once the
Soviets had increased their ICBM lead substantially, they would then attack
the United States and its allies.

The missile gap myth was formally articulated in the Gaither Report, a
study of the state of America's defense establishment, which was completed

shortly after Sputnik was launched. (The committee that issued the report was named after its chairman, Rowan Gaither, who was also the chairman of the Ford Foundation.) The report warned that, even if the United States launched a crash missile-building program, which it strongly recommended, it would not be possible to erase the Soviet lead until the early 1960s. Until that time, the report recommended a number of steps—including the dispersal of Strategic Air Command (SAC) bombers, the creation of an airborne alert system, and the construction of fallout shelters on a nationwide basis—to deal with the possibility of a Soviet attack. To finance these and related programs, the report strongly recommended an annual defense spending increase of between $8 billion and $13 billion.[32]

Eisenhower was appalled by the conclusions of the Gaither Report. He found the vast majority of its recommendations not only unnecessary but exorbitant. In coming to that conclusion, the President relied on a source of information not available to the Gaither Committee, the U-2 photo-reconnaissance plane. Even before the first Soviet ICBM test in August 1957, U-2 aircraft had found the Soviet test site at Tyura Tam, near the Aral Sea. On the basis of U-2 photos taken of that installation as well as the Trans-Siberian Railway, along which it was believed the Soviets would be deploying the bulk of their ICBMs, Eisenhower concluded that the Soviets were not in the process of a massive ICBM deployment.[33]

Displaying a degree of realistic analysis not found in the Gaither Report, Eisenhower concluded that, while the Soviets might have a small lead in missile development, America's strategic forces were far more than adequate to deter a Soviet first-strike. The United States, he knew, had many more bombers than the Soviets had missiles, and he realized that the bombers would have to be destroyed before the Soviets could hope to escape a devastating American retaliation. On the other hand, he knew that the United States could destroy every conceivable Soviet target with as few as 300 of its 4,000 nuclear-capable aircraft. The Soviet Union, in short, would have far more to lose in a nuclear war than the United States, even if, as Eisenhower initially believed, the Soviets had a small lead in ballistic missiles.[34]

As American intelligence data accumulated, even Eisenhower's initial belief of a Soviet missile lead proved false. On the basis of U-2 photos, as well as Soviet missile test data obtained by CIA radar and electronic eavesdropping installations in Turkey and Iran, Department of Defense intelligence spokesmen stated that as late as September 1959, when the first U.S. Atlas missiles became operational, the Soviets had still deployed no operational missiles. But it was not until early in 1961, after Kennedy entered the White House, that a National Intelligence Estimate stated with any degree of certainty that the Soviets possessed a small number of operational ICBMs. In short, during the Eisenhower years, there never was a missile gap, at least not one favoring the Soviet Union. While the Soviets deployed more operational IRBMs in Europe than the United States, the Soviets until the early 1970s never had more operational ICBMs than the Americans.[35]

Sputnik Diplomacy

Why did the Soviets build far fewer ICBMs than the administration initially believed? One reason appears to have been the huge costs involved. Instead of expending scarce economic resources on the Soviet Union's unreliable, first generation of ICBMs, it is quite probable that Khrushchev decided to deploy them in very limited numbers while concentrating on the development of a more reliable, second generation of missiles.[36]

At the same time, however, Khrushchev could not ignore the opportunities provided by the American inclination to minimize their own strength, particularly as it related to the Soviet Union. He repeatedly played on American insecurities, minimizing the technological achievements of the United States while exaggerating those of the Soviet Union, in order to disguise the real, relative inferiority of his nation's strategic forces. The refusal of the United States to intervene in the Hungarian Revolution and America's decision to side with the Soviet Union in condemning the 1956 Franco-British-Israeli invasion of Egypt probably convinced Khrushchev that the Soviet Union, even with a far weaker nuclear arsenal than that of the United States, could use the threat of nuclear war as an instrument of foreign policy.[37]

The result was a campaign of Soviet "rocket-rattling" that seemed to confirm the worst fears of Americans. During the Suez War, Khrushchev warned both Britain and France that in the nuclear age it was dangerous for them to wage war with Egypt. During the Taiwan Straits crisis of 1958, he sent Eisenhower a letter stating that the Soviet Union would "do everything" to defend China if the United States attacked the Chinese mainland. He went out of his way to warn the West Germans that in the event of war they would have "no chance of survival," even promising that the distant American homeland would not escape nuclear devastation.[38] On November 27, 1958 Khrushchev gave an ultimatum to the Western powers to end their occupation of West Berlin, reminding them that Soviet IRBMs were targeted on Western Europe. Eisenhower reponded by reiterating his policy that the United States would not fight a ground war in Europe. If necessary, the United States would use nuclear weapons to preserve the Western status in Berlin. He had no illusions that any recourse to nuclear weapons in Central Europe would inevitably lead to all-out nuclear war. "Possibly we were risking the very fate of civilization on the premise that the Soviets would back down from the deadline," Eisenhower recalled. "Yet this . . . was not really gambling, for if we were not willing to take this risk, we would be certain to lose." Fortunately for civilization, diplomacy averted a nuclear showdown. Khrushchev withdrew his ultimatum after the United States agreed to discuss Berlin in a face-saving, but as it turned out, inconclusive foreign ministers' conference.[39]

Khrushchev's rocket-rattling was ultimately counterproductive. The short-term gain it may have produced in enhancing Soviet diplomatic leverage was more than offset by the boost it gave the American missile program. As a result of the expansion of the American nuclear arsenal, the Soviet Union, by the end of the 1950s, was more vulnerable to an American first-strike than it was before the Soviet leader embarked on his rocket-rattling campaign. Rather than disguising the Soviet Union's strategic military inferiority, Khrushchev only aggravated it. The growing imbalance in Soviet-American strategic power, some historians believe, was the major factor behind Khrushchev's decision to place ballistic missiles in Cuba in 1962. That decision nearly produced a nuclear war.[40]

The Military-Industrial Complex

After the findings of the Gaither Report were leaked to the press, apparently by those who favored increased defense spending, the American people reacted with almost the same degree of anxiety that they had displayed after the announcement of Sputnik. The military establishment, as a whole, capitalized on the hysteria by attacking the ceilings on defense spending that were imposed by the administration. The Air Force, in particular, saw the "missile gap" as a useful technique with which to spur its own missile programs while cementing its role, in the face of Army and Navy opposition, as the primary instrument of the administration's massive retaliation strategy. The Air Force went to almost any length to achieve these ends, even allowing its spokesmen to exaggerate the projections of Soviet missile production made by the administration's intelligence sources. Key journalists were told that the Soviet Union would have as many as 1,000 ICBMs by the end of 1961—a six-to-one advantage over the United States.[41]

In its battle for more missiles, the Air Force was assisted by powerful allies in Congress, who saw the alleged missile gap as a way to embarrass the administration while serving their own vested interests. Senator Henry Jackson, a Democrat from Washington, a state with major aircraft industries, asserted that the United States lagged far behind the Soviet Union in numbers of missiles. He called for restoration of budget reductions for missile development and increased bomber production until the United States caught up.[42]

The efforts of Jackson and other congressmen with close ties to the Air Force were augmented by the lobbyists of corporations that did business with that service. The breadth of the relationship between the Air Force and the defense industry can be grasped by looking at work done on the Atlas, Titan, and Thor missiles, which involved 18,000 scientists and technicians affiliated with universities and industry, and 70,000 others working in 22 industries that included 17 prime contractors, over 200 subcontractors, and innumerable small suppliers. Among the top one hundred corporations that divided three-quarters of total expenditures for military contracts, three of the largest were aircraft manufacturers: General Dynamics, Lockheed, and Boeing.[43]

Eisenhower feared that the scare tactics used by special-interest groups would lead to an unwarranted expansion of the nation's military establishment. He argued that "everybody with any sense" could see that unless military spending were restricted to what was necessary, America would not only go bankrupt, it could easily become "a garrison state." In his 1961 farewell address to the nation, he warned that "we must guard against the acquisition of unwarranted influence, whether sought or unsought, by the military-industrial complex. . . . We must never let the weight of this combination endanger our liberties or democratic processes."[44]

The military-industrial complex was not alone in encouraging the myth of the missile gap. One interest group after another—business associations, organized labor, educators, and scientists—used the Sputniks and the "missile gap" to press the administration for more federal spending in their particular fields. Walter Reuther, president of the United Auto Workers, demanded that the administration mobilize American industry and resources in an all-out effort to catch up with the Soviets. Even the U.S. Chamber of Congress, which had condemned Eisenhower's fiscal 1958 budget as excessive, called for greater defense spending in fiscal 1959.[45]

Yet of all the groups that perpetuated the missile gap myth, the news media was perhaps the most important. With the exception of *The Nation,* every major mass periodical accepted the "missile gap" as fact at one time or another between 1957 and 1961. Most of the nation's leading newspapers, including *The New York Times,* also considered the missile gap a reality. Portraying America as endangered by a "barbarism armed with Sputniks," *The New York Times* depicted the Soviets as eager to take over the world. The paper attacked the Eisenhower administration repeatedly for allowing the Soviets to seize a lead in missile development by its excessive adherence to budget-balancing. The net effects of the press' belief in a missile gap were not only to frighten the American people but also to galvanize the administration's critics into attacking its defense policies.[46]

With a presidential election approaching in 1960, the Democrats needed little encouragement from the press to attack the administration's defense program. Democratic Senate Majority Leader Lyndon Johnson's Subcommittee on Preparedness released a report blaming the missile gap squarely on the administration's "record of underestimation of [Soviet] military progress." And another Democrat, Senator John F. Kennedy of Massachusetts, running for the presidency in 1960, charged that as a result of the administration's "complacent miscalculations, penny-pinching, budget cut-backs, incredibly confused management, and wasteful rivalries and jealousies," the United States had become the second strongest military power in the world. Kennedy called for nothing less than a crash program to overcome the alleged Soviet lead in missiles.[47]

It was not until a year after Kennedy's election victory in November 1961 that he gave any public indication that the Republicans had been right, that indeed there had never been a missile gap. Kennedy then revealed, on the basis of what he called "a lucky intelligence break"—probably the data received from the first American reconnaissance satellites—that the Soviets had not deployed as many ICBMs as American intelligence had initially thought. Not surprisingly, considering the short memory span of the American electorate, neither Kennedy nor any other politician, including Johnson and Jackson, who had capitalized on the missile gap myth suffered lasting political damage for the hysteria they had helped to produce.[48]

Aftermath of the Missile Gap Scare

The hysteria generated by the missile gap scare ultimately resulted in the production of hundreds of unneeded ballistic missiles. In response to public and congressional pressure, the administration increased its ICBM program from its originally planned level for 1962 of 80 Atlases and Titans to a new level of 255. The Minuteman program was expanded from a planned force level of 400 missiles to 450. An additional 90 mobile Minutemen were added in the administration's last budget for fiscal 1962. The Polaris program was also expanded to a planned force of 304 missiles on 19 submarines with an additional 8 missiles on the nuclear-powered cruiser *Long Beach*.[49]

To reassure America's European allies that their defense was intimately bound to the defense of the United States, both the Army's Jupiter and the Air Force's Thor IRBMs were ordered into production, while negotiations were begun with the NATO allies to secure missile launch sites in Europe. In the end, Britain accepted the Thors, and Italy and Turkey the Jupiters. France, however, refused to accept any American missiles until the United States agreed to assist the French nuclear weapon program. French President Charles de Gaulle was angered by the apparent eagerness of the United States to provide Britain with nuclear weapons materials and technology (which was made possible by a 1958 amendment to the Atomic Energy Act) but not his own country. Turned down by the United States, the French developed nuclear weapons on their own. With the explosion of the first French atomic bomb in 1960, the number of nuclear weapon states reached four.[50]

6

Eisenhower and Nuclear Arms Control, 1953–1961

Eisenhower and the Nuclear Peril

President Eisenhower stated repeatedly that the main concern of his administration was the unprecedented peril of nuclear war. On the day Lewis Strauss was sworn in as chairman of the Atomic Energy Commission, Eisenhower told him that "my chief concern and your first assignment is to find some approach to the *dis*arming of atomic energy. . . . The world simply must not go on living in the fear of the terrible consequences of nuclear war." Nevertheless, after eight years in the White House, Eisenhower was compelled to admit that the results of his own effort to end the nuclear arms race were, in his words, "meager, almost negligible."[1]

The Atoms-for-Peace Plan

In his first attempt to bring nuclear weapons under control, Eisenhower emphasized the potential benefits of nuclear energy rather than its destructive potential. Early in his term, he called on the nuclear powers to provide nuclear materials and technology to an international atomic energy "bank" that, in turn, would provide nuclear assistance to nations promising to forgo the production of nuclear weapons. The result of this so-called Atoms-for-Peace Plan, Eisenhower promised, would be to "strip" nuclear energy of "its military casing and adapt it to the arts of peace." He also hoped it would stimulate an international market for commercial nuclear power that would spur domestic development. The Soviets, however, ignored the President's proposal, believing that any reduction in their stockpile of fissionable materials would only widen the American lead.[2]

In spite of the negative response of the Soviet Union, Eisenhower went ahead with the Atoms-for-Peace Plan. In 1954 a new Atomic Energy Act was enacted that enabled the United States to participate in an "international atomic pool." Two years later, 40,000 kilograms of U-235 for nonmilitary nuclear research were made available to friendly governments. In an attempt to ensure that nuclear materials and technology would not be diverted to military projects, the United States in 1957 helped to create the International Atomic Energy Agency (IAEA). Its primary function was to establish and monitor a system of international atomic energy controls. And yet the nuclear activities of the nuclear weapon states—the United States, Britain, and the Soviet Union— were excluded from the IAEA controls, and their exclusion was not overlooked. France and China, in the process of developing or contemplating the development of nuclear weapons, refused to submit to IAEA controls and eventually constructed their own nuclear weapons. Unfortunately, the safeguard system set up by the IAEA proved to be inadequate even for states that agreed to accept them. In 1974 India, which submitted partially to the IAEA controls, used nuclear materials and technology provided by the United States and Canada for civilian purposes to produce and to detonate a nuclear device.[3]

The Threat of Radioactive Fallout

In the 1950s, however, most Americans were not concerned about the prospects of nuclear proliferation. They were more upset about the possibility of radioactive contamination from the fallout produced by the atmospheric testing of nuclear weapons. Public concern about radioactive fallout was first aroused by a nuclear test series, code-named CASTLE, which the United States conducted in the Bikini Atoll in early 1954. In one of the tests, called BRAVO, a thermonuclear device was detonated with an energy yield of fifteen million tons of TNT—the largest thermonuclear explosion ever produced by the United States.[4]

Tragically, there was an unanticipated result of the BRAVO test. An unexpected change in wind direction carried radioactive fallout from the explosion over a Japanese tuna trawler, inappropriately named the *Lucky Dragon,* which was sailing some eighty miles from the test site. The crew soon suffered the effects of severe radiation exposure: nausea, fever, and bleeding gums. One crewman, Aikichi Kuboyama, subsequently died. In Japan the shock produced by the *Lucky Dragon* tragedy was compounded by reports that fish contaminated by the CASTLE tests, and by Soviet nuclear tests in Siberia, were being sold in Japanese marketplaces. The Japanese, the first victims of an atomic weapon, were now to be the first victimized by thermonuclear weapons.[5]

A worldwide wave of protest followed in the wake of the *Lucky Dragon* tragedy. Highly respected world leaders, including Jawaharlal Nehru, the Indian prime minister, Albert Schweitzer, the famed physician and philosopher, Albert Einstein, and Pope Pius XII, called for a cessation of nuclear testing. And while the uproar prompted Eisenhower to admit "that this time something must have happened that we have never experienced before," AEC chairman Strauss tried to assure the nation that "no civilian has ever been injured as a result of these tests." Nevertheless, the U.S. government paid Aikichi Kuboyama's widow an indemnity of $2,800.[6]

Almost a year later, in February 1955, the AEC released a report describing the effects of the BRAVO test. It stated that the radioactive debris from the test's thermonuclear explosion covered a cigar-shaped area 200 miles in length and 40 miles at its widest point. It also estimated that a similar explosion on a city would kill at least half of the people 160 miles downwind. Moreover, the report found the most dangerous long-term by-product of thermonuclear explosions to be radioactive strontium-90. With a half-life of twenty-eight years, strontium-90 could find its way into human bones by way of the food chain— that is, it could travel from the soil on which it fell, to plants and animals, and ultimately to humans. The AEC report attempted to assure the public that the amount of strontium-90 produced by the CASTLE tests was insignificant and not a health hazard to Americans.[7]

More than a few scientists disagreed, some vehemently, with the AEC's report. Physicist Ralph Lapp, a veteran of the Manhattan Project, stated that the report ignored many issues, such as the lingering nature of local fallout and the dangerous consequences of breathing fallout particles. Many geneticists stated that, while the danger posed to an individual by radioactive fallout from a single explosion might be slight, the cumulative effects of several tests were more likely to be harmful. This, they believed, should be considered when weighing the military necessity of continuing the atmospheric testing of nuclear weapons.[8]

That recommendation, however, had little impact on Strauss. He argued that continued testing was imperative if the United States were going to maintain its technological lead over the Soviet Union. He also pointed out that a nuclear war would produce far more lethal quantities of fallout than could result from the continuation of nuclear testing. The tests, he said, must continue until there was a workable international agreement on disarmament, something he did not expect in the foreseeable future.[9]

Disarmament Talks, 1954–1955

While Eisenhower could not disagree, at least not initially, with Strauss' arguments in favor of continued nuclear testing, he could not adopt the same cavalier attitude toward disarmament that seemed to govern his AEC chairman. In the spring of 1954 the administration used the opportunity provided by the convening of the new U.N. Subcommittee on Disarmament to try to revive the long-stalled talks on the international control of atomic energy. The Soviets, however, continued to consider any plan for international inspection of nuclear facilities, which the Eisenhower administration insisted was a necessary ingredient of effective nuclear arms control, a disguise for American espionage.[10]

On May 10, 1955 the Soviets responded to the American proposal with a disarmament plan of their own. In order to "create the requisite conditions for . . . a broad disarmament program," the Soviet plan called for a preliminary agreement that would liquidate all foreign military bases, require the withdrawal of all foreign military forces from Germany, and resolve all outstanding problems in the Far East.[11]

After this was accomplished, the Soviet plan called for phased reductions in conventional forces that would have left the Soviet Union, China, and the United States with between one million and one-and-a-half million troops apiece. Control posts to guard against surprise attack would be established at large ports, railway junctions, airports, and on major highways in the territories of the participating countries. They would be supplemented by a single control organ that would have an unlimited right to inspect these facilities.[12]

Finally, the Soviet plan addressed the problem of nuclear weapons. It proposed, for the first time by either superpower, a nuclear test ban as the initial step toward nuclear disarmament. In what appeared a major concession on the inspection issue, the plan stated that the test ban should be supervised by an international commission subject to the U.N. Security Council. Later, however, the Soviets reversed themselves and denied that inspection was a prerequisite for an effective test ban.[13]

But there were other problems besides the inspection issue that would deny a favorable response from the West on the Soviet plan. For one, neither the United States nor its allies were prepared to accept the Soviet proposal for the liquidation of all foreign military bases. The West considered foreign bases vital to the defense of its worldwide interests. Also unacceptable was the Soviet condition that all measures in response to violations would be the responsibility of the veto-empowered U.N. Security Council. In spite of these objections— and they were significant objections—the gap that separated the Soviet and Western disarmament positions seemed to have narrowed, leading some to believe that an East-West agreement on nuclear arms control would eventually be possible.[14]

Khrushchev and Disarmament, 1955–1956

A number of reasons can be found for the Soviets' adoption of a more realistic approach to the nuclear weapons issue. For one, the development of Soviet nuclear weapons, and the means to deliver them on American targets, provided the Soviets with a degree of security they had previously lacked when the United States enjoyed a nuclear monopoly. With the American nuclear monopoly gone, the Soviets felt freer to express an appreciation of the destructive potential of nuclear weapons. At the twentieth congress of the Soviet Communist Party in 1956, Khrushchev reversed Stalin's public stance that nuclear weapons were inconsequential and announced that their awesome destructive power made peaceful coexistence between the East and West a necessity. Moreover, Khrushchev hoped that peaceful coexistence would provide the Soviet Union a breathing space in its struggle with the West that would enable him to concentrate on pressing domestic problems, particularly a stagnant agricultural sector.[15]

At the same time, the arms concessions made by the Soviets in their May 1955 plan posed little military risk to the Soviet Union while they offered the possibility of considerable diplomatic gain. Moscow's willingness to reduce conventional forces, a step it took unilaterally in 1955–1956 when Soviet ground forces were slashed by about two million men, was a reflection of the fact that, with nuclear weapons now at its disposal, it could afford to cut its conventional forces. Furthermore, when combined with the withdrawal of Soviet troops from Austria in 1955, Khrushchev apparently believed that an offer to reduce the size of the Soviet military presence in Central Europe might have the effect of preventing the rearmament of West Germany, a major priority of Dulles' policy.[16]

The Soviets probably realized that there was little possibility of a test ban in the near future. Neither the American people nor the Eisenhower administration appeared ready to support a cessation of nuclear weapon tests. And Khrushchev's hold on the Kremlin was probably not sufficiently strong in 1955 to permit Politburo ratification of a test ban agreement, particularly with the Soviet nuclear arsenal still in its embryonic phase. Nevertheless, Khrushchev appreciated the propaganda value of the Soviet test ban proposal and he exploited it fully. However, instead of testing the sincerity of the Soviet offer, if only to deflate its propaganda value, the West chose to ignore it. Over Soviet objections, the deliberations of the U.N. Disarmament Subcommittee were suspended until after the heads of government met in a July summit meeting in Geneva, Switzerland.[17]

The Open Skies Plan, 1955

The Geneva Conference, July 18–22, 1955, was a result of a British effort, first by Churchill, then by his successor, Anthony Eden, to ease international tensions and, not incidentally, to help the Conservative Party win an approaching parliamentary election. Eisenhower went along with the British gambit primarily because he believed that a new leadership group in the Kremlin provided an opportunity for a Cold War thaw.[18]

The summit agenda covered proposals on disarmament, European security, Germany, and economic exchange programs. With respect to disarmament, Soviet Premier Nicolai Bulganin offered a modified version of his country's May 10 Plan. The new version no longer considered a settlement of Cold War differences a prerequisite to a disarmament agreement. Gone, too, was the Soviet demand for the liquidation of foreign bases. But the altered Soviet plan substituted new obstacles for the ones they removed. One snag was the demand for a ceiling of 150,000 to 200,000 troops on the armed forces of smaller nations, an obvious attempt to scuttle NATO's plan to create a 500,000-man West German army. Another was a call for a ban on the first use of nuclear weapons, an option NATO wanted to retain to deter a Soviet invasion of Western Europe. The Soviets also renewed their earlier proposal for the creation of an all-European security pact that would have replaced both the NATO alliance and the Warsaw Pact; it would have created a reunited, but neutralized, German state. However, the Western powers turned down the security pact proposal because they refused to contemplate either the dissolution of NATO or the neutralization of West Germany, both of which were considered vital to the preservation of democracy in Western Europe.[19]

The Western powers once again refused to consider the Soviet May 10 Plan either in its original or its modified form. Instead of the comprehensive approach to disarmament suggested by the Soviet plan, the Western powers offered a number of interim disarmament steps. Eden proposed the creation of an experimental zone of arms limitation and inspection in Central Europe. French Premier Edgar Faure suggested a system of budgetary controls on nuclear weapons development. But it was Eisenhower's "Open Skies" proposal, which called for aerial inspection of both the Soviet Union and the United States, as well as the exchange of blueprints of American and Soviet military installations, that created the greatest excitement.[20]

The Open Skies proposal was the work of a committee under the general supervision of Harold Stassen, who was appointed Special Assistant on Disarmament by Eisenhower on March 19, 1955. The proposal reflected the administration's new view that it was futile to try to implement disarmament programs as ambitious as the Baruch Plan or, for that matter, the Soviet May 10 Plan. The administration now believed that instead of trying to negotiate a comprehensive disarmament plan it would be wiser to begin with so-called confidence-building measures, like the Open Skies Plan, that might pave the way for general disarmament some time in the future. In effect, the new approach favored by the United States would attempt to control nuclear weapons rather than abolish them. The new approach was prompted at least partially by the administration's massive retaliation strategy. Because the defense of the West had come to rest so heavily on nuclear weapons, their complete elimination was considered too dangerous to implement.[21]

For a number of reasons, the Soviets rejected the Open Skies Plan. For one, it offered the Soviets little beyond what they probably already knew about American military and nuclear installations or could find out through conventional espionage tactics, or even by being receptive to the American news media. On the other hand, by opening the Soviet Union to aerial inspection, the plan would have revealed the location of Soviet nuclear and military installations, thereby guaranteeing that American nuclear bombs would be targeted on them. After Eisenhower presented the Open Skies Plan, Khrushchev called it "a very transparent espionage device," telling the President, "you could hardly expect us to take this seriously." Little did the Soviets suspect, however, that within a year of the Geneva Conference, the ingenious Americans would begin a unilateral, and secret, version of the Open Skies Plan when U-2 photo-reconnaissance missions began over Soviet territory.[22]

The Test Ban Issue in 1956

The year 1956 was no more productive in dealing with the nuclear weapons problem than the preceding year. On March 1, Eisenhower wrote Bulganin a letter in which he proposed a mutual halt to nuclear weapons production. He also suggested that both countries transfer a certain amount of fissionable materials from their nuclear stockpiles to nonmilitary nuclear projects. The Soviets rejected the President's suggestions not only because they would have required on-site inspection, but also because a production cutoff would have frozen the Soviet nuclear arsenal in a distinctly inferior position. Instead of accepting the latest American offer, the Soviets countered with proposals calling for the denuclearization of both East and West Germany, the reduction of military budgets by fifteen percent, and the cessation of nuclear testing. Surprisingly, the Soviets also accepted, for the first time, the concept of aerial inspection, first introduced in the Open Skies Plan, although only on a very limited scale. They indicated that they were willing to permit aerial inspection of a zone 800 kilometers deep on either side of the line dividing NATO and Warsaw Pact forces in Central Europe.[23]

The United States refused to consider the first two Soviet proposals—the denuclearization of Germany and reductions in military budgets—for essentially the same reasons that they were rejected when they were a part of the Soviet comprehensive plan of May 10, 1955. The Eisenhower administration tied the defense of West Germany and all of Western Europe to nuclear weapons and did not want to have restricted by treaty the amount of funds that could be spent for them. It rejected a test ban agreement primarily because the Pentagon feared that the Soviets had established a lead not only in bombers, but also in the development of ballistic missiles. Without continued testing, the Pentagon insisted, the United States could not hope to catch up to the Soviets.[24]

Yet the administration found it necessary at least to respond to the latest Soviet bid for a nuclear test ban. Public concern about the effects of atmospheric nuclear testing was increasing worldwide, especially in the aftermath of the explosion of a Soviet hydrogen device on November 22, 1955 which produced highly radioactive rain over Japan. Three weeks later on December 13, Khrushchev and Bulganin responded to the outrage their test produced by calling for an "unconditional prohibition of the production, use or experimentation of nuclear and thermonuclear weapons." Their proposal was followed by Pope Pius XII's Christmas Eve message linking a test ban with agreements to eliminate nuclear weapons. In response to this pressure, Secretary of State Dulles could only say that the United States had not yet found "a formula" for a test ban that would be "both dependable and in the interest of the United States."[25]

While the Eisenhower administration refused to consider a nuclear test ban in 1956, Adlai Stevenson attempted to make it a key issue in his bid to win the White House. Stevenson's support for a test ban infuriated Eisenhower, who believed that the nuclear weapons issue transcended partisan politics. The President indirectly accused his Democratic opponent of advocating "disarmament without inspection" and stated that continued testing was vital to the nation's security. However, after the Soviets invaded Hungary on November 4, the test ban became a dead issue, at least so far as the election was concerned. Any thought of negotiating with the Soviets on any issue, let alone one as sensitive as a nuclear test ban, was out of the question in the immediate aftermath of the Soviet Union's brutal suppression of the Hungarian revolutionaries.[26]

The Revival of the Test Ban Issue in 1957

In spite of Eisenhower's public defense of continued testing, he was still bothered by the intensification of the nuclear arms race. On October 12, during his election campaign, he said to an aide, "My God, we have to simply figure a way out of this situation. There's just no point in talking about 'winning' a nuclear war." For this reason, he told Dulles that he did not want his hands unalterably tied on the test ban issue. Yet at the same time, he could see no way around the Pentagon's insistence that continued testing was a military necessity. Less than two months after his election to a second term, on December 28, he agreed to permit a new test series on the condition that they would be restricted to the Nevada test range, where fallout could be kept to a minimum.[27]

The Soviets, on the other hand, displayed considerably less sensitivity in resuming their nuclear testing program. After conducting five tests within a two-week period during April 1957, all of which produced heavy fallout, Khrushchev boasted that the Soviet Union had developed a hydrogen bomb too powerful to test. He said it "could melt the Arctic ice cap and send the oceans spilling all over the world."[28]

The resumption of American and Soviet nuclear testing during the spring of 1957, and the explosion of Britain's first hydrogen bomb in May, produced a chorus of worldwide protest. Albert Schweitzer, from the hospital he administered in French Equatorial Africa, warned that the continued testing of nuclear weapons would produce significant genetic damage. In West Germany, Otto Hahn, the scientist who split the uranium atom in 1938, joined eighteen of his colleagues in informing their government that they would have nothing to do with the production, testing, or use of nuclear weapons. The American biochemist and Nobel laureate Linus Pauling circulated a petition calling for a nuclear test ban that was signed by more than 9,000 scientists from 43 countries. As a result of anti-test publicity like this, public opinion on nuclear testing was slowly transformed. In April 1957, sixty-three percent of the people who participated in a Gallup poll supported a total ban on nuclear testing as long as it applied to all nations; only six months earlier, fifty-six percent of those polled by Gallup opposed the proposal.[29]

As letters, telegrams, and petitions calling for a test ban poured into the White House, the administration began to consider the fallout problem as important as the military's argument that continued testing was vital to the nation's security. Even Dulles, who had repeatedly defended nuclear testing as a military necessity, began to appreciate the negative impact on world opinion of the continued American testing. Nevertheless, Strauss and the Pentagon, realizing they were losing ground on the test isssue, were able to persuade Eisenhower that a test ban could only follow a halt to all nuclear weapons production. They argued that prolonged negotiations on a test ban without a production cutoff could give Congress and the American people a false sense of security that would have a disruptive impact on the administration's defense programs.[30]

In May 1957 at the London disarmament talks, Harold Stassen, the chief American delegate, presented the administration's offer tying a test ban to a halt to nuclear weapons production. The Soviet negotiator, Valerian Zorin, responded to the American offer by proposing an immediate, but temporary, ban on testing, without inspection or a weapon production cutoff. Although Stassen realized that a test ban without inspection would never be acceptable to the United States, he nevertheless urged the administration not to abandon the effort to ban testing.[31]

Stassen's effort aroused the angry opposition of the Pentagon. Admiral Arthur Radford, chairman of the Joint Chiefs of Staff, stated, "We cannot trust the Russians on this or anything. . . . The Communists have broken their word with every country with which they ever had an agreeement." Radford's remarks angered Eisenhower. He reportedly said, in a meeting with the admiral on May 25, 1957, "This is a question of survival and we must put our minds at it until we can find some way of making progress." In an attempt to mollify both Radford and Stassen, Eisenhower decided to offer a brief suspension of testing in the hope of securing future restrictions on the production of nuclear weapons.[32]

American opinion reacted favorably to the administration's temporary test suspension offer. But the British and French did not. Prime Minister Harold Macmillan, who had succeeded Eden in early 1957, feared that any test ban, even if temporary, would retard deployment of British nuclear weapons and leave his nation's nuclear arsenal at a very low, and inadequate, level. The proposal was even less accepable to the French, whose nuclear weapons program was still only in the developmental stage.[33]

Yet the Soviet pressure for a test ban eclipsed the Anglo-French objections. On June 14 the Soviets restated their proposal for a temporary suspension of testing and suggested a moratorium of two or three years duration. What was startling about the latest Soviet proposal was their apparent willingness to tie a test ban to the American condition for on-site inspection. In this spirit, they suggested the establishment of control posts on Soviet, American, and British territory to monitor compliance with the test ban. Overwhelmed by the dramatic transformation of the Soviet attitude on testing, Eisenhower told a press conference on June 19: "I would be perfectly delighted to make satisfactory arrangement for temporary suspension of tests."[34]

The President's announcement, however, did not sit well with those who believed that continued testing was vital to the security of the nation. On June 24 Edward Teller, Ernest Lawrence, and Mark Mills, all scientists associated with the development of nuclear weapons at the Livermore Laboratory in California, visited the White House and told the President that they opposed the temporary test suspension for two basic reasons. First, a test ban would hinder their effort to develop a "clean" bomb, a nuclear weapon ninety-six percent free of radioactivity. Second, they were convinced that it would be relatively easy for the Soviets to continue testing in secret after agreeing to a test suspension. Eisenhower was absolutely fascinated by the prospect of a "clean" nuclear weapon. He saw it as the way out of the dilemma of continuing nuclear testing as a military necessity or cancelling testing to avoid radioactive contamination.[35]

Upon hearing the news of Eisenhower's enthusiastic reaction to the possibility of a "clean" nuclear weapon, more than a few scientists were amazed by the President's gullibility. It was common knowledge to nuclear physicists that some radioactive debris would always result from the explosion of a hydrogen bomb's uranium trigger. *The New Republic* felt that "Strauss and his technicians" were simply pleading "for a continuation of the arms race." James Reston of *The New York Times* suggested that what Eisenhower needed to counter the heavy emphasis on the military applications of nuclear energy provided by Strauss and Teller was an independent, objective source of scientific advice. The appointment of James Killian as the President's special science adviser later in the year served that purpose. It was only then that Eisenhower began to feel more secure in his scientific knowledge of nuclear weapons and of the feasibility of a test ban.[36]

Dulles and the Test Ban

With Harold Stassen's drive for a test ban stalled by the phantom of Teller's "clean" bomb, Dulles moved to regain control of the administration's policy on nuclear arms control. In an attempt to allay public fears about fallout, without endangering American security, the secretary of state reestablished a weapon production cutoff as an essential condition of a temporary suspension of nuclear testing. As presented by Stassen to the Disarmament Subcommittee in London, the formal American proposal offered a test suspension of twelve months, with a pledge to renew for a second year, contingent on the implementation of a weapon production cutoff and an effective inspection system.[37]

Although the American proposal was close to a Soviet offer proposing a two to three year suspension of testing, the conditions the United States tied to its test ban proposal were unacceptable to the Soviets. The American offer, the Soviets said, made no provision for the reduction of nuclear arsenals and permitted the continued production of new nuclear weapons from existing stocks of fissionable materials. They insisted that a production cutoff would only be effective when it was "indissolubly linked to the prohibition of atomic weapons, their elimination from the armaments of states, and the destruction of atomic weapons stockpiles." It was highly unlikely, however, that the Soviets were about to freeze their nuclear arsenal in a state of permanent inferiority by agreeing to a halt to nuclear weapons production. Nor were they about to accept an inspection system as extended as the one demanded by the United States. The Soviet Union was not about to become an open society simply to halt the testing of nuclear weapons. They apparently considered the risks of nuclear contamination less dangerous to the Soviet system than the risks of Western snooping.[38]

HARDTACK

As a result of the Soviet rejection of the American test suspension offer in September 1957, Eisenhower believed he had no choice but to approve Strauss' request for additional tests. But the President based his approval on the condition that the tests would be held to the absolute minimum required. To Strauss, the absolute minimum involved the explosion of twenty-five thermonuclear devices, which occurred during the fall of 1957 in a series of tests code-named HARDTACK. Including the nuclear devices detonated by the British and the Soviets, the American tests brought the total number of weapons tested in 1957 to forty-two, compared to the combined total of nineteen during the preceding year.[39]

As the amount of radioactive fallout increased, so, too, did the activities of the test ban movement. During the summer of 1957 the American Friends Service Committee and the Women's International League for Peace and Freedom circulated petitions calling on the United States to cease all thermonuclear weapons tests. They were joined in their effort by the World Council of Churches and the Lutheran World Federation. Prompted by Bertrand Russell, the British philosopher and mathematician, and financed by the American industrialist Cyrus Eaton, a conference of scientists met in Pugwash, Nova Scotia, in July 1957 to consider the threat posed by the nuclear arms race. A report released by the conference asserted that the primary danger facing humanity was not radiactive fallout, which the report estimated increased the risk of cancer deaths and genetic mutations by only one percent, but rather the increasing probability of a nuclear war.[40]

Perhaps even more effective than any of the anti-nuclear weapons organizations in arousing public awareness of the perils of the nuclear arms race was a novel, *On the Beach,* written by Nevil Shute, which appeared in 1957. The story described how fallout produced by a nuclear war fought in 1961 gradually wiped out life on the planet, even in the Southern Hemisphere, which was not directly involved in the nuclear exchange. But while *On the Beach* may have frightened many people, it had no observable impact on the Atomic Energy Commission's attitude toward nuclear weapons testing.[41]

The Test Ban Issue in 1958

The nuclear testing issue dominated the arms control negotiations during 1958. On December 10, 1957, flushed with the Sputnik successes, Soviet Premier Bulganin proposed a two to three year suspension of testing beginning January 1, 1958. The Eisenhower administration, in the wake of Sputnik not really anxious for a test suspension that would prevent the United States from closing the alleged missile gap, responded to the Soviet initiative on January 12, 1958 by renewing its offer of the preceding August, which linked a test suspension to a weapon production cutoff. In all probability, the administration was not surprised, nor terribly disappointed, when the Soviets again rejected that offer. The administration was still free to close the presumed missile gap, while it had gained some satisfaction from its attempt at placating negative world opinion.[42]

Harold Stassen wanted to do more. In early 1958, he suggested that to get a test suspension, it would be necessary to drop the weapon cutoff condition. Predictably, his suggestion ran into the stone-wall opposition of the Pentagon and the AEC, which were supported by Dulles, who increasingly viewed Stassen as a threat to his authority. Eisenhower, not wanting to lose his secretary of state, sided with him; the President requested Stassen's resignation, and received it on February 7, 1958.[43]

But Stassen did not give up the fight after he left the administration. On February 28, before Senator Hubert Humphrey's Senate Disarmament Subcommittee, he proposed a two-year suspension of nuclear testing, the creation of a UN supervising agency, an inspection system consisting of seismological stations in both the Soviet Union and the United States, and additional negotiations to pinpoint more extensive disarmament agreements. He stated that twelve seismological stations in the Soviet Union, each manned by inspection teams empowered to investigate dubious earth tremors, would be more than adequate to prevent Soviet cheating. In response to the argument that continued testing was vital to American security, Stassen said that there was no security in "an all-out arms race," which would only continue if testing were not halted.[44]

The use of seismological inspection posts was based on the proven ability of seismologists in Fairbanks, Alaska, who were able to detect an underground test (code-named RANIER) of a 1.7-kiloton nuclear device which was conducted in Nevada in September 1957. On the basis of the RANIER test, a panel of scientists, headed by physicist Hans Bethe and appointed by James Killian, the President's new science adviser, concluded that seismological inspection posts were a feasible method of detecting Soviet underground tests. The conclusions of the Bethe Panel, however, created considerable controversy in the scientific community. Edward Teller vigorously attacked the panel's findings, saying that the Soviets could and would cheat on a test ban no matter how many seismographs were emplaced in the Soviet Union.[45]

While scientists debated the feasibility of inspecting a test ban, Dulles on March 24 warned the President that Khrushchev was preparing to announce within a few days a unilateral suspension of Soviet nuclear testing. To offset the propaganda advantage the Soviets would gain from that step, the secretary of state recommended that Eisenhower immediately announce that the United States would suspend testing at the conclusion of the HARDTACK series. Dulles' proposal was immediately and vigorously opposed by Strauss and the Pentagon. They insisted that further testing was necessary to develop warheads for an anti-missile missile, the Polaris SLBM, and other advanced weapon systems. While Eisenhower informed Dulles that he was determined to bring the arms race to an end, he again sided with the military and the AEC. Quite clearly, despite the President's desire to end the nuclear arms race, he did not know how or when it could be accomplished.[46]

Just as Dulles had predicted, the Soviets announced a unilateral suspension of testing on March 31. In making the announcement, Soviet Foreign Minister Andrei Gromyko added that the Soviet Union would have no recourse but to resume testing if the United States did not follow suit. Although the Soviet test suspension was somewhat hypocritical, since the Soviets had just completed their own test series, they nevertheless scored propaganda points in the forum of world opinion. Although Eisenhower was furious with the Soviets for the position they had placed him in, he had no one to blame but himself. For months he had considered suspending the American test program, only to allow Dulles, Strauss, and the Pentagon to talk him out of taking that step. Now, as the pressure applied by the Soviets forced him to follow his own inclination to suspend nuclear testing, he was angry that he had not done so first. Nevertheless, on April 8 Eisenhower sent Khrushchev a letter inviting the Soviets to participate in a conference of technical experts who would examine the feasibility of verifying a nuclear test ban agreement. In effect, he finally informed the Soviets that the United States was prepared to unhinge a test suspension agreement from a cutoff of nuclear weapons production.[47]

On May 9 Khrushchev agreed to allow Soviet scientists to participate in the technical conference proposed by the President. The conference began in Geneva on July 1. It concluded its work on August 21 with a report which stated that seismological inspection was a feasible way to verify a nuclear test ban. The report stated that if sufficient numbers of seismological stations were established, it would be possible to detect all nuclear explosions above a yield of five kilotons. It recommended the construction of 170 to 180 seismological stations, 100 to 110 of which were to be located on continental areas, while the remainder would be placed on oceanic islands. The seismological stations would be supplemented by aerial inspection and by teams of ground inspectors who would be sent to investigate suspicious tremors. Unfortunately, the "Geneva system," as the report's recommendations came to be called, contained a number of glaring omissions. Nothing was stated about the nationality of those who would staff the inspection posts. Nor was any determination made relative to who would decide when on-site inspection became necessary. In addition, little was said about the system's inability to detect explosions under the five-kiloton threshold or in outer space, above an altitude of fifty kilometers.[48]

Ignoring the omissions, or perhaps unaware of them, Eisenhower accepted the conclusions of the Geneva report and decided, over the objections of the AEC and the Pentagon, to begin negotiations for a test ban treaty. On August 22 he issued a statement indicating that he had invited the Soviets to begin the negotiations on October 31. As a sign of American good intentions, he offered to suspend testing for one year as of that date provided the Soviets also refrained from conducting tests. He also said he was prepared to extend the test moratorium beyond one year if progress were made in the negotiations. In response, the Soviets accepted the American invitation to begin negotiations on a test ban, but they did not make any commitment to continue their test suspension while the negotiations were being conducted.[49]

A Binge of Nuclear Weapon Tests

Not surprisingly, the American call for a test moratorium that would take effect on October 31 produced a spate of last-minute nuclear testing. The British exploded four nuclear devices before the moratorium began, the Soviets fourteen, and the Americans nineteen. The American tests, code-named HARD-TACK II, were conducted at the Yucca Flats test site in Nevada and produced the highest level of fallout ever observed in the United States. Because of an unexpected shift in wind direction, the fallout produced by one test in late October was carried over Los Angeles, raising that city's radioactivity to a level 120 times its normal reading. The Soviet tests which resumed on September 30 were even dirtier.[50]

Finally, when the Soviets completed their tests on November 3, three days after the American moratorium began, thirteen years of nuclear testing came to an end, albeit only a temporary end. In that interval, the United States had detonated 174 nuclear and thermonuclear devices, while the Soviet Union and Great Britain had exploded 53 and 21, respectively. In 1958 alone, these three countries had exploded almost half as many nuclear devices as they had detonated in the previous twelve years. While the test moratorium which began in November 1958 would last only three years, that period constituted the longest abstention from nuclear weapon testing in the history of the nuclear arms race.[51]

The Geneva Test Ban Conference, 1958–1959

The Conference on the Discontinuance of Nuclear Weapons Tests, as the test ban negotiations were officially called, opened as scheduled on October 31, 1958 in the Palace of Nations in Geneva. The talks, however, were soon deadlocked. A controversial aspect of the inspection problem that had long plagued nuclear arms control talks surfaced again. The Soviets asserted that each of the nuclear powers was entitled to a veto over the activities of the control commission that would police the test ban. The Americans and British disagreed. They feared that a Soviet veto would make the inspection system meaningless.[52]

As if the inspection issue were not enough of a problem facing the Geneva talks, the Americans raised another. It concerned the feasibility of detecting underground tests. On the basis of data gathered from the HARDTACK II nuclear weapon test series conducted during the fall of 1958, some American scientists began to question the conclusion of the Geneva Technical Conference that all underground tests above a five-kiloton yield could be differentiated from earthquakes. The new data indicated that the Geneva system would not be able to detect explosions having less than a twenty-kiloton yield, perhaps not even those with a fifty-kiloton yield. For the Geneva system, this meant that many more control posts would be required to make inspection effective, and many more would have to be located within the Soviet Union than the previously anticipated 180 posts.[53]

The presentation of the new HARDTACK II data by the American delegation at Geneva, recalled James J. Wadsworth, the head of the delegation, produced "the most violent reaction imaginable" from his Soviet counterparts. According to Wadsworth, "the Soviets were convinced that the United States was deliberately sabotaging the conference and was simply seeking a pretext to resume testing." If the Soviet Union were to accept the new data, chief Soviet delegate Semyen Tsarapkin predicted, "inspection group after inspection group will be roaming all over the country." Needless to say, the Soviets not only rejected the new data, they even refused to give it an objective hearing.[54]

In addition to the HARDTACK II results, another verification problem was raised by data gleaned from a series of tests, code-named Project Argus, that was conducted by the Navy in September 1958. In the Argus series, three nuclear devices were exploded some 300 miles above the surface of the South Atlantic Ocean. Not only did the explosions, in the low-kiloton range, fail to produce any fallout, they also were undetectable by ground observers. On the basis of the Argus data, Edward Teller asserted that outer space provided a potential violator with an opening large enough through which "to drive a herd of elephants." Even though, as Stanford University physicist Wolfgang Panofsky pointed out, the "incentives were not plausible for tests to be conducted under such extreme conditions," the possibility that nuclear tests could be conducted in space without being detected was, for some opponents of a test ban, sufficient reason to reject any test ban agreement.[55]

On the other hand, those who favored a test ban believed that there was a far greater danger than the possibility of Soviet cheating: a nuclear war arising from the failure to restrain the nuclear arms race. Senator Hubert Humphrey believed that the main obstacle in the way of a test ban agreement was the reluctance of its opponents to take some risks. They wanted, he said, an "absolutely perfect, absolutely foolproof"—and absolutely unattainable—inspection system. As a more realistic alternative, Humphrey proposed an inspection system based on the deterrence principle; that is, a system that would "deter a violator because he can never be sure that he won't be caught." What was needed, he believed, was a number of annual inspections sufficiently large to deter Soviet cheating, but small enough to be politically acceptable to the Soviets.[56]

Rather than attacking the test ban idea directly, however, some of its critics attempted to reduce its scope. John McCone, who succeeded Strauss as chairman of the AEC in July 1958, advanced an idea first proposed by Democratic Senator Albert Gore of Tennessee. The United States, Gore had suggested, should resume testing, but only beneath the surface of the earth. Underground testing, he argued, would not only reduce fallout and enable the United States to develop new nuclear weapons, it might also make it possible to develop improved test detection techniques that eventually could make a total test ban feasible.[57]

Dismayed by the impasse in the negotations produced by the inspection issue, Eisenhower leaned toward the approach proposed by Senator Gore. On April 13, 1959 he wrote Khrushchev a letter proposing an agreement that would ban only those activities that could be monitored, beginning with tests beneath the surface of the oceans and in the atmosphere below an altitude of fifty kilometers. Eisenhower's offer was rejected by Khrushchev. He called it a "dishonest deal" because it would permit the United States to continue to develop its nuclear arsenal with underground tests.[58]

Khrushchev then tried to break the inspection impasse by proposing an approach that was first suggested to him by British Prime Minister Harold Macmillan in February 1959. It was the same approach favored by Senator Humphrey—simply a small number of annual inspections. Eisenhower accepted the Soviet proposal contingent on Khrushchev's willingness to modify his position on the veto in the control commission and his acceptance of the need for further technical talks on the problem of detecting high-altitude tests. On April 27 a breakthrough was achieved when Tsarapkin stated that his government would drop its demand for a veto over on-site inspections provided that the West agreed to a fixed number of annual inspections. On May 14 Khrushchev agreed to continued Soviet participation in the technical talks, but said nothing about the veto issue. Eisenhower ignored the omission, and the technical talks resumed in July 1959.[59]

No sooner did the talks resume, however, than they were confronted with another problem, another one raised by Edward Teller. Soon after the Geneva Technical Conference had produced its report during the summer of 1958, Teller had urged the Rand Corporation to punch a hole in its findings. Albert Latter, working for Rand, succeeded in doing just that. Latter asserted, in what came to be called his "decoupling" theory, that it would be possible to conduct nuclear tests in underground caverns large enough to muffle, and thereby conceal, the seismic impact of the explosions. A 300-kiloton explosion, Latter explained, could be decoupled in that it would register on seismographs as only a one-kiloton explosion. Even Hans Bethe conceded that Latter's decoupling theory was possible, but, he pointed out, building an underground cavern large enough to muffle underground explosions would be difficult, costly, and easy to detect. James Killian, Eisenhower's science adviser, believed that the "incentives were not plausible for tests to be conducted under such extreme conditions." Privately to Killian, Latter admitted that no matter what advances might be made in the detection techniques, "Teller would find a technical way to circumvent and discredit them."[60]

Killian attempted to circumvent Teller's obstructions with another scientific panel, chaired by Lloyd Berkner, president of Associated Universities. The Berkner Panel concluded in a report presented to the Geneva Conference in June 1959 that the use of "deep-hole" seismometers, and the addition to the

Geneva system of unmanned seismic stations at 175-kilometer intervals in seismic areas could make possible the identification of ninety-eight percent of the seismic events as small as a one-kiloton equivalent. Even with the suggested improvements, however, the Berkner Panel estimated that there would be 300 worldwide continental earthquakes above a five-kiloton equivalent that could not be identified as such, compared to the twenty to one hundred undetectable earthquakes originally estimated by the Geneva Technical Conference.[61]

Despite Teller's efforts to scuttle the test ban talks, the Technical Conference worked through the summer and fall of 1959 to save them. One group of experts, the so-called Technical Group I, worked on the problem of detecting high-altitude tests. In July it concluded that the problem could be overcome by placing five or six satellites in orbit 18,000 miles above the earth. In November another working group, "Technical Group II," was established to consider the HARDTACK II data, the decoupling theory, and a list of possible grounds for dispatching on-site inspection teams. While the Soviets accepted the possibility that underground explosions could be decoupled, they still refused to consider the HARDTACK II data and set the conditions for on-site inspection so rigidly that the likelihood of effective inspection was precluded. As a result, the Geneva Technical Conference was unable to come to an agreement on the inspection problems. It ended its work on December 18 in an atmosphere of scientific hostility unprecedented in the postwar history of arms control talks.[62]

The only bright spot in what otherwise was a dismal December for arms control was the signing on December 1 of the Antarctic Treaty. The agreement, which was signed by twelve countries—including Britain, France, the Soviet Union, and the United States—demilitarized the Antarctic, provided for full multilateral inspection, and prohibited the dumping of radioactive wastes there.[63]

The sense of satisfaction that accompanied the signing of the Antarctic Treaty could not, however, overcome the gloom that resulted from the failure of the Geneva Technical Conference. In fact, Eisenhower was so aggravated by the demise of the conference that on December 29 he threatened to resume testing in the new year. The Soviets responded by stating that they would refrain from testing so long as the United States continued to do so. Unwilling to suffer the opprobrium of world opinion by being the first to resume testing, and increasingly sensitive to the danger of fallout, Eisenhower decided to continue the test moratorium on a day-to-day basis. And although the test ban was continued, the failure of the Geneva Conference after fourteen months of negotiation marked a new low point in the effort to conclude a test ban that would be comprehensive in nature—that is, covering all nuclear weapon tests.[64]

The Administration's Last Attempt at a Test Ban, 1960

Faced with the inability to work out a comprehensive test ban agreement, the Eisenhower administration returned to its earlier bid for a partial ban on testing. On February 11, 1960 it proposed, in a further modification of its April 1959 offer, a phased approach to the inspection problem. In the first stage of the agreement, only those tests would be prohibited that could be monitored by the Geneva system. In later stages, as detection capabilities improved, the prohibitions could be extended. The administration suggested that all atmospheric and underwater tests would be prohibited in the first stage, as well as tests in space above an unspecified height, where detection procedures were considered unreliable. Underground tests that produced seismic signals greater than 4.75 on the Richter scale would also be prohibited. This threshold was selected because it corresponded to the Western evaluation of the detection and identification capabilities of the Geneva system. The administration also made an important concession by indicating its willingness to accept a limited number of annual inspections. It now considered a quota of twenty annual on-site inspections sufficient to deter Soviet cheating.[65]

The Soviets, at first, responded negatively to the administration's latest offer. They believed that the only acceptable solution was "a plan to end all tests." But it did not take the Soviets long to realize the futility of pursuing a complete ban on tests. On March 19 they stated that they would consider the American offer. However, they also indicated that at least three modifications of the plan would be necessary before it became acceptable to the Soviet Union. First, all tests in space would have to be prohibited, regardless of whether they could be detected. Second, a four- to five-year moratorium on all underground tests below the 4.75 Richter-scale threshold would have to be observed. Third, the number of annual on-site inspections would have to be treated as primarily a political, rather than a scientific, issue. What the Soviets wanted in this regard was a small number of "symbolic" inspections that would simply express the good intentions of the parties to uphold a test ban treaty and not a rigorous inspection that would determine with scientific preciseness whether or not violations had occurred.[66]

Hubert Humphrey was elated by the Soviet offer. He believed that it had brought a comprehensive test ban "in sight." He thought the gap that still separated the two sides could be closed if the Soviets would only compromise on the number of annual inspections and if the West would accept a continued moratorium on small underground tests. On the other hand, hardliners, like John McCone, believed a moratorium below the 4.75 threshold would enable the Soviets to cheat and to continue to develop their nuclear arsenal. McCone denounced the Soviet counterproposal and strongly recommended that the President reject it. The vehemence of McCone's rejection upset Eisenhower. He angrily told the AEC chairman that he was out of line, that he was only "an operator, not a foreign policy maker." Eisenhower insisted that it was absolutely essential to halt the "terrific" arms race, and that he was determined to do so before he left office, at least by stopping nuclear testing in the atmosphere.[67]

Alarmed by the uproar that the Soviet counterproposal had caused within the Eisenhower administation, Prime Minister Macmillan decided to travel to the United States at the end of March 1960 to confer with the President. After a day and a half of discussions at Camp David, the two leaders agreed to accept the Soviet conditions. They would accept a continuation of the moratorium on underground tests beneath the 4.75 threshold, but for only one or two years, provided that the Soviets agreed to sign a treaty barring all verifiable tests and creating a coordinated seismic research program. When Tsarapkin reacted favorably to the Western proposition, the general anticipation was that a test ban treaty could be concluded at the Paris summit meeting that both Eisenhower and Khrushchev had agreed to attend in May.[68]

Just as the superpowers appeared to be moving toward a comprehensive test ban, an event occurred with fatal impact on the agreement. On May 7 Khrushchev announced that six days earlier the Soviet Union had shot down an American U-2 aircraft deep inside Soviet territory. The aircraft's pilot, Francis Gary Powers, survived the crash and was being held in Soviet captivity. The incident seemed to confirm Eisenhower's earlier prediction that "some day one of these machines [the U-2] is going to be caught, and we're going to have a storm."[69]

Khrushchev, at first, tried to give the President a face-saving way out of the administration's embarrassment. He stated that he was prepared to accept that Eisenhower knew nothing about the U-2's mission, but he also wanted the President's assurance that similar flights would not be repeated. Eisenhower, however, refused to evade responsibility for the incident. To do so, he believed, would be an admission on his part that he was not fully aware of his nation's military activities, especially one as sensitive as the U-2 flights over Soviet territory. Turned down by the President, Khrushchev then demanded an apology. Eisenhower angrily refused. He told De Gaulle that he was not about to "crawl" on his knees to the Soviet leader. Khrushchev responded by angrily denouncing the President and by canceling the invitation he had extended to him to visit the Soviet Union. Obviously, in the atmosphere of acute superpower hostility produced by the U-2 affair, the Paris summit meeting had little chance of success. On May 19, only two days after the summit began, Eisenhower left Paris. Although the Geneva test ban talks continued until the end of the year, it was clear that the summit represented the last opportunity to finalize a test ban agreement during the Eisenhower presidency.[70]

The collapse of the Paris summit not only doomed the test ban, it also contributed to a further deterioration of Soviet-American relations. A crisis in the Congo erupted in July 1960. Communist insurgency escalated in Indochina. American relations with Fidel Castro's Cuba reached the breaking point, and preparations for an American-backed invasion of that island nation began. Alarmed by the deteriorating world situation, Eisenhower was prepared to resume nuclear weapon tests, and would have done so had Nixon been elected in the November election. After Kennedy won, Eisenhower advised the president-elect to resume testing as soon as possible.[71]

7

Kennedy, Nuclear Weapons, and the Limited Test Ban Treaty, 1961–1963

An Opportunity for Arms Control

Jerome Wiesner, President John F. Kennedy's special assistant for science and technology, believed that an excellent opportunity for halting the nuclear arms race existed at the beginning of 1961. "Soon after President Kennedy took office," Wiesner recalled, "we learned that the Soviet missile force was substantially smaller than earlier estimates which provided the basis for the so-called missile gap." Wiesner and others in the administration proposed restraint in developing the American nuclear arsenal, hoping and expecting the Soviets would follow suit.[1]

Indications at the beginning of the Kennedy presidency suggested that the Soviets might be prepared to curtail their nuclear arms program and accept a thaw in the Cold War. On Kennedy's inauguration day, Khrushchev sent the new President a warm letter of congratulations. A few days later, the Soviets released two American Air Force officers whose RB-47 reconnaissance plane had been shot down over Soviet territory the preceding July. However, shortly before Kennedy took office, Soviet Deputy Foreign Minister Vasily Kuznetsov had warned Wiesner personally that the Soviet Union would not stand by idly if the new American administration embarked on another arms buildup.[2]

Kennedy responded to Khrushchev's gestures by removing restrictions on the importation of Soviet crab meat and by proposing agreements for more consulates and scientific and cultural exchanges. But he also ignored Kuznetsov's warning and began what his close aide and biographer, Theodore Sorensen, described as the most rapid and large-scale military buildup in American peacetime history until that time. By mid-1964, the number of nuclear warheads in America's arsenal had increased by 150 percent and its total megatonnage had more than doubled since Kennedy took office.[3]

Kennedy and the "Missile Gap"

Why did Kennedy ignore the opportunity for halting the nuclear arms race?

One explanation offered by administration partisans held that Kennedy was not fully aware of the extent, or even the existence, of American ICBM superiority until the fall of 1961, when the administration's review of the American-Soviet military equation was completed. According to this interpretation, Kennedy was under the impression that the United States had to close a missile gap that he believed favored the Soviet Union before he could seriously undertake negotiations with the Soviet leaders.[4]

Others have argued that Kennedy realized very early in his presidency, if not before his inauguration, that a missile gap favorable to the Soviet Union did not exist. According to historian Henry Trewitt, Kennedy was informed by a Navy intelligence officer during the Presidential campaign that the national intelligence estimates did not support the figures concerning Soviet missile strength that he was using. "Kennedy would not accept the [intelligence] figures," Trewitt wrote; "he was, after all, a politician just short of the greatest prize the nation could offer."[5]

Both Kennedy and his defense secretary, Robert S. McNamara, in fact were briefed on the American-Soviet strategic relationship by the CIA before the inauguration, if not before the election. McNamara seemed to reflect the knowledge he gained from the CIA in a February 6, 1961 background press briefing. He told reporters there was "no missile gap." Yet the next day Kennedy stated that McNamara had meant no such thing. He also refused to comment further on the missile gap until the administration had completed its review of the American and Soviet strategic relationship.[6]

It was only after the review was completed, in the fall of 1961, that the administration officially admitted that the Soviets were not pursuing a massive ICBM buildup. On November 8 Kennedy told a press conference that American military power was second to none. However, it took another year before the administration offered an explanation of why it had accepted the validity of the missile gap in the first place. "The myth" of the missile gap, McNamara said in late 1962, "was the result of incomplete intelligence; although it was created by intelligence analysts acting in good faith, it was a myth all the same."[7]

Although American intelligence was at least partly responsible for the creation of the missile gap myth, it was nevertheless possible, if not likely, that Kennedy refused to accept the political embarrassment that would have resulted from an admission that he had been wrong, or untruthful, about Soviet missile strength, particularly after Congress had approved supplementary defense funding he had insisted was necessary. By claiming that the missile gap issue was under review during most of 1961, Kennedy may have wanted to divert public and congressional attention from the nonexistent gap until it was no longer an issue, or until after he had obtained appropriations for his defense program. For whatever reason, the end result was still the same, another massive buildup of the American nuclear arsenal.[8]

Flexible Response

Why did Kennedy initiate this unprecedented nuclear buildup? Part of the explanation is related to the fact that the new President was boxed in by his past and frequent criticisms of Eisenhower's defense policies. Under Eisenhower, Kennedy charged that "we tailored our strategy and military requirements to fit our budget—instead of fitting our budget to our military requirements." Unlike Eisenhower, he believed the United States had sufficient wealth to finance both increased domestic and military spending.[9]

As a senator, Kennedy had also frequently castigated Eisenhower's "dangerous dependence" on America's nuclear deterrent forces. Massive retaliation, he said, could deal with the threat of an all-out nuclear attack, but it could not deal effectively with communist aggression below the nuclear threshold. Nuclear weapons, Kennedy insisted, were not usable without destroying the area one intended to defend. America, he stated, needed a strategy that would enable her to meet the full spectrum of communist challenges, from nuclear war to guerrilla infiltration.[10]

The "flexible response" strategy championed by General Maxwell Taylor, whom Kennedy made chairman of the Joint Chiefs of Staff, fit the new President's requirements. It would enable the United States "to respond anywhere, anytime, with weapons and forces appropriate to the situation," Taylor asserted. But the new strategy also required a major expansion of America's conventional forces. It was a requirement Kennedy went a long way toward fulfilling. During his administration, the number of Army divisions was increased from eleven to sixteen, the number of tactical air squadrons was increased by thirty percent, and the number of ships in the fleet was doubled. One would think that with more conventional forces the administration could have reduced the size of the nation's nuclear arsenal. Obviously, that did not happen.[11]

A Damage Limitation Strategy

The Kennedy administration wanted more nuclear weapons partly because it wanted greater flexibility and control in using them. In June 1962 McNamara explained that a more flexible nuclear strategy was needed to implement a "damage limitation" strategy. In the event of a Soviet attack on America's allies, or one limited to American military installations, as McNamara explained the new strategy, the United States would attempt to avoid damage to its own cities by retaliating initially against only Soviet military installations (a counterforce attack), rather than Soviet cities (a countercity attack). By contrast, the Eisenhower targeting strategy, contained in the so-called Single Integrated Operational Plan (SIOP), had called for simultaneous nuclear strikes against Soviet cities and military installations as well as those of Communist

China. Superior American nuclear forces were required to implement a damage limitation strategy, McNamara believed, in order to give the United States an "assured destruction" capability—that is, sufficient forces, even after a Soviet first-strike, to retaliate effectively against Soviet military installations and, if necessary, Soviet cities as well.[12]

McNamara's damage limitation strategy was subjected to severe criticism almost as soon as it was announced. Some critics stated flatly that it would not be possible to limit damage in a nuclear exchange. Because fallout would be widespread, millions would be killed even if cities were not directly attacked. Some critics wondered how the Soviets, who lacked counterforce-capable forces in the 1960s, could refrain from attacking American cities in return for restraint by the United States. Moreover, even if a President kept his wits sufficiently calm to launch a limited retaliatory attack, the critics asked, how effective could it be? The primary military targets, Soviets ICBMs, would have already been expended in the Soviet first-strike, leaving only empty silos for American missiles to destroy. Primarily, for this reason, many critics felt that the only value in having a counterforce capability was to enable the United States to strike first, before Soviet ICBMs were launched. An American first-strike capability, they warned, would be inherently destabilizing. It would tempt the Soviets to strike before their ICBMs were destroyed by an American preemptive attack. Moreover, it would require the Soviets to build more ICBMs, as they eventually did, in order to have a retaliatory capability sufficient to deter an American first-strike.[13]

Criticism of McNamara's damage limitation strategy was not restricted to the United States. West Europeans feared that an American desire to limit damage to the Soviet Union following a Soviet attack on a NATO country would weaken deterrence and thereby make Soviet aggression more likely. Some Europeans suspected that the flexible response strategy, both in its nonnuclear and nuclear components, was designed to limit a war to Europe in order to save American cities from nuclear destruction. By reducing European confidence in the American nuclear deterrent, some critics charged, McNamara's damage limitation strategy reinforced the determination of the French to increase the size of their own nuclear arsenal.[14]

In the face of these criticisms, the administration took a number of steps in attempting to reestablish European confidence in the American nuclear deterrent. The number of American tactical nuclear missiles deployed on the continent was increased by more than sixty percent, even though the administration considered them vulnerable to enemy attack or seizure, and impossible to use in battle without causing extensive collateral damage to populated areas in Central Europe. In addition, the administration agreed to sell Polaris missiles to the British after Kennedy had scrapped, primarily for reasons of cost-effectiveness, the Skybolt air-to-ground missile, which he had promised to the British for use on their Vulcan bombers. Partly to ward off the development of an independent, NATO nuclear force, as well as to "couple" the nuclear defense of Europe to the United States, the administration in May 1963 agreed to create an inter-allied nuclear force consisting of the British Vulcans and five American

Polaris submarines. Simultaneously, the administration began negotiations with the allies to create a NATO, mixed-manned fleet of ships armed with Polaris missiles. The creation of this so-called multilateral nuclear force, however, proved to be unworkable primarily because neither the British nor the French were willing to give up their own independent nuclear forces. In the end, the administration's measures only marginally increased European confidence in the American nuclear deterrent. They did, however, substantially increase the number of American nuclear weapons deployed in Western Europe—a development that did not go unnoticed by the Soviets.[15]

To no small extent, the ease with which Kennedy and McNamara accepted the damage limitation strategy and the Soviet first-strike scenario, on which it was based, was directly related to their own limited experience with national security affairs. The President admitted reading only one contemporary military book, while McNamara's experience with national security affairs before he became secretary of defense was limited to a three-year hitch in the Air Force during World War II. While it can be argued that every president and every secretary of defense have to go through an on-the-job training period, it is also true that all of the major Kennedy-McNamara defense decisions were made during that period.[16]

Although the Soviet first-strike scenario stretched credibility beyond its limits in the early 1960s, considering the fact that the Soviets had less than a hundred ICBMs at that time, and although the administration had no evidence to believe that the Soviets were about to engage in a massive ICBM deployment program, McNamara still felt that it was necessary to plan for the day when the Soviets would have sufficient ICBMs to make a Soviet first-strike a plausible possibility. Rather than attempting to persuade the Soviets to restrain their nuclear programs by engaging in a policy of American nuclear self-restraint, the administration, in effect, attempted to intimidate the Soviets with a massive American nuclear buildup. The Soviet first-strike scenario came much closer to becoming a reality after the Soviets responded to the American nuclear buildup with one of their own.[17]

Nuclear Superiority

According to one Kennedy adviser, physicist Herbert York, worst-case analysis during the Kennedy administration was used primarily as an attempt to justify the maintenance of American nuclear superiority rather than as a planning device. Without a doubt, the Congress, the Pentagon, the defense industries, and a great majority of the American people expected the administration to maintain America's wide nuclear lead over the Soviet Union. Since the advent of the atomic age, not only America's security, but also her national prestige, had been intimately tied to her overwhelming superiority in nuclear weapons. In addition, without strong nuclear forces, the administration feared, Europe would not accept the American program to strengthen the alliance's conventional forces, one of the key components of the flexible response strategy. While the Kennedy administration wanted strong conventional forces to defend Western Europe in case deterrence failed, the Europeans wanted strong nuclear forces to ensure that deterrence would continue to succeed.[18]

Not everyone, however, believed that nuclear superiority was either necessary or wise: General Taylor, an advocate of the so-called finite deterrence school, believed that the purpose of nuclear weapons was to deter a nuclear war and not to fight one. Taylor believed that a "few hundred reliable and accurate missiles" would be sufficient to deter a Soviet attack on the United States. Jerome Wiesner, Kennedy's science adviser, indicated that the United States could deter a Soviet attack with as few as twenty delivery vehicles. But Wiesner also realized that while a small nuclear deterrent force might be a sound military proposition, it was a political impossibility; the pressure to continue nuclear weapons production was enormous. In August 1963 Senator George McGovern (Democrat, South Dakota) discovered as much when he introduced an amendment that would have cut the strategic weapons budget for fiscal 1964 in half. The amendment was soundly defeated by a roll-call vote of 74 to 2. "How many times," McGovern asked in vain, "is it necessary to kill a man or a nation?"[19]

As in the past, much of the pressure to continue nuclear weapons production came from the Air Force, which insisted that the nation needed at least 3,000 ICBMs. McNamara considered that figure excessive. He recommended the construction of 950 ICBMs (a number he subsequently rounded off at 1,000), not because he believed that that many missiles were needed for effective deterrence, but because he thought that it was the smallest number the Congress would accept. I. F. Stone, a persistent critic of the military-industrial complex, found it "impossible to believe [that] decisions which cost so many billions— and were so profitable to the arms industry—were merely the result of intramural bureaucratic bargaining and appeasement."[20]

Kennedy and the Cold War

Kennedy also may have been limited in his ability and desire to wind down the nuclear arms race by the image of an ardent Cold Warrior he had successfully projected during his years as a member of Congress, and particularly during his successful campaign for the presidency. What he told an audience on September 23, 1960 was typical of the statements on communism that he had made for some time: "The enemy is the communist system itself—implacable, insatiable, uneasy in its drive for world domination." It is quite possible, even probable, as Kennedy's intimates have argued, that his public pronouncements on the communist menace were often nothing more than campaign rhetoric; that, in reality, Kennedy's policy toward the Soviet Union was less ideologically motivated than the Soviet policy of the Eisenhower administration. But it is also true that Kennedy's anti-communist rhetoric, by aggravating, rather than allaying, American public hostility toward the Soviet Union, may have precluded the possibility of cultivating popular and congressional support for arms control negotiations with the Soviet Union early in his administration.[21]

Kennedy biographer and confidant Arthur M. Schlesinger, Jr. was one who believed that the President was far more realistic than his campaign rhetoric implied. In the Third World, for example, Schlesinger argued, Kennedy often expressed sympathy for the efforts of the developing nations to free themselves from the last vestiges of colonial rule, and even asserted that Kennedy had no quarrel with communist regimes that had come to power by means of democratic processes. According to Schlesinger, what bothered him was the possibility—indeed, what he believed was the probability—that the Soviet Union would use Third World revolutions to alter a balance of power that definitely favored the United States. Indeed, in a speech on January 6, 1961 Khrushchev declared the support of his country for "wars of national liberation." It was this speech, Schlesinger wrote, that the President would hold up as the "authoritative exposition of Soviet intentions." These words of the Soviet leader "alarmed Kennedy more than Moscow's amiable signals assuaged him."[22]

Revisionist historians, however, have been much less charitable in interpreting Kennedy's Cold War policies. In his very first month in office, according to Bruce Miroff, Kennedy launched an anti-communist crusade which, despite his widely acclaimed realism, was far more ambitious, and dangerous, than anything ever attempted by his predecessor. Miroff wrote that "despite all of the talk of 'brinksmanship' in the Eisenhower-Dulles era of foreign policy, there was really nothing in that era comparable to the Berlin crisis of 1961 and the Cuban missile crisis of 1962," both of which represented the closest approaches to nuclear war made by the superpowers during the Cold War.[23]

According to Miroff, Kennedy's approach to the Soviets revealed not a particular ideological bent, but a psychological flaw. Kennedy, he asserted, suffered from an acute inferiority complex which was manifested in a perverse need to prove his leadership capabilities. Kennedy was aware, however, as he wrote in his book *Profiles in Courage*, that "great crises produce great men and great deeds of courage." He also realized that great crises rarely occurred in the domestic arena, and thus gave relatively little attention to his domestic programs. What alone could satisfy his desire for greatness, what alone could prove the paramount test of his courage, Miroff asserted, was the global struggle with the Soviet Union. Rather than ignoring Khrushchev's taunts and threats, as Eisenhower usually did, Kennedy welcomed them, absorbed them, almost reveled in them. "The call to an intensified Cold War struggle," Miroff wrote, "supplied his presidency with the aura of purpose and passion it otherwise lacked."[24]

Miroff went so far as to say that Kennedy's obsessive need to prove himself led him to create crises where crises need not have existed. In his first State of the Union address, only ten days after his inauguration, Kennedy stated: "No man entering upon this office . . . could fail to be staggered upon learning—even in this brief ten day period—the harsh enormity of the trials through which we must pass in the next few years." Why should Kennedy be so horrified? To be sure, the communists were active in Indochina and in the Congo, and Fidel Castro was moving toward an alliance with the Soviet Union. But, Miroff asked, could these setbacks be legitimately considered crises capable of placing the United States in jeopardy?[25]

For whatever reasons, whether they were primarily ideological, political, or psychological, in formulating his initial response to the Soviet Union, Kennedy initially chose to emphasize Khruschchev's more bellicose actions rather than his friendly gestures. Although he was inclined to negotiate an end to the Cold War, the Third World challenge which Khrushchev threw at him would have to be dealt with first. Apparently, only after Kennedy had proved to the Soviet leader and to his domestic critics that he was not "soft" on communism would diplomacy make any headway during his presidency.[26]

The Bay of Pigs Invasion

Although Kennedy had steeled the nation and himself, his first foray into Third World confrontation with the Soviet Union was a disaster. Shortly after taking office, he permitted the CIA to proceed with plans for an invasion of Cuba that Eisenhower had initiated in March 1960. It would be carried out by 1,500 CIA-trained Cuban exiles who would land on the south coast of Cuba at the Bay of Pigs. The invaders, the CIA anticipated, would serve as the nucleus of a popular uprising that would lead to the overthrow of Castro's communist regime. The invasion, which began on April 18, 1961, ended in abject failure. Kennedy, at the last hour, refused to permit any direct American military involvement, and as a result Castro's forces had little trouble isolating the invaders' bridgehead before counterattacking. After three days of fighting, the invaders surrendered and were imprisoned until Kennedy bought their release in December 1962 for $53 million worth of supplies.[27]

Kennedy, embittered by his humiliation in Cuba, lashed out at the CIA, the press, and especially the communists. On April 20, when it had become clear that the invasion had failed, he stated, "let the record show that our restraint is not inexhaustible." That same day he authorized American advisers in Laos, who had dressed as civilians, to don their military uniforms. On May 5 he announced that he was considering an expansion of American military aid to Vietnam. On May 25, for the second time since taking office, he asked Congress for a supplemental increase in defense appropriations. "There is much to suggest," wrote historian Louise FitzSimons about the Bay of Pigs, "that John F. Kennedy, consciously or unconsciously, spent the rest of . . . his life trying to recover from and make up for that initial colossal error."[28]

The Berlin Crisis of 1961

In the wake of the Bay of Pigs fiasco, Kennedy agreed to meet with Khrushchev in Vienna on June 3–4, 1961. The Soviet leader, who was probably still angered by Kennedy's military buildup, attempted invasion of Cuba, and rejection of Soviet peace feelers, tried to take advantage of the President's embarrassment over Cuba and his apparent belief in a missile gap favoring the Soviet Union with all the bombast with which he was capable. At Vienna, Khrushchev tried to impress upon Kennedy the Soviet view that the victory of communism in the Third World was inevitable. He also revived the Soviet demand for a German peace treaty, one that would end the Western presence in Berlin. If the West refused to comply by December, Khrushchev threatened, the Soviet Union would sign a separate peace treaty with the East Germans. If the Western allies attempted to maintain their garrisons in West Berlin after that treaty was concluded, force would be met with force, the Soviet leader promised, and the risk of a nuclear war would result. Kennedy responded to Khrushchev's Berlin threat by saying: "It will be a cold winter."[29]

In the two months following the Vienna summit, Khrushchev backed up his Berlin threat with action. On July 8, partly to increase the pressure on Berlin and partly as a response to Kennedy's military buildup, the Soviet leader suspended ongoing reductions in the size of the Soviet army and ordered an increase in military spending. To halt the outflow of East Germans to the West (estimated at 1,000 people per day), Khrushchev permitted the East German government to begin erecting the infamous Berlin Wall on August 13.[30]

Kennedy considered Khrushchev's Berlin challenge not only a threat to the freedom of West Berlin, but also a test of the American commitment to defend the entire free world. He feared that if the United States wavered in its determination to maintain a presence in West Berlin, the Soviets would be encouraged to attack American interests elsewhere. However, unlike Eisenhower's reaction, Kennedy's initial response was military rather than diplomatic. As a show of American determination, 1,500 U.S. soldiers were sent down the *Autobahn* in armored vehicles, and another 150,000 Army reservists were called to active duty. If necessary, Kennedy was prepared to go further. He told Schlesinger that he believed "that there was one chance out of five for a nuclear exchange." When he requested a $207 million appropriation from Congress for an augmented civil defense program, many Americans began to build fallout shelters in their backyards.[31]

Fortunately, the Berlin crisis abated on August 21, after Kennedy decided to accept Khrushchev's "feelers" for a negotiated settlement of the crisis. Although the talks, which began in September, proved inconclusive, Kennedy gave Khrushchev a face-saving way to back down from a dangerous confrontation with the West. By the end of the summer, the Berlin crisis was over.[32]

The Resumption of Nuclear Testing

Yet the waning of the Berlin crisis did not produce an immediate improvement in East-West relations. Quite to the contrary, on August 30, 1961 Khrushchev announced the resumption of Soviet nuclear testing. In the next sixty days, the Soviets conducted over fifty atmospheric nuclear tests. The total megatonnage of the exploded devices exceeded the yields of all previous tests combined.

The highlight of the Soviet test series came on October 30, when the Soviets detonated the most powerful thermonuclear device ever tested, with a yield of fifty-eight megatons.[33]

The Soviet decision to resume testing, it appears, was motivated primarily for military reasons. In all probability, Khrushchev was under considerable pressure from his military leaders to resume testing before the Americans could develop the first-strike capability Kennedy's nuclear buildup seemed to indicate the United States was pursuing. With the United States showing little restraint in deploying new nuclear weapons, Khrushchev probably found it impossible to restrain his own military. The resumption of testing, in the end, enabled the Soviets to make significant improvements in their nuclear forces, including reductions in sizes of warheads and increases in warhead yields. But the resumption of Soviet nuclear testing also served a political purpose. Khrushchev admitted that the tests were designed to shock the West into negotiating a settlement of the Berlin issue as well as to motivate the Americans to halt their nuclear buildup. Kennedy obviously was not alone in believing that military power is a prerequisite of effective diplomacy.[34]

Obviously, the resumption of nuclear testing by the Soviets placed enormous pressure on Kennedy to follow suit. Since February, the Joint Chiefs of Staff had been pushing for a renewal of the American test program. At first, Kennedy resisted. He wanted the full force of world opprobrium for resuming testing to fall on the Soviets. On September 3 he joined Prime Minister Macmillan in calling for an immediate cessation of atmospheric testing and excluded for the first time a demand for internal inspection. The proposal was scornfully rejected by Khrushchev.[35]

After the third Soviet test, Kennedy's patience evaporated. On September 5 he ordered the resumption of American testing, but limited the new tests to underground detonations. "What choice did we have?" Kennedy asked U.N. ambassador Adlai Stevenson. "They [the Soviets] had spit in our eye three times. We couldn't possibly sit back and do nothing at all." In the wake of the Bay of Pigs fiasco and the construction of the Berlin Wall, Kennedy was certain that Khrushchev "wants to give out the feeling that he has us on the run. . . . Anyway, the decision has been made. I'm not saying it was the right decision. Who the hell knows?"[36]

On September 15, 1961 the first American test since October 30, 1958 was conducted beneath the surface of the earth. It was not until March 1962 that Kennedy decided to resume atmospheric testing, primarily because the Soviets were testing in the atmosphere. McNamara, the Joint Chiefs of Staff, and hardline scientists like Edward Teller were afraid that the Soviets would use atmospheric tests to establish a nuclear lead, particularly in the development of an anti-ballistic missile system (ABM). But what really mattered in resuming American atmospheric testing, the President told a nationwide television and radio audience, was the image of the United States. "Should we fail to follow the dictates of our own security," he said, "they [the Soviets] will chalk it up, not to goodwill, but to a failure of will."[37]

The Cuban Missile Crisis

While the superpowers contaminated the atmosphere with unprecedented levels of radioactive fallout during the summer and fall of 1962, another threat to humanity began to develop in the Caribbean. In early 1962, the Soviets decided to deploy in Cuba forty-eight medium-range ballistic missiles with a range of 1,000 nautical miles and twenty-four intermediate-range ballistic missiles with a range of 2,200 nautical miles. In his memoirs, Khrushchev said that the deployments were made in order to deter another American invasion of that island-nation. Moreover, the United States had deployed IRBMs in Italy and Turkey ostensibly for defensive purposes; Khrushchev had no qualms about trying to do the same thing in Cuba. It is also possible that Khrushchev wanted a dramatic means of achieving a breakthrough on the Berlin problem, perhaps expecting that the successful deployment of missiles in Cuba would do much to neutralize American nuclear superiority, thereby enabling him to increase Soviet pressure on that beleaguered city.[38]

However, considering the risk Khrushchev ran in attempting to place Soviet missiles so close to the American homeland, some analysts believe that the Soviet leader's move was prompted primarily by a desire to defend the Soviet Union, rather than to protect Cuba or to pressure West Berlin. According to this view, the Soviet missile deployments in Cuba were designed to redress the growing threat to the Soviet homeland posed by Kennedy's rapid expansion of the American nuclear arsenal. Soviet fears about American nuclear superiority probably turned to nightmares when Deputy Secretary of Defense Roswell Gilpatric in October 1961 announced publicly that America not only enjoyed missile superiority over the Soviet Union, but that the United States was aware of how small—and vulnerable—the Soviet ICBM force actually was.[39]

When combined with McNamara's stated intention to build a counterforce capability, the news that the Americans realized the extent of Soviet ICBM inferiority probably only increased Soviet fears of an American first-strike. And yet, Khrushchev, for technological or economic reasons, was unable or unwilling to initiate an immediate counterbuildup of Soviet strategic nuclear forces. By placing in Cuba medium-range and intermediate-range missiles, which the Soviets had in larger numbers, the Soviet leader was effectively and inexpensively deterring an American first-strike against the Soviet homeland. Khrushchev admitted later that the Soviet missile deployments in Cuba "would have equalized what the West likes to call the balance of power."[40]

Kennedy, however, refused to allow Khrushchev to redress an imbalance of strategic power that favored the United States. After a U-2 photo-reconnaissance plane first spotted the Soviet missiles in Cuba on October 14, the President's closest advisers, the so-called Executive Committee, met to decide the nature of the American response. Not all the members of the committee favored a drastic American reaction. McNamara believed that the Soviet missiles, while reducing the warning time before American targets could be hit, did not significantly alter the Soviet-American strategic equation. The Soviets simply could not have enough missiles in Cuba to prevent the United States from delivering a massive retaliatory attack against the Soviet Union, even if they struck first. "A missile is a missile," he reportedly said. "It makes no great difference whether you are killed by a missile fired from the Soviet Union or from Cuba." He argued that the best response to the Soviet missile deployments was simply to ignore them, and thereby give them no unwarranted significance.[41]

Yet Kennedy could not ignore the Soviet missiles. Perhaps it was because of past humiliation suffered in Cuba, or maybe a result of playing nuclear poker with Khrushchev in Berlin, but Kennedy came to view the Cuban missile crisis as a personal test of his ability to lead the nation and the free world. He insisted that even though the missiles in Cuba did not "substantially" alter the strategic balance, they nevertheless created "in appearance" a Soviet strategic advantage. And appearances, Kennedy believed, contributed to reality. If he did not stop this Soviet adventure, others were sure to follow, and under circumstances much less favorable to the United States. Therefore, the Soviet missiles in Cuba would have to be withdrawn.[42]

How could the Soviets be persuaded to withdraw their missiles? U.N. ambassador Adlai Stevenson recommended a diplomatic approach to the crisis. The United States, he suggested, should offer to dismantle its obsolete Jupiter IRBMs in Italy and Turkey in exchange for the withdrawal of Soviet missiles from Cuba. Kennedy rejected Stevenson's recommendation immediately. The President, Schlesinger recalled, "felt strongly that the thought of negotiations at this point would be taken as an admission of the moral weakness of our case and the military weakness of our posture."[43]

At the same time, however, Kennedy refused to approve the opposite approach—direct American military action against the Soviet missile bases in Cuba. Over the objections of the Joint Chiefs of Staff, who wanted to blot out the Soviet missiles with air strikes, Kennedy decided on a naval quarantine of Cuba. After the Navy took its positions around Cuba, the President in a nationwide television address called upon Khrushchev "to halt and eliminate this clandestine, reckless and provocative threat to world peace." Kennedy backed up his words with other strategic moves. American forces in Florida began preparations for an invasion of Cuba in the event that one became necessary. More ominously, he prepared the armed forces for the possibility of nuclear war. One hundred and fifty-six ICBMs were readied for firing. SAC's B-47 bombers were dispersed to airports around the nation, while its B-52s were placed on alert—the most threatening alert in the history of the Air Force.[44]

Finally on October 28, after days of anguished waiting, Khrushchev agreed to withdraw the Soviet missiles. He had little choice, short of war with the United States, a war neither the Soviet Union nor the United States had any chance of winning. In return for meeting Kennedy's demand for the withdrawal of the missiles, Khrushchev exacted a pledge from the President that the United States would not again invade Cuba; there was also an unpublicized promise that, after the crisis had ended, the Jupiters would be withdrawn from Italy and Turkey. The American concessions enabled the Soviet leader to salvage some meager semblance of face.[45]

Aftermath of the Cuban Missile Crisis

Kennedy was praised nationwide, even by his Republican critics, for the masterful way he handled the Cuban missile crisis—a response characterized by Schlesinger as a "combination of toughness and restraint." Few discussed the possible implications had he failed. "The essential, the terrifying question about the missile crisis," I. F. Stone asked, "is what would have happened if Khrushchev had not backed down?" In the same vein, historian Louise FitzSimons wrote that Kennedy's "restraint in not invading or bombing Cuba has been praised, but many have forgotten, in the flush of success and relief from danger, that Kennedy was determined to force Khrushchev's total capitulation— no matter the cost." To remove Soviet missiles from Cuba, missiles which even Kennedy admitted affected the strategic balance only marginally, he was prepared to wage nuclear war. The President himself during the height of the crisis placed the likelihood of disaster at "somewhere between one out of three and even," and lamented the possibility that all the world's innocent children might never have the chance for full lives.[46]

Was it worth the potential risk of destroying the Northern Hemisphere, if not both hemispheres, to get the Soviet missiles removed from Cuba? Some did not think so. In FitzSimons words, "the only conceivable time when national leadership can be justified in risking a nuclear holocaust is in the face of an immediate and overwhelming threat to the very existence of the nation. . . . If all other threats are less than mortal, than the means to counter them must be primarily political; they should be removed by negotiation." Yet Kennedy during the Cuban missile crisis, as he did during the Berlin crisis, turned to negotiations only after military actions had been emphasized first. Journalist Walter Lippmann was among those who lamented the suspension of diplomacy early in the crisis. A private message to Khrushchev at that point, Lippmann wrote, would have been less dangerous than a public confrontation, and would have given the Soviet leader a chance to back down without humiliation.[47]

Revisionist historians have argued that image, not security, was the major factor behind Kennedy's actions during the Cuban missile crisis. Wrote Henry Pachter, Kennedy had been longing for an opportunity to correct Khrushchev's perception of his leadership abilities, a perception that was largely formed in the wake of the Bay of Pigs fiasco. "That this opportunity came in Cuba may have given him additional satisfaction," Pachter believed. "Here was a chance to cancel out last year's humiliation" in the very country where it had occurred.[48]

Another revisionist, Ronald Steel, has gone so far as to state that the real hero of the crisis was not Kennedy, but rather Nikita Khrushchev. Steel conceded that Khrushchev was unnecessarily bellicose, pompous, and insensitive to Kennedy's domestic situation. He was also foolish enough to believe that Kennedy would tolerate the presence of Soviet missiles in Cuba, the site of the President's greatest humiliation. Yet it was Khrushchev, not Kennedy, who backed down and saved the world from the scourge of nuclear war. To Steel, "Kennedy showed his skill in throwing down the gauntlet, but it required greater courage for Khrushchev to refuse to pick it up."[49]

Ironically, but perhaps not surprisingly considering the history of the nuclear arms race, the enhanced short-term prestige that Kennedy experienced in the wake of the Cuban missile crisis only translated into a greater long-term insecurity for the United States. The humiliation Khrushchev suffered at the hands of Kennedy during the missile crisis contributed to his ousting from power in October 1964. The new Soviet leadership, headed by Leonid Brezhnev, was determined to avoid a repetition of the humiliation Khrushchev experienced. Beginning in early 1965 the Kremlin embarked on a massive expansion of the Soviet nuclear arsenal, one that would enable the Soviet Union to achieve rough nuclear parity with the United States by the end of the decade.[50]

But the Cuban missile crisis also had beneficial consequences. For one, it helped to create a climate for productive arms control negotiations that had not existed since the abortive Paris summit conference of May 1960. Though both superpowers had made attempts to build momentum toward a nuclear test ban during Kennedy's first two years in office, the American and Soviet military buildups, the Bay of Pigs and Berlin crises, and the increased Soviet-American competition in the Third World, particularly in Indochina, had hindered those efforts. But the close brush with nuclear war experienced by the superpowers seemed to have deeply affected both Khrushchev and Kennedy. According to Schlesinger, Kennedy's "feelings underwent a qualitative change after Cuba: a world in which nations threatened each other with nuclear weapons seemed to him not just as irrational but an intolerable and impossible world." Moreover, the successful handling of the crisis had proved that Kennedy was not "soft" on communism. Bathing in the adulation he received in the wake of the Cuban missile crisis, he was able to exorcise from his mind the deadweight of the Bay of Pigs fiasco. He was free, at last, to demonstrate his capabilities as a statesman, which, in the last year of his presidency, proved to be considerable.[51]

The Test Ban Talks Resume

Although Kennedy's arms control policy built on the foundation laid by his predecessor, the new President enjoyed a number of advantages that Eisenhower had lacked, advantages that enabled him to achieve a major arms control agreement in an area where his predecessor had failed. One was Kennedy's conception of presidential power. Eisenhower had delegated much power and influence to subordinates, one result of which was to permit the likes of Dulles, Lewis Strauss, and John McCone to obstruct his efforts to achieve a test ban agreement. Kennedy, on the other hand, believed that presidential leadership should be centralized. He was not only less willing to delegate authority to subordinates, he made sure that opponents of arms control did not hold prominent positions in his administration. Both Secretary of State Dean Rusk and Defense Secretary McNamara saw the necessity for restraining the nuclear arms race. And so, too, did Glenn Seaborg, who replaced the hardline McCone as head of the Atomic Energy Commission, and William Foster, head of the new Arms Control and Disarmament Agency (ACDA), which was established by Congress in September 1961 to conduct arms control research and manage American participation in arms control negotiations and agreements.[52]

In addition, Kennedy came to see, to a much greater extent than Eisenhower, the collateral benefits that could result from the negotiation of meaningful arms control agreements. A test ban agreement would not only lessen the risks of a nuclear war by discouraging nonnuclear weapon states from acquiring nuclear arsenals, it could also serve as the basis for agreement on other issues, such as the wheat sale which the administration concluded with the Soviets in 1963. Moreover, a test ban would serve to widen the growing Sino-Soviet split by causing the Soviets to take a position on testing that was diametrically opposed to China's policy.[53]

Kennedy also benefitted from a number of technical developments that had occurred since Eisenhower left office, developments which increased his confidence that Soviet compliance with a comprehensive test ban could be verified. One was the ability of American seismologists to detect, with seismographs in the United States and on ships in the oceans, an underground test of a thirty- to fifty-kiloton nuclear weapon that France had conducted in the Sahara Desert on May 1, 1962. The experiment opened up the possibility of long-range seismic detection of nuclear explosions.[54]

In spite of enhanced verification capabilities that developed during Kennedy's presidency, the United States was still not willing to abandon on-site inspection. Yet the administration was willing to reduce the number of on-site inspections and to consider a smaller number of detection stations. Although the number of annual on-site inspections was left undetermined in a draft treaty presented to the Soviets on August 27, 1962, the American and British delegates said that the number would be less than the ten to twelve originally anticipated. Furthermore, if there were no unidentifiable seismic events, there would be no inspections. As an alternative to a comprehensive ban, the administration offered the Soviets another draft treaty on August 27. It provided for a partial test ban, permitting only underground testing, but requiring no international inspection. The partial ban would serve as a first phase to a comprehensive agreement, and the administration urged the Soviets to accept it immediately.[55]

But the Soviets rejected both draft treaties. In regard to the comprehensive plan, they argued that the improved prospects for verification made on-site, seismic inspection stations no longer necessary. They rejected the partial ban, on the other hand, because it was not comprehensive and because it "legalized" underground testing. The Soviets argued that the continuation of the nuclear arms race, which underground testing would make possible, was far more dangerous than the elimination of radioactive fallout. In effect, the Soviets threatened to continue atmospheric testing if the West would not stop its underground tests.[56]

It took the Cuban missile crisis to bring about the dramatic change in the Soviet position that made possible the conclusion of a test ban agreement. On December 19, 1962 Khrushchev sent Kennedy a personal letter in which he expressed his willingness to accept a limit of two or three annual on-site inspections, "when it was considered necessary," in seismic areas of the Soviet Union. The Khrushchev initiative was prompted by discussions between American and Soviet scientists on the possibility that the use of automatic recording stations, called "black boxes," would necessitate very few on-site inspections. In addition, on the basis of discussions between Arthur Dean, the administration's chief test-ban negotiator, and Soviet Deputy Foreign Minister Kuznetsov, Khrushchev apparently received the impression that three inspections would be sufficient to gain Senate ratification of a comprehensive test ban agreement. Yet in responding to the Soviet leader, Kennedy stated that Khrushchev had misunderstood the American position. The President informed him that three inspections were insufficient, not to mollify the Senate, but rather to guarantee American national security. Privately, Kennedy told his negotiators that he would accept no fewer than six annual inspections, although he authorized an initial offer of eight.[57]

While national security helped Kennedy determine that six annual inspections was the "rock bottom" number he would accept, the Senate also strongly influenced his decision. Kennedy had to contend with organized opposition from the Republican Conference Committee on Nuclear Testing, which had called a comprehensive nuclear test ban a "risky, unenforceable pact." Leading Democrats, too, were cool to a comprehensive agreement. Senators Thomas Dodd, Henry Jackson, and Stuart Symington stated that the administration was too willing to cut the number of inspections to satisfy the Soviets, while getting nothing in return.[58]

Khrushchev was not pleased by Kennedy's counteroffer. He said to journalist Norman Cousins, "I notified the U.S. I would accept three inspections. Back came the American rejection. They now wanted—not three inspections or even six. They wanted eight. And so once again I was made to look foolish. But I can tell you this: it won't happen again."[59]

More than a few analysts believed that the difference between the American and Soviet positions on the number of inspections was too small to prevent an eventual agreement. But, as AEC chairman Glenn Seaborg pointed out, there was more to the Soviet-American stalemate over the test ban than just their divergent positions on annual inspections. The Soviets flatly refused to discuss the technical details demanded by the Americans to implement on-site inspection until after an agreement had been reached on the number of inspections. It is also possible, Seaborg believed, that Khrushchev did not have sufficient support within the Soviet Politburo for a test ban agreement. The procedures for aerial inspection alone that were demanded by the Americans, assuming six inspections per year, would have opened for inspection 3,000 square kilometers of Soviet territory, including much of the strategically sensitive zone in central Asia. The United States, on the other hand, emphasized that it could not agree on a specific number of inspections until it had first determined, through a discussion of the technical details of the inspection process, how many annual inspections would be needed. The failure to reach agreement on the inspection issue more than any other factor was responsible for the ultimate inability of Kennedy and Khrushchev to conclude a comprehensive test ban treaty.[60]

Negotiating a Limited Test Ban Agreement

Although a comprehensive test ban seemed an impossible objective, both sides attempted to break the stalemate. On June 8, 1963 Khrushchev invited the British and Americans to Moscow to participate in a conference designed to negotiate a nuclear test ban treaty. Two days later, in a commencement address at American University, Kennedy accepted Khrushchev's invitation. He called on Americans to reexamine their attitudes toward peace and the Soviet Union. Peace without competition between the United States and the Soviet Union, he said, was not possible; however, the prevention of a nuclear war was. "In the final analysis," he told his audience, "we all inhabit this small planet. . . . And we are all mortal." Kennedy called upon the American people to support his effort to conclude a nuclear test ban agreement. It would be, he said, an initial step toward preserving the life of the planet. As a sign of American good faith in the ultimate success of the test ban talks, Kennedy announced that the United States would not conduct atmospheric nuclear tests as long as the Soviet Union employed similar restraint.[61]

The American University speech had an extremely favorable impact on the Soviet leadership. Khrushchev told Averell Harriman that it was the best speech delivered by an American president since Franklin Roosevelt. In response to the address, the Soviets on June 20 signed the "Hot Line" agreement. It established a direct teletype link between Moscow and Washington designed to reduce the risks of an accidental nuclear war as well as to ease tensions during international crises.[62]

Soon after the Moscow Conference convened on July 15, it became apparent that there was no chance of negotiating a comprehensive test ban agreement. Khrushchev in his opening remarks rejected any and all on-site inspections. Harriman, who headed the American delegation, responded by urging President Kennedy to abandon the quest for a comprehensive agreement and instead concentrate on getting one that was limited in scope. Since Khrushchev on July 2 had said he would support a limited agreement, Kennedy went along with Harriman's recommendation. Although he was pleased with the President's decision, Harriman nevertheless lamented the demise of the comprehensive ban idea. Years later he wrote: "When you stop to think of what the advantages were to us of stopping all testing in the early 1960s, when we were still ahead of the Soviets, it's really appalling to realize what a missed opportunity we had."[63]

AEC chairman Glenn Seaborg believed that the main obstacle to a comprehensive ban was the Soviet refusal to accept a system of on-site inspection. But also responsible, he admitted, was the American fear of Soviet cheating. That fear, he believed, was exaggerated. "Certainly," he wrote, "the Soviets did not wish to be caught cheating because of the great political embarrassment this would have caused." Moreover, he believed it was "doubtful that the clandestine tests the USSR might have undertaken in violation of a comprehensive treaty would have been militarily significant in the aggregate." Nevertheless, because the Joint Chiefs of Staff believed the Soviets would cheat, they informed the President that they could not support a comprehensive test ban. Without the support of the Joint Chiefs, Kennedy realized, a comprehensive agreement stood little chance of ratification. American as well as Soviet fear in the end blocked agreement on a comprehensive ban.[64]

With the comprehensive test ban idea summarily dismissed, the negotiations on a limited agreement were quickly concluded. On July 25 Britain, the Soviet Union, and the United States signed an agreement that prohibited the testing of nuclear weapons in the atmosphere, in outer space, and beneath the surface of the seas. At the insistence of the United States, the treaty included a clause providing for the right of withdrawal upon three-months' notice if a party found that "extraordinary events" had jeopardized its interests. At Soviet insistence, the treaty also prohibited peaceful nuclear explosions; the Soviets feared that they would be indistinguishable from nuclear tests for military purposes.[65]

Ratification of the Treaty

On August 8, 1963 President Kennedy placed the Limited Test Ban Treaty before the Senate for its advice and consent. He stated that, while the agreement would not guarantee world peace, it was a "first step toward limiting the nuclear arms race." Furthermore, he promised, the treaty would not only "curb the pollution of our atmosphere," it would help get the nuclear "genie back into the bottle" by prohibiting nuclear assistance to nations engaged in nuclear testing in the prohibited environments.[66]

Even though the Limited Test Ban Treaty was a much diluted version of the comprehensive plan, it still attracted considerable and often vehement opposition. The most effective testimony against the treaty that was heard in Senate hearings was, not surprisingly, that of Edward Teller. As so often in the past, when Teller had acted to block a test ban agreement, he asserted that any prohibition on testing would prevent the development of new nuclear weapons he considered vital to America's security. He insisted that the Limited Test Ban Treaty would interfere with the development of an anti-ballistic missile (ABM) system by prohibiting nuclear explosions in the atmosphere. He also argued that the treaty would prevent the United States from developing nuclear explosives for peaceful purposes, one of Teller's pet projects.[67]

The administration attempted to counter Teller's testimony with a bevy of pro-test ban scientists. One of them, Harold Brown, director of Defense Research and Engineering (and later Jimmy Carter's secretary of defense), argued that Soviet and American anti-ballistic missile research was about even and that, in any event, an ABM system was not feasible primarily because there would always be potentially more incoming warheads to inundate ABM sites than there could be ABMs to counter them. Norris Bradbury, director of the Los Alamos Scientific Laboratory, urged the senators, before they became "too bemused with megatons and multimegatons, . . . to look again at the pictures of Hiroshima and Nagasaki in 1945 after fifteen or twenty kilotons, not megatons."[68]

Among the military services, the Air Force provided the most vehement opposition to the treaty. General Thomas Power, the SAC commander, told the Senate: "I think it is a mistake to maintain any posture other than one of overwhelming military superiority," which, he said, could only be maintained by continued and unbridled nuclear testing. Initially, in varying degrees, the Joint Chiefs of Staff also opposed the treaty. In general, they felt the treaty required that Americans trust the Soviets on issues too vital to defense, offering little in return.[69]

Kennedy realized that the opinions of the Joint Chiefs carried considerable weight in the Congress, and that their support, or at least their neutrality, was essential to the ratification of the treaty. To get it, he agreed to four conditions that the Joint Chiefs offered him as the price of their support. The first required the administration to conduct comprehensive, aggressive, and continued underground nuclear testing. The second required maintenance of modern nuclear laboratories and programs to ensure the continuation of nuclear weapons research. The third condition called for the maintenance of the facilities to resume nuclear testing in the atmosphere should the United States deem the resumption of atmospheric testing essential to the national security or should the Soviets abrogate the treaty. The last condition required the administration to take steps to enhance America's capabilities to detect and monitor Soviet and Chinese nuclear activities.[70]

Kennedy also put considerable personal effort into getting public support for the treaty. He enlisted the assistance of prominent businessmen, scientists, clergy, farmers, educators, and labor leaders. He helped to organize, and then supervised, the activities of the Citizen's Committee for a Nuclear Test Ban. He personally targeted key senators, such as Henry Jackson, for special-interest group attention. He obtained Eisenhower's support by giving the former President the same assurances he had given the Joint Chiefs of Staff.[71]

The intensity of Kennedy's involvement in the administration's effort to secure ratification of the treaty paid off. Harris public opinion polls showed that unqualified support for the treaty jumped from forty-seven percent in late July to eighty-one percent by September 1. While Kennedy should be given much of the credit for bringing about this dramatic change in public opinion, it was also a result of ineffective opposition to the treaty, particularly by the Republican Party, which saw nothing to gain from opposing a treaty endorsed by Eisenhower. On September 24, 1963 the Limited Test Ban Treaty received an overwhelmingly favorable vote in the Senate—80 to 14—fourteen votes more than the required two-thirds majority. According to Theodore Sorensen,

Kennedy considered ratification of the treaty his greatest achievement. On October 11, 1963 the treaty went into effect.[72]

Aftermath of Ratification

The Limited Test Ban Treaty was in many ways a major step in the effort to control the nuclear arms race. By the end of 1980, 125 countries had given their assent to its provisions. With the ratification of the treaty, the atmospheric testing of nuclear weapons by the superpowers and Britain officially ended and, after the treaty went into effect, the problem of radioactive fallout was greatly diminished. However, the Limited Test Ban Treaty in no way ended the nuclear arms race. It simply drove it, quite literally, underground. Between 1945 and 1980 the number of announced American nuclear detonations totaled 638. More than half that number occurred after the Limited Test Ban Treaty was signed. In fact, on the very day of the treaty signing, the Senate passed the largest peacetime military appropriations bill up to that time in American history.[73]

Historian Richard Walton considered Kennedy's successful effort to gain ratification of the treaty a Pyrrhic victory. "With all the harm that has come from underground testing," in still allowing the nuclear arms race to continue, Walton wrote, "it might have been better to have had no treaty at all for a while longer in the hope that continued public pressure would have resulted in a comprehensive treaty." He lamented that Kennedy was "prepared to take far greater risks toward war"—an allusion to the President's handling of the Cuban and Berlin crises—"than he was toward peace."[74]

On the other hand, historian Bernard Firestone has argued that for Kennedy to have achieved "more than an incremental amelioration of the Cold War would have required a remarkable and fortuitous confluence of international and domestic changes," timely changes that did not occur in the early 1960s. Given the almost universal American distrust of the Soviets, which Kennedy only late in his presidency tried to diminish, given the intensity with which he pursued the American military buildup, as well as an interventionist approach in the Third World, it is understandable that he could go no further than the Limited Test Ban Treaty. Nevertheless, as Firestone pointed out, by producing a nuclear arms agreement, even one so limited as the Nuclear Test Ban Treaty, Kennedy helped to institutionalize the collaborative aspects of the Soviet-American relationship. In so doing, Firestone wrote, Kennedy "paved the way for the future integration of arms control considerations into defense policy—something that Eisenhower, despite his good intentions, could not or would not do." Indeed, Kennedy's successors would accept the necessity for strategic arms control largely because of the spade work he did during his short presidency.[75]

8

Johnson, Nuclear Weapons, and the Pursuit of SALT, 1963–1969

Johnson's World View

As a politician in Texas and in the U.S. Congress, both as a representative and as a senator, Lyndon Baines Johnson built a reputation as a "wheeler-dealer," as a man who got things done. Johnson brought this reputation with him to the White House. In terms of the quantity of domestic legislation, no president equaled Johnson before or since with the exception of Franklin D. Roosevelt. But while Johnson was a master of domestic politics, he knew little about the world outside the United States. "Foreigners," he once remarked, only half-jokingly, "are not like the folks I am used to." Johnson's inability to understand the conditions in the Third World, in places like Vietnam, would cost him and the nation dearly before he left office in 1969.[1]

Yet in spite of the failure of Johnson's Vietnam policy, he made notable progress in relaxing tensions with the Soviet Union, a policy that the media would soon call détente. During Johnson's presidency, the United States and the Soviet Union concluded a cultural exchange agreement, a consular convention, a commercial air-service agreement, a fishing agreement, and a treaty guaranteeing the return of astronauts or space vehicles that descended accidently on each other's territory. More important, progress was made in the area of arms control. In 1967 both superpowers signed a treaty banning weapons in outer space and successfully negotiated a nuclear nonproliferation treaty designed to prevent the spread of nuclear weapons to the nonweapon states of the world. Yet the Soviet-American agreement that Johnson wanted most, a strategic arms limitation agreement, eluded him while he was in the White House.[2]

Nuclear Doctrine

Johnson's inexperience with international relations was matched by the shallowness of his understanding of military strategy and particularly its relationship to diplomacy. More than Kennedy, Johnson was compelled to rely on the judgment of Secretary of Defense Robert S. McNamara when questions involving nuclear weapons arose.[3]

During the Johnson presidency the objectives of McNamara's nuclear strategy remained essentially what they had been since they were first defined in 1962: assured destruction and damage limitation. By 1967, however, the foundation on which they were based, continued overwhelming American nuclear superiority, was beginning to erode. As a result of the expansion of the Soviet nuclear arsenal which began in 1964, Soviet ICBM strength stood at about 720 in late 1967. By the end of 1969, the number of Soviet ICBMs reached 1,060, giving the Soviet Union for the first time a slightly larger ICBM force than the United States, with 1,054.[4]

Although in 1967 the United States continued to maintain overwhelming superiority in numbers of delivery systems (2,400 for the United States compared to 900 for the Soviet Union) and nuclear warheads (4,500 and 1,000, respectively), McNamara realized that because of the rapid expansion of the Soviet missile program the United States could no longer hope to escape massive destruction in the event of a nuclear war. "The fact is," he admitted in a San Francisco speech in September 1967, "neither the Soviet Union nor the United States can attack the other without being destroyed in retaliation." The assured destruction capability that once was an American monopoly had now become a mutual phenomenon, mutual assured destruction (or MAD, as its critics love to call it). No matter which superpower struck first and no matter how massive its attack, the other would still be able to launch a devastating retaliatory attack.[5]

McNamara also realized that any attempt by the United States to stay ahead of the Soviets in numbers of missile launchers or warheads would be futile. The Soviets sooner or later would match the American increments. In his San Francisco speech he publicly acknowledged for the first time the validity of the so-called action-reaction cycle. He conceded that the Soviet nuclear buildup was "in part a reaction to our own buildup since the beginning of this decade. Soviet planners undoubtably reasoned that if our buildup were to continue at its accelerated pace, we might conceivably reach in time a credible first-strike capability against the Soviet Union." While McNamara insisted that the acquisition of a first-strike capability was not the administration's goal, he also admitted that the Soviets "could not read our intentions with any greater accuracy than we could read theirs." As a result, both sides built up their forces "to a point that far exceeds a credible second-strike capabilty."[6]

McNamara also came to realize that a counterforce capability could be not only ineffective but extremely destabilizing. In a period of acute Soviet-American tension, the Soviets might be tempted to launch their missiles first, before they were destroyed by America's superior nuclear forces. As a result, he explained, "in a second-strike situation, we would be attacking, for the most part, empty sites from which the missiles had already been fired." Also disconcerting was the realization that the Soviets refused to accept the damage limiting strategy, contrary to McNamara's expectation. They continued to assert that any nuclear war would be an all-out nuclear war. The fact is the Soviets did not have the capability to implement a damage limiting strategy until the mid-1970s. Up to then, Soviet ICBMs and SLBMs were armed with high-yield, single warheads, primarily because Soviet guidance systems were too inaccurate to permit the use of smaller warheads. Obviously, the use of large-yield warheads in a Soviet nuclear attack on the United States, even one limited to American ICBM air and submarine bases, would cause extensive damage to American cities.[7]

McNamara also realized that any attempt on the part of the United States to maintain quantitative nuclear superiority was bound to be economically counterproductive. Such an attempt, he believed, would inevitably fall prey to the law of diminishing returns. In other words, significant increases in the number of American warheads would not significantly increase the capability of the United States to destroy the Soviet Union, even if the Soviets attacked first. Doubling the number of one-megaton warheads from 400 to 800, for example, would "only" kill nine percent more Soviet people and destroy "only" one percent more of the Soviet Union's industrial capacity, because fewer large cities would remain after the initial number of warheads were expended.[8]

In an attempt to break the action-reaction cycle, McNamara was now prepared to allow the Soviet Union a rough parity in numbers of missile launchers. He decided in late 1964, a few days *after* Johnson's election victory, to maintain the number of American ICBMs at 1,054 and SLBMs at a total of 656, a level that was achieved in October 1967. To diminish Soviet fears of an American first-strike, McNamara publicly downplayed the counterforce option on which his damage limitation strategy was based. In fact, by the time he left office in 1968, damage limitation was rarely mentioned as a goal of American nuclear strategy.[9]

Multiple Warheads

Yet even after the counterforce strategy was downgraded in the administration's rhetoric, counterforce options remained an integral part of the American targeting doctrine. And even while McNamara admitted that "our current numerical superiority over the Soviet Union in reliable, accurate and effective warheads is both greater than we originally planned and in fact more than we require," the Johnson administration approved the development and deployment of additional nuclear weapon systems to enhance the American counterforce capabilty.[10]

In 1964 the administration began equipping Polaris A-3 SLBMs with multiple reentry vehicles (MRVs)—that is, multiple warheads (three warheads on each missile), all of which could be delivered in a close pattern to the same target. Even more sophisticated were two additional weapons, the Minuteman III ICBM and the Poseidon SLBM, whose development the administration also initiated in 1964. Each would be equipped with multiple "independently targetable" reentry vehicles (MIRVs)—multiple warheads each of which could be internally guided to a different target. MIRVs were designed to be carried on a "bus" equipped with its own internal guidance system and with motors that would enable it to alter its course and velocity. When the bus reached a programmed position, it would release its warheads, which would then fall on their individual targets. The Minuteman III would carry three MIRVed warheads and the Poseidon ten warheads. As a result of MIRV deployments, more than 3,000 warheads were added to the American strategic arsenal by the mid-1970s.[11]

Considering that McNamara publicly admitted that additional warheads were unnecessary and a counterforce capability was potentially destabilizing, why then did the Johnson administration decide to develop and deploy additional counterforce weapons?

Part of the explanation is related to what American intelligence believed was the first step in the deployment of a nationwide Soviet anti-ballistic missile system. The first Soviet ABMs were deployed around Leningrad in 1962. Two years later, the Soviets began to deploy "Galosh" anti-ballistic missiles around Moscow. These deployments were augmented by the so-called Tallinn Line, a system of Soviet radar installations across the northwestern approach to the Soviet Union, which American intelligence believed was designed to implement a general Soviet ABM capability. American military planners feared that the Soviets eventually would be able to launch a first-strike attack with their ICBMs, SLBMs, and bombers, and then prevent effective American retaliation with their ABMs. MIRVed missiles, the Pentagon argued, would not only ensure that more American warheads would survive a Soviet first-strike, they would also give the United States the ability to overwhelm any Soviet ABM system with far more warheads than the Soviets could ever hope to shoot down.[12]

But the desire to counter Soviet ABM deployments was not the only factor motivating the American decision to deploy MIRVs. By placing more warheads on each American missile, stated John S. Foster, Jr., director of Defense Research and Engineering at the Pentagon, the United States would gain the capacity to destroy far more targets than could be destroyed by single-warhead missiles. What Foster and the Pentagon feared was the possibility that there would not be enough American warheads surviving a Soviet first-strike to retaliate against the some 1,500 Soviet targets on the American targeting list. Without sufficient surviving warheads, the Pentagon feared, the United States would have no alternative but to retaliate against Soviet cities.[13]

McNamara, in fact if not in rhetoric, never intended to abandon a counterforce strategy should nuclear war erupt. Even though he realized that the amount of damage limitation in a nuclear exchange would continue to decrease as the Soviets added more high-yield warheads to their nuclear arsenal, he still wanted to provide the Soviets with an incentive to refrain from directly striking American urban areas. Moreover, he believed, a counterforce capability was far more useful for diplomatic and military purposes than a targeting doctrine limited to Soviet cities. A flexible targeting doctrine and force posture would not only contribute to deterring Soviet probes in Western Europe and elsewhere, it would also give the United States what he regarded as usable nuclear power if a conventional conflict escalated into a nuclear war. Without the capability to use nuclear weapons, McNamara emphasized, America's nuclear deterrent would no longer be credible.[14]

An important consideration in McNamara's decision to develop MIRVs, then, was the degree of targeting flexibility they provided. The increasing accuracy of American guidance systems made it possible for the United States to rely on multiple, smaller-yield strategic warheads, like the Mark 12 (with a yield of 170 kilotons). If used in a counterforce attack on Soviet military installations, McNamara believed, the Mark 12 would reduce the collateral damage to surrounding areas, thereby providing the Soviets with incentive to refrain from attacking American cities. Even the Air Force was eventually persuaded that it would be better off with a mixed force of large-yield warheads and low-yield MIRVs than with only large warheads. With a mixed force, low-yield MIRVs could be used against Soviet "soft" targets, such as air and naval bases, while large warheads could be employed against "hardened" Soviet missile silos and command posts. In fact, by 1968 the Pentagon had become so satisfied with the capabilities of the Mark 12 MIRV that General Earle Wheeler, chairman of the Joint Chiefs, could state that an increase in the Minuteman force was no longer necessary.[15]

The MIRV decision was also prompted by economic considerations. With expenditures for the Vietnam War escalating from $12 billion to $25 billion annually, and with the President determined to simultaneously finance his Great Society domestic programs without raising taxes, the administration was hard-pressed to get funds for its strategic programs. McNamara considered MIRVs, at a cost in the hundreds of millions of dollars, a far more economical way to deal with the anticipated vulnerability of Minuteman than a full ABM system, at an estimated cost of $24 billion; or the 200 additional ballistic missiles, for $1.3 billion; and the B-70 long-range bomber that the Air Force badly wanted, at an estimated cost of between $8.9 billion and $11.5 billion for development and deployment over a five-year period. McNamara used the anticipated savings from MIRV deployments to justify his decisions to cancel the B-70 project and to retire 345 older B-52s and two squadrons of B-58s.[16]

McNamara had other noneconomic reasons for reducing the nation's bomber force. To the defense secretary, the bomber was no longer more important than the ballistic missile. Bombers on the ground were more vulnerable to destruction than missiles, would take longer to reach their targets (up to twelve hours, compared to ten to thirty minutes for SLBMs and ICBMs), and due to Soviet air defense systems were likely to be less effective than missiles against fixed targets, such as military installations. At the same time, however, McNamara did not plan to totally eliminate bombers; they remained a key ingredient of the "triad" of American strategic forces, which also included ICBMs and SLBMs. Without a bomber force, McNamara feared, the Soviets could attack America's ICBMs and, if they deployed a nationwide ABM system, could hope to escape unacceptable devastation from a weakened American retaliatory response. But since he expected the role of the bomber to be limited to "mopping up" operations in a nuclear war, he did not think a large force of bombers necessary.[17]

The decision to deploy MIRVs helped the defense secretary dampen the outcry of the Air Force and its allies against his bomber decisions and his plan to halt Minuteman deployments at the 1,000 level. Adding MIRVs to the Minuteman ICBMs would allow the Air Force to meet the demands of its target list without as many bombers as before and without building additional ICBMs. By not building these systems and instead relying on MIRVs, McNamara believed, the administration would not only save money, it could also take a step toward breaking the action-reaction cycle; he hoped that this would then set the stage for meaningful strategic arms limitation talks.[18]

There were also purely political reasons for developing the MIRV. Johnson had been attacked by Barry Goldwater, the President's Republican opponent in the 1964 election, for not adding any new weapon systems to the American nuclear arsenal. Johnson did not want to be criticized during the 1968 presidential campaign for the same reason. Moreover, by enabling the United States to retain superiority in numbers of warheads at a time that it was conceding the Soviets parity in numbers of launchers, Johnson expected the MIRV program to enable him to escape the political flak that would result from surrendering superiority in this important area of the nuclear arms race.[19]

ABM

The MIRV program served still another purpose. The administration believed that a decision to proceed with MIRVs would make possible a delay in the deployment of an American ABM system it did not want, while it tried to get the Soviets to agree to an ABM ban. The decision to deploy MIRVs, however, did little to satisfy the proponents of an American ABM program. They argued that, in spite of the planned MIRV deployments, ABMs would still be needed to protect American cities or to prevent them from becoming hostages to Soviet nuclear blackmail following a crippling Soviet first-strike on America's ICBM fields. They argued, moreover, that an American ABM system would protect the nation against an accidental nuclear launch or against an attack by China's embryonic nuclear force or, for that matter, from a nuclear attack from any other quarter. But the most effective argument of those who favored the ABM was the fact that the Soviet Union had one. If the administration allowed the Soviets to retain an ABM monopoly, it would give the Soviets an enormous psychological, technical, and political advantage.[20]

McNamara at first refused to budge on the ABM issue. He opposed an American ABM system primarily because he believed that one sufficiently effective to avoid massive destruction from an all-out Soviet nuclear attack was an impossibility. Even if the United States were able to ring every conceivable American target with ABMs, the Soviets would still have the capability to overwhelm them simply by constructing more missiles and warheads. Moreover, even if it were possible to destroy most incoming Soviet ICBM warheads, the United States would still have to contend with Soviet SLBM warheads as well as nuclear weapons delivered by Soviet bombers. While effective air defenses could conceivably be built against Soviet bombers, defense against a Soviet SLBM attack was virtually impossible.[21]

Moreover, as McNamara pointed out, any defense program, even an ABM system, would have to be supplemented by a nationwide fallout shelter system if American fatalities from a Soviet nuclear attack were to be kept below the 100 million to 150 million figure he anticipated without an adequate shelter system. Yet the Congress persistently refused to finance a large-scale shelter system. Like West Europeans, most Americans preferred to have the government prevent the outbreak of a nuclear war than attempt to prepare people to survive one. If America's deterrent forces were not adequate to prevent a nuclear holocaust, most Americans seemed to believe, neither would a nationwide system of shelters.[22]

McNamara opposed deployment of an ABM system for other reasons, too. Mutual ABM deployments by the United States and the Soviet Union, he feared, would destabilize the strategic balance and increase the risks of a nuclear war. By protecting cities from nuclear destruction, an ABM network would reduce the psychological inhibitions against launching a nuclear attack. Moreover, by promoting what McNamara referred to as the "mad momentum" of the nuclear arms race, ABM deployments would make it difficult if not impossible to conclude a strategic arms limitation agreement. As a result, McNamara believed that the appropriate response to the Soviet ABM was not to build an ineffective and extremely expensive American ABM system, but rather to increase the capability of the United States to penetrate Soviet ABM sites by adding multiple warheads to America's ballistic missiles.[23]

Although the ABM idea was blatantly unworkable, McNamara still experienced overwhelming pressure to deploy it. In addition to the military, the ABM project had the strong support of the Pentagon's civilian technicians, engineers, and scientists. The Pentagon's John Foster believed that what was at stake in the ABM decision was not only the security of the United States, but also the continued effectiveness of the weapons laboratories and the scientific research teams of America's defense industries. He feared that the morale of those who participated in the development of the ABM would be adversely affected if it were not deployed.[24]

The pressure exerted on the administration by the Pentagon was compounded by the Congress. Many congressmen could not understand how or why McNamara could oppose what they considered a purely defensive weapon system whose major objective was the preservation of American lives. "Any defense against an enemy weapon can be better than no defense at all," stated Congressman Robert L. F. Sikes (Democrat, Florida). More than a few congressmen also realized that a decision to deploy an ABM system would mean substantial contracts for defense industries and jobs for defense workers in districts they represented. It was estimated that over 15,000 companies and one million people in 172 congressional districts in forty-two states would be involved in the production of an ABM system.[25]

In addition to the pressure from the military and the Congress, McNamara had to deal with the realization that President Johnson was adverse to challenging either on the ABM issue. Johnson had sided with McNamara on a number of issues opposed by the Joint Chiefs, including the retirement of B-52 and B-58 bombers and the decision to forgo development of the B-70 bomber. The President was not inclined to challenge the Joint Chiefs on the ABM, especially after key senators on the Armed Services Committee who were close to Johnson politically and personally, such as Richard Russell, John Stennis, and Henry Jackson, strongly supported the ABM. Furthermore, by 1967 Richard Nixon and other Republicans were increasing the pressure for an American ABM. With another presidential election approaching, Johnson did not want to be on the short end of an ABM "gap," one he believed the Republicans were sure to exploit if he rejected all ABM expenditures. He also did not want to compound his problems by losing his secretary of defense on the issue. Johnson wanted to cap his presidency with a strategic arms limitation agreement, and he accepted McNamara's argument that a massive ABM deployment would diminish his chances of concluding one.[26]

McNamara was able to help Johnson escape his predicament by presenting him with two ABM options. The first called for the procurement of only those ABM components that required a long lead-time to produce. This option would provide enough time to determine if the Soviets were interested in agreeing to a mutual prohibition on ABM deployments. The second option proposed to begin deployment of a "thin" ABM system; that is, one designed to ward off only a Chinese nuclear attack, thereby preserving the Soviet assured destruction capability. Johnson accepted McNamara's first option. He told the Joint Chiefs and congressional leaders that he had approved the procurement of ABM components, but that he would postpone the actual deployment of the system until he had first determined if the Soviets were willing to restrict their ABM deployments.[27]

The Soviets, however, were not willing, at least not initially, to curb their ABM program. At a hastily arranged summit conference at Glassboro, New Jersey, on June 23 and 25, Johnson and McNamara tried in vain to persuade Soviet Premier Alexei Kosygin that an ABM race would be the height of folly. Kosygin insisted that the Soviet ABM system was purely defensive in nature and was no threat to the United States. Nor could he accept the American argument that ABMs were inherently destabilizing. Bowing to the inevitable, McNamara in his September 1967 San Francisco speech indicated that the United States had no choice but to go ahead with the second ABM option he had presented Johnson, the construction of a thin ABM system labeled "Sentinel." The President in his January 1968 budget message requested the appropriation of $1.2 billion for its production and deployment.[28]

Like so many other strategic weapon systems deployed by the United States, worst-case analyses were used to attempt to justify the decision to deploy an ABM system. McNamara asserted that the inability of the United States to determine Soviet intentions and capabilities left the administration no alternative but to hedge "against what was then only a theoretically possible Soviet buildup." Yet when one considers the overwhelming nuclear superiority the United States possessed during the Johnson years, the American ability by means of satellite and radar reconnaissance to detect any massive Soviet missile construction program, and the capability of American industry to build additional missiles quickly to counter any massive Soviet buildup, one is hard-pressed to accept the rationale behind McNamara's ABM decision.[29]

It is even more difficult to justify the continuation of the MIRV and ABM programs after the administration had learned that the Soviets were not building a massive ABM system. McNamara in his January 1968 defense posture statement told Congress that, "although construction of the Galosh ABM system around Moscow is proceeding at a modest pace, no effort has been made during the last year to expand that system or extend it to other cities." By then the Soviets had built four sites around their capital, with a total of sixty-four missiles, but they added no more after that. In addition, both McNamara and Air Force Secretary Harold Brown admitted that the Tallinn Line had almost no ABM capability, but rather was a radar system designed for air defense. The Leningrad ABM, first observed in 1962, was dismantled in 1964. Belatedly, it appears, the Soviets had accepted McNamara's argument that ABM deployments were potentially destabilizing.[30]

Nor did the Chinese become as great a threat as the administration had anticipated. In 1968 McNamara also admitted that "it is now clear [the Chinese] failed to conduct either a space or a long-range ballistic missile launching before the end of 1967, as we thought possible last year." The admission prompted Jerome Wiesner, Kennedy's science adviser, to call Sentinel "a bad joke perpetrated on us by Mr. McNamara and Mr. Johnson in an election year. It seems to me," Wiesner said, "that their very rationalization—that it was to defend us against the Chinese, but we would stop building it if the Russians agreed not to build one—demonstrates this well enough."[31]

Curbing the Worldwide Dissemination of Nuclear Weapons

While the superpowers proceeded with the expansion of their nuclear arsenals, they—and many other nations besides—were becoming more alarmed by the prospect of nuclear weapons dissemination to the nonnuclear weapon states of the world.

Many states cooperated in an attempt to diminish the spread of nuclear technology. On February 14, 1967 fourteen Latin American nations signed the Treaty of Tlatelolco, which prohibited nuclear weapons in their countries. The treaty, which went into effect in 1968, was signed by every Latin American state except Guyana and Cuba. (Argentina, Brazil, Chile, Colombia, Panama, and Trinidad and Tobago signed the treaty, but did not ratify it.) A protocol to the treaty called on the nuclear powers to respect the denuclearization of Latin America and to refrain from using or threatening to use nuclear weapons against any of the treaty's contracting parties. The protocol was ratified by all the nuclear powers, including France and China.[32]

Shortly before the signing of the Treaty of Tlatelolco, the superpowers concluded negotiations on an agreement prohibiting the installation of military bases and fortifications, the testing of weapons, and the conduct of military maneuvers on the moon and other celestial bodies. Compliance with the treaty could be checked through inspection. Of more immediate importance, the treaty prohibited the placing into orbit around the earth, or the stationing in outer space or on celestial bodies, of weapons of mass destruction. Since neither superpower saw much need to deploy nuclear weapons in these environs, the negotiation of the agreement, which began in 1965, proved relatively easy. The Outer Space Treaty was approved by the Senate in an 88 to 0 vote in April 1967 and went into effect the following October.[33]

The international effort that attracted the greatest interest, however, was one aimed at preventing the manufacture or acquisition of nuclear weapons by nonweapon states. Although discussions between the United States and the Soviet Union on a nuclear nonproliferation treaty (NPT) had begun as early as 1962, several obstacles delayed the conclusion of the negotiations until 1967. One was the plan initiated by the Kennedy administration for a NATO multilateral nuclear force (MLF). The Soviets vigorously opposed any nonproliferation agreement that would permit West Germany to have access to nuclear weapons, a key feature of the MLF proposal. The United States, on the other hand, insisted that the MLF did not constitute a case of nuclear proliferation because, while nuclear weapons in West Germany would be placed under MLF control, the United States would retain a veto over their use.[34]

Another obstacle to the early conclusion of a nuclear nonproliferation agreement was created by the six member states (France, West Germany, Italy, Holland, Belgium, and Luxembourg) of the European Atomic Energy Agency (EURATOM). They objected to the inspection of their nuclear facilities by the International Atomic Energy Agency (IAEA), as called for in the draft NPT. They indicated that they would sign a nonproliferation agreement, but only on the condition that inspection of their nuclear installations would be conducted by EURATOM. It was a condition the Soviets, who were members of the IAEA but not EURATOM, initially refused to accept.[35]

Other problems were created by the nonnuclear weapon states. Many of them insisted that their adherence to the NPT would be conditioned by the willingness of the superpowers to reduce their nuclear arsenals, which the nonweapon states considered the greatest threat to the continued existence of humanity. The nonweapon states also demanded as a condition for their participation in the NPT continued assistance for their own nonmilitary nuclear programs. Some nonweapon states, like India, which feared a Chinese nuclear attack, made their participation contingent on a pledge from the superpowers that they would be protected against nuclear attack or intimidation.[36]

In the end, the obstacles that blocked the conclusion of the NPT agreement were removed to the satisfaction of an overwhelming majority of nuclear weapon and nonweapon states. The MLF obstruction disappeared when the MLF plan failed to win sufficient NATO support. Soviet objections to a nuclear weapon role for West Germany were apparently satisfied by the inclusion in the NPT of a provision (in Article I) prohibiting the transfer to nonweapon states of nuclear weapons or nuclear explosive devices as well as by a provision (in

Article II) prohibiting nonweapon states from manufacturing or procuring nuclear weapons or nuclear explosives from any source whatsoever. The inspection imbroglio was resolved by a formula (in Article III) which allowed EURATOM to conduct inspections provided that the IAEA could satisfy itself that nuclear materials could not be diverted from civilian to military projects. The security issue was settled to the satisfaction of most countries by a resolution drafted by Britain, the Soviet Union, and the United States which promised prompt action by the U.N. Security Council in case of "aggression with nuclear weapons." The treaty also contained a pledge (in Articles IV and V) by the weapon states that they would provide nonmilitary nuclear assistance to nonweapon states that ratified the treaty and thereby forswore the acquisition of nuclear weapons. One of the major obstacles to the successful completion of the negotiations was removed when the nuclear powers promised (in Article VI) that they would engage in a serious effort to reduce the size of their respective nuclear arsenals, and as a way of providing a method of checking their compliance with that pledge, agreed (in Article VIII) to participate in a review conference five years after the treaty went into effect. The Treaty on the Non-proliferation of Nuclear Weapons, as the NPT was formally entitled, was signed in Washington, Moscow, and London on July 1, 1968.[37]

Unlike the Limited Test Ban Treaty, the NPT attracted little opposition in the United States, because it affected states contemplating the acquisition of nuclear weapons and left intact the American nuclear arsenal. The Senate approved ratification of the treaty on March 16, 1969 by a vote of 83 to 15. It was also ratified by Britain and the Soviet Union. By the time the treaty went into effect on March 5, 1970, 97 states had signed it and 47 states had ratified it. By 1975, the date of the first Treaty Review Conference, there were 111 signatories, including 96 full parties. Although France refused to ratify the treaty, it did declare that it would abide by the treaty's provisions. The Chinese, on the other hand, not only refused to ratify the treaty, they also declared that further nuclear weapon proliferation among the "revolutionary" states would be a positive development because, they believed, it would contribute to the breakup of the "Soviet-American nuclear monopoly."[38]

India was one of the nonnuclear weapon states refusing to sign the NPT. Not only dissatisfied with the security guarantees of the nuclear powers, the Indians also felt that the treaty was inherently discriminatory against the nonweapon states. In return for surrendering the sovereign right to build nuclear weapons, the Indians believed the nonweapon states received little more than what they could obtain, or were already obtaining, in the way of nonmilitary nuclear assistance. The nuclear powers, on the other hand, were permitted to retain nuclear weapons, were not required to submit to IAEA inspection of their nuclear facilities, and were only required to make what the Indians considered a faint-hearted pledge to reduce the size of their nuclear arsenals. As a way of removing what the Indians considered the discriminatory features of the NPT, they proposed a two-stage alternate treaty requiring the nuclear weapon states to reduce their nuclear arsenals prior to the acceptance by the nonweapon states of any obligation requiring them to refrain from the acquisition of nuclear weapons. The Indian proposal was unacceptable to both the Soviets and the Americans. While the superpowers expressed a willingness to engage in talks to restrict the size of their nuclear arsenals, they were not prepared to allow the Third World nations to set the conditions under which they would do so.[39]

SALT

Despite India's objections, the Nuclear Nonproliferation Treaty did provide an impetus for further superpower arms talks. On the day Lyndon Johnson signed the NPT, he announced that an agreement had been reached with the Soviet Union to begin strategic arms limitation talks (SALT).

A strategic arms limitation agreement had been an early goal of the President. In his first State of the Union message in January 1964, he proposed as a first step toward restricting strategic weapons a mutual Soviet-American reduction in the production rate of fissionable materials for use in nuclear weapons. In the same address, Johnson attempted to spur implementation of his proposal by announcing a unilateral twenty-five percent reduction of American uranium production. Additional cutbacks the following April resulted in a forty percent reduction in American uranium output and a twenty percent reduction in plutonium production. Khrushchev responded favorably to the Johnson initiative by announcing that the Soviet Union was halting the construction of two nuclear reactors whose primary purpose would have been the production of plutonium; the Soviets also promised to reduce the production of uranium for nuclear weapons substantially and to allocate a higher percentage of fissionable materials for nonmilitary purposes.[40]

In January 1964 Johnson also proposed that both superpowers engage in a "bomber bonfire." The proposal called for the destruction of equal numbers of American B-47 bombers and Soviet TU-16 bombers, about 500 bombers for each side, at the rate of twenty bombers a month over a two-year period. However, since both bombers were obsolete and scheduled for scrapping with or without a mutual agreement, the "bonfire" idea was recognized by the Soviets for what it really was, a meaningless gesture. What the Soviets favored was the destruction of all strategic bombers, a proposal which, because it would have removed one leg of the American nuclear triad, was unacceptable to the United States.[41]

More significant was another Johnson proposal made on January 21, 1964 calling for a "verified freeze on the number and characteristics of strategic, nuclear offensive and defensive launch vehicles." Under the proposed freeze agreement, production of these weapons would be halted, except to provide for maintenance or loss by accident, for training and "confidence" firings, and for outer space programs. While new weapon development programs would be permitted under the Johnson plan, they would be closely restricted. The Soviets opposed the Johnson freeze not only because it included a provision for on-site inspection, but also because it excluded American forces in Europe that were capable of delivering nuclear weapons on the Soviet Union. But perhaps the most important factor behind the Soviet rejection was the fact that the Soviets were not about to agree to a freeze that would have left them in a permanent condition of nuclear inferiority.[42]

In January 1964 Johnson also called for a verifiable agreement to ban underground nuclear testing. But the bid made no progress during his presidency for essentially the same reason that it failed during the Kennedy administration. The Americans insisted that on-site inspections would be necessary to determine if seismologically unidentifiable events were due to nuclear explosions or natural causes, such as an earthquake. But the Soviets continued to insist that so-called national means of verification, primarily photo-reconnaissance satellites or electronic eavesdropping gear, were more than adequate for that purpose.[43]

The superpowers made little progress toward closing the gap that separated their negotiating positions until 1967. In February of that year Kosygin responded favorably to a letter from Johnson suggesting the possibility of initiating ongoing talks to discuss "means of limiting the arms race in offensive and defensive nuclear missiles." The United States immediately followed up with a proposal to limit offensive nuclear weapons—ICBMs, SLBMs and bombers, as well a defensive ABMs. In spite of this auspicious beginning, however, the effort to get SALT started languished until very late in the Johnson presidency. The delay was partly the result of the Soviet's reluctance to negotiate with the United States on an issue as important as strategic weapons at a time when the Johnson administration was trying to bomb North Vietnam into submission. But it was also partly the fact that the Soviets were still in the process of achieving parity with the United States in numbers of strategic nuclear delivery vehicles and were obviously reluctant to restrict their strategic programs until they reached that goal.[44]

Yet Johnson did not give up on SALT. He saw the talks as a way of diverting public attention away from the war in Vietnam, and after March 31, when he announced that he would not seek another term in office, a way to cap his presidency. His decision to restrict the bombing of North Vietnam and the willingness of the North Vietnamese to begin peace talks in Paris no doubt made it easier for the Soviets to begin the negotiations. But probably more important as an explanation of Soviet motivation was the continued buildup of the Soviet strategic arsenal. With 800 ICBMs and 130 SLBMs by 1968, the Soviets probably felt that they could deal with the United States from a position of strength. As a result, in May 1968, they informed Johnson that upon completion of the NPT negotiations in July they would be prepared to begin SALT.[45]

While waiting for the NPT negotiations to conclude, the Johnson administration laid out a framework for SALT. In the first stage of the talks, the administration envisaged an exchange of technical papers in which both sides would set forth their respective positions on limiting both offensive and defensive strategic weapons. It was the consensus of Johnson's advisers that a SALT agreement should limit Soviet and American SLBMs, ICBMs, and ABMs at some specific number, but should in no way restrict MIRV or bomber deployments. In the second stage of the talks, a summit conference would be held for the purpose of reaching an agreement on a broad agenda designed to guide subsequent negotiations. In the final stage, the guidelines drafted at the summit would direct the talks, but would be under the control of Johnson's successor.[46]

On August 19, 1968 Soviet Ambassador Anatoly Dobrynin informed the administration that the Soviet Union was ready to begin the strategic arms talks. They would open with a summit meeting attended by Johnson and Kosygin on September 30. Johnson, however, was not fated to end his presidency with a SALT triumph. On August 20 Soviet tanks and troops stormed into Czechoslovakia as a part of a successful move to overthrow the government of Alexander Dubcek, whose reforms the Soviets had come to consider "anti-socialist." In the wake of the Soviet invasion of Czechoslovakia, Johnson was compelled to cancel the September 30 summit.[47]

Although Johnson canceled the summit meeting, he did not abandon his hope of initiating the arms talks before he left office. Four days after the Soviet invasion of Czechoslovakia, the administration concluded work on a SALT proposal. On October 26 Under Secretary of State Nicholas Katzenbach said that SALT must proceed in spite of Czechoslovakia. On November 12, after Richard Nixon won the presidential election, the Soviets responded to the Johnson feelers by announcing that they were ready "without delay to undertake a serious exchange of views on this question." In early December, Johnson told Senator William Fulbright, chairman of the Foreign Relations Committee, that he intended to launch SALT at a summit meeting before he left office. While Fulbright thought that this was "not an orthodox action for a man leaving office," he saw no reason to object. Perhaps Fulbright realized what Johnson did not: that Nixon would refuse to be bound by any agreements concluded by the outgoing President. And when the President-elect so informed the Soviets, they withdrew their offer to meet with Johnson. So ended Johnson's SALT effort.[48]

9

Nixon and SALT I, 1969–1972

The Pragmatic Cold Warrior

The presidency of Richard M. Nixon marked a high point in the Soviet-American effort to control the strategic arms race. In 1972 Nixon succeeded in achieving what had eluded his predecessor, Lyndon Johnson: agreements limiting the deployment of both defensive and offensive nuclear weapon systems. "It is an irony of American history," wrote Gerard Smith, Nixon's chief SALT I negotiator, "that the only man who ever resigned the presidential office was the first President who succeeded in making significant arms limitation agreements." What is even more ironic is the fact that this achievement was made by a man who had built his political career on an ardent Cold Warrior reputation.[1]

One factor that accounts for Nixon's success in SALT was the flexibility of the man himself. Beneath a veneer of ardent anti-communism was a hard core of realistic pragmatism. If cooperation with a communist state advanced American interests, he was prepared to bend. In fact, Nixon believed his anti-communist reputation was a major asset in dealing with the Soviet as well as the Chinese Communists. He thought he could end decades of Sino-American hostility and pursue arms control agreements with the Soviets without incurring the criticism that he was "soft" on communism.[2]

Why was Nixon so interested in improving America's relationship with the Soviet Union? For one reason, he believed that a thaw in the Cold War would help relieve the growing weariness of the American people with the war in Vietnam. In fact, he even believed that the Soviets could be persuaded to help end that conflict on terms satisfactory to the United States. Nixon also realized that with the Soviets about to achieve parity with the United States in numbers of ICBMs and SLBMs, and with no new American strategic programs under development, it was politically and economically impractical for the United States to attempt to maintain nuclear superiority. With Congress pressing for a reduction in American troop strength in Europe, and reluctant to support additional defense spending while the war in Vietnam was raging, Nixon surmised that SALT was the only feasible way to restrain the Soviet strategic buildup. Indeed, to Nixon, the prospect that the Soviets would build a nation-wide ABM system seemed to make SALT an urgent necessity.[3]

Henry Kissinger

While Nixon was the source of power behind the new American approach to the Soviet Union, Henry Kissinger was its architect. In several books and articles written before he joined the administration, as national security adviser and beginning in 1973 as secretary of state, Kissinger criticized the traditional American approach to the communist world, with its heavy emphasis on military rather than diplomatic solutions to the problems of the Cold War. He believed that, if the United States were going to play an effective role in preserving peace, it would have to learn how to blend military power with diplomacy so that both worked in tandem to achieve realistic goals. Kissinger's primary goal was the creation of a new framework of international relations, one in which the United States and the Soviet Union would attempt to resolve the issues that had perpetuated the Cold War as well as to take the steps necessary to reduce confrontations that could trigger a nuclear war. Like the President, Kissinger also apparently believed in SALT. Without some restraint on the deployment of nuclear weapons, he stressed, diplomats would lack the sense of security they would need to make accommodations in the negotiating process. In short, without arms control, Kissinger did not think lasting détente with the communist world was either possible or desirable.[4]

The Soviet Approach to SALT

Auspiciously, the Soviet Union seemed to be even more eager than the United States to get on with SALT. On the very day Nixon was inaugurated, the Soviet government sent the President a note inviting him to resume the talks as soon as possible. The Soviets were eager for a number of reasons. Economic factors were important if not primary. By the end of the sixties, the Soviet economy was nowhere near Khrushchev's proclaimed goal of surpassing the United States by 1980. The inability of the Soviet Union's collectivized farms to feed that nation was compounded by a declining rate of industrial growth. The relative backwardness of Soviet technology was apparent in the way the Soviet space successes in the late fifties and early sixties gave way to a series of space failures late in the decade, failures that were made all the more galling by America's success in putting a man on the moon in 1969—a feat yet unaccomplished by the Soviet Union. The Soviet leadership undoubtably saw a relaxation of tensions with the United States as the prerequisite for obtaining

Western assistance in overcoming the shortcomings of the Soviet economic system.[5]

For the Soviet Union, the necessity of a Cold War thaw was motivated as much by external threats as by internal problems. The Soviet Union's problems with its European satellites were only compounded by its deteriorating relationship with Communist China. A one-time ally of Moscow, Beijing had split with the Soviets in the late fifties, partly because of ideological differences that centered on Khrushchev's doctrine of peaceful coexistence but primarily because of long-standing national animosities that were reflected in the Soviet refusal to help the Chinese develop nuclear weapons. By 1969, when a border dispute led to several exchanges of gunfire between Soviet and Chinese troops along the Ussuri River, which separates Chinese Manchuria and Soviet Siberia, all-out war seemed a distinct possibility. As the Nixon administration took office, the Soviets seemed prepared to reduce tensions with the United States if only to be free to deal with the seemingly unreasonable Chinese.[6]

The Soviets also had compelling military reasons for resuming the quest for a SALT agreement. Having achieved rough nuclear parity with the United States, they were more than eager to preserve it. What they had gained in quantities of missile launchers they were about to lose to the Americans in numbers of nuclear warheads. It is probable that they hoped that SALT could provide a way of halting the American MIRV deployments, at least long enough for them to deploy their own. But the absurdity of deploying additional nuclear weapons did not escape them. The attempt of one superpower "to achieve military superiority over its rival," Soviet Foreign Minister Andrei Gromyko warned, "would compel the latter to mobilize still more national resources for the arms race. And so on, *ad infinitum*." In short, the Soviet leadership, like the American, appreciated the possible benefits as well as the necessity of restricting the growth of nuclear weapon arsenals.[7]

The Nixon-Kissinger Approach to SALT

While both Nixon and Kissinger favored the resumption of SALT, they were not about to be rushed into the negotiations by the Soviets. Months of preparation would be required before the administration would be ready to begin the talks. Kissinger especially wanted to move slowly toward SALT to determine if the Soviets were interested in the talks only for their propaganda value.[8]

In his capacity as chairman of the National Security Council's Verification Panel, Kissinger dominated the preparations. According to Kissinger, the panel not only studied various arms control proposals, but also tried to determine their impact on current and projected American military programs, as well as those of the Soviet Union. In addition, it made detailed analyses of the American capability to verify compliance with each proposal. The work of the panel, Kissinger believed, enabled the American SALT delegation to respond meaningfully and quickly to Soviet counterproposals. However, Gerard Smith, head of the U.S. delegation, had a much different assessment of the panel's work.

He believed that most of its meetings were meaningless since most of the decisions had already been made by Kissinger and the President. "A serious flaw" in the system, he believed, "was that Kissinger . . . had too much influence."[9]

One of the most controversial aspects of the Nixon-Kissinger SALT strategy was the administration's intention of linking the talks to other issues that divided the United States and the Soviet Union—particularly the status of Berlin, the Arab-Israeli conflict, and the war in Vietnam. Smith, as well as others, disagreed entirely with this approach. The threat of thermonuclear destruction, he emphasized, took precedence over all other issues, including Berlin, the Arab-Israeli conflict, and Vietnam.[10]

Not surprisingly, considering the history of previous administrations, Kissinger saw no inconsistency in building additional nuclear weapons while simultaneously seeking their limitation. There was nothing in history, he wrote, to support the view that "all arms races caused tensions; arms buildups, historically, were more often a reflection rather than a cause of political conflicts and distrust." Moreover, Kissinger believed that in the absence of a SALT agreement the United States would have to take action to maintain the credibility of its deterrent forces. And, like all previous administrations, he believed that arms control talks had to be pursued from a position of overwhelming strength. Only ongoing weapon programs or the threat to initiate new weapon programs, he believed, would provide the Soviets with the incentive to reach an agreement. In other words, to Kissinger, American weapon programs were needed as "bargaining chips" at the SALT table.[11]

The Nixon Doctrine

In reviewing the military component of its SALT strategy, the Nixon administration was faced with a situation unique in the nuclear arms race: the Soviets had achieved rough strategic parity with the United States in numbers of missile launchers. What was worse, from the perspective of American military planners, there seemed to be no end in sight for the Soviet strategic buildup. By 1975 the Soviets would have almost 1,600 ICBMs while the number of American ICBMs remained the same. In addition, the Soviets were about to surpass the United States in numbers of missile-launching submarines. In 1973 they would have tested their first MIRV.[12]

The Soviet strategic buildup should have seemed inevitable to American diplomats. For one reason, it represented a sustained reaction to America's nuclear activities. The United States had an ongoing MIRV program that would bring the total number of American warheads to almost 10,000 by the end of the new decade. While proceeding with their own MIRV program, the Soviets hoped to compensate for their own inferiority in numbers of warheads by building more ICBMs than the United States. The larger number of Soviet SLBMs was seen by them as a way of compensating for America's Poseidon SLBMs, which were not only being MIRVed but were far superior in accuracy and range to those of the Soviet Union. Moreover, like the Americans, the Soviets were not opposed to the idea of negotiating from a position of strength. Numerical superiority in ICBMs and SLBMs would not only give the Soviets greater bargaining power in SALT, it would give them greater leverage in any superpower confrontation.[13]

Finally, after six months of intensive study, the administration's new military strategy was completed and announced by the President during his visit to Guam in July 1969. For as new as the so-called Nixon Doctrine supposedly was, it is interesting how much it borrowed from its predecessors. With respect to conventional forces, it resembled the Kennedy-Johnson "flexible response" strategy; however, instead of the two and one-half war strategy followed by Kennedy and Johnson (two major wars, one in Europe and one in Asia, and a minor conflict elsewhere), the Nixon administration adopted a one and one-half war strategy, in effect, stating that the United States no longer considered China a threat. The Nixon Doctrine emulated the Eisenhower strategy by insisting that America's allies must assume the primary burden of engaging an aggressor in ground combat, while America would confine her role to one of providing supplementary conventional and, if needed, nuclear assistance. American conventional forces, partly as a result of the Nixon Doctrine, and partly as a result of the winding down of American participation in the Vietnam War, would be reduced appreciably in the first half of the decade. Between 1970 and 1977, deployable American ground forces were cut by 207,000 men. Between 1969 and 1975, Air Force squadrons declined from 169 to 110, Army and Marine divisions from 23 to 16, and combat ships from 976 to 495.[14]

Like Eisenhower's massive retaliation strategy, the Nixon Doctrine placed its emphasis on the deterrent value of strategic nuclear forces. Also like Eisenhower, Nixon called the nuclear component of his strategic doctrine "sufficiency." It reflected the Eisenhower assumption, one that was also shared by McNamara, that it would be prohibitively and unnecessarily costly to maintain America's nuclear superiority in numbers of strategic missiles. What the Nixon administration wanted was enough nuclear force to inflict an unacceptable level of damage on a potential aggressor—in other words, McNamara's assured destruction capability. Yet, also like McNamara, both Nixon and Kissinger believed that a targeting doctrine based exclusively on an assured destruction capability was no better than the Eisenhower massive retaliation strategy. To retaliate only against Soviet cities, they insisted, would be suicidal, for the Soviets would then retaliate against American cities. What Nixon and Kissinger wanted was further development of an existing American capability to retaliate against Soviet military installations as well as cities—in other words, an enhanced counterforce capability.[15]

Unlike McNamara, however, the Nixon administration decided to emphasize, rather than downplay, its counterforce option both for domestic and diplomatic purposes. Counterforce weapons were seen by Nixon and Kissinger as the only feasible way that the United States could attempt to maintain nuclear superiority, at least in numbers of warheads, at a time that it was conceding the Soviets numerical superiority in launch vehicles. To strengthen the counterforce option, in late 1971 the administration announced that it was accelerating the development of the Trident submarine. Designed to carry twenty-four SLBMs, each with up to fourteen warheads and with a guidance system much more accurate than Poseidon, Trident would be a potent, counterforce-capable weapon system. At the same time, the administration decided to go ahead with a project abandoned by McNamara, a new long-range bomber, eventually dubbed the B-1. While Trident and the B-1 were designed to give the President an enhanced counterforce option, they were also initiated because they would provide the administration with "bargaining chips" that could be used in SALT. Without them, Nixon and Kissinger believed, the Soviet Union would have no incentive to negotiate restrictions on their own nuclear weapon programs.[16]

Safeguard

Not surprisingly, the new emphasis on a counterforce strategy brought howls of protest from its critics. But even more controversial was Nixon's decision to proceed with an American ABM, renamed "Safeguard." Unlike the ABM proposed by the Johnson administration, Sentinel, Safeguard was not designed to protect American cities against a possible Chinese attack; instead, it was designed to protect Minuteman silos against a Soviet attack, particularly if SALT failed to curb Soviet offensive and defensive missile deployments. Moreover, the administration wanted some defense, even a limited one, against the possibility of an accidental launch or an attack by a future nuclear power.[17]

The administration's critics, on the other hand, considered Safeguard not only unworkable but unnecessary and dangerous. It was unworkable, they argued, because the Soviets would soon have the capability to overwhelm any American ABM site with scores of warheads ejected by MIRVed vehicles. Distinguished scientists, including Hans Bethe, Herbert York, George Kistiakowsky, and James Killian, asserted that there was no way to match the number of Safeguard interceptor missiles with the number of warheads the Soviets would be capable of launching at them. The critics argued that Safeguard was unnecessary because, even in the unlikely event that America's ICBMs were destroyed by a Soviet first-strike, the United States would have more than enough almost invulnerable SLBMs to devastate the Soviet homeland. In short, Safeguard would add nothing to America's existing deterrent capability. And in fact, Safeguard could be dangerous because the Soviets would respond to the deployment of American ABMs by augmenting their own ABM arsenal, thus adding a new menace to an already escalating nuclear arms race.[18] It would appear that Safeguard was anything but a safeguard.

When the program was debated in Congress, the strength of these criticisms proved more and more difficult for the administration to try to overcome. The point is, however, Safeguard was not desired primarily because of its intrinsic value, but rather because it could be used as a bargaining chip in SALT. The Soviets, Kissinger wrote, regarded Safeguard as "a harbinger of future American superiority. . . . Offering to limit our ABM would become the major Soviet incentive for a SALT agreement."[19]

It was by only the narrowest of margins, a 50–50 tie broken by the vote of Vice President Spiro Agnew, that the Senate voted to appropriate Safeguard on August 6, 1970. In subsequent years, however, its opponents were able to whittle away at the appropriations until the original twelve-site program proposed by the administration in 1969 was reduced to three sites in 1972. In the end, only one site was built, and even that one was shut down in 1975 after $5 billion had been spent on its construction.[20]

SALT I: Round One, November–December 1969

On November 17, 1969 the first round of SALT finally got under way in Helsinki, Finland, which, along with Vienna, would serve as the site of the negotiations. Neither side made any formal proposals in this first round, but disagreement resulted anyway. Two questions were the basic sources of contention. First, should both offensive and defensive limitations be included in the talks? Second, exactly what constituted strategic forces?[21]

As far as the first issue was concerned, the Soviets shocked the Americans by stating that they were willing to discuss the limitation of ABM systems, a reversal of their earlier position. However, they were unwilling to consider any restrictions on offensive systems, even apparently MIRVs, where the Americans enjoyed a substantial lead. The Americans, on the other hand, insisted that there would be no agreement unless both offensive and defensive systems were restricted. Both Nixon and Kissinger feared that a limitation on defensive systems alone would result in a continuation of the ominous Soviet offensive buildup. They also believed that, if only offensive weapons were limited, defensive systems might multiply to a point where the United States would be unable to retaliate effectively against Soviet targets.[22]

The second issue, the definition of strategic forces, was equally troublesome. The Soviets wanted to include in the definition of strategic forces all nuclear delivery systems capable of attacking Soviet territory. This meant that the so-called forward-based systems (FBS), nuclear-capable aircraft deployed by the United States in Western Europe and on aircraft carriers around the Soviet littoral, were to be considered strategic forces since they had the capacity to strike Soviet targets. The Americans disagreed and refused to consider inclusion of FBS in the strategic definition. American forces in Western Europe, the administration insisted, were designed to counter Soviet conventional forces in Eastern Europe as well as the approximately 600 intermediate- and medium-range Soviet missiles aimed at Western Europe. The United States also argued that its FBS were integral parts of defense agreements with America's NATO allies and consequently could not be dealt with in bilateral negotiations with the Soviets. The FBS issue proved impossible to resolve and ultimately had to be dropped from the SALT agenda.[23]

Pressure for a MIRV Ban

While the administration planned its strategy for round two of SALT, it experienced considerable pressure from those who believed that the best way to curb the nuclear arms race was to build less nuclear weapons, and that a good place to start would be by prohibiting the deployment of MIRVs. In July 1969, 113 members of the House endorsed a resolution calling for a negotiated ban on MIRV deployments. A Senate resolution calling for a freeze on MIRV deployments passed 73 to 6 on April 9, 1970. The vote was a clear indication that Congress was prepared to go much further on the MIRV issue than the administration desired.[24]

Gerard Smith was the chief spokesman of those within the administration who supported inclusion of a MIRV ban in the administration's SALT package. He argued that American willingness to halt MIRV deployments might be the only way to get the Soviets to halt their ongoing deployments of ICBMs and SLBMs. More important, he felt that it was not in the long-term interests of either side to deploy MIRVs. Multiplying the number of deployed and ever more accurate warheads, he argued, would enhance each side's first-strike capability and thereby increase the vulnerability of both side's ICBMs. MIRVs, in other words, would destabilize the mutual deterrent balance. If they were not restricted, Smith predicted quite accurately, America's strategic nuclear arsenal would increase from 2,600 warheads in 1969 to between 7,000 and 9,500 by 1977. "This," he stressed, "is hardly a way to limit strategic arms."[25]

Smith's effort to get an administration proposal for a MIRV ban, however, ran straight into the stone wall of Pentagon opposition. Without MIRVs, the Pentagon argued, neither Trident nor a counterforce strategy would have much meaning. MIRVs were also necessary, Defense Secretary Melvin Laird insisted, to counter the potential deployment of a nationwide Soviet ABM system, as well as to provide a greater number of surviving warheads in the event America's ICBM complex were attacked by Soviet missiles armed with MIRVs. Smith was surprised that Laird refused to admit that the best way to deal with the future threat of Soviet MIRVs was a ban on all MIRV deployments. But Laird and the Pentagon had little faith in a negotiated agreement to halt MIRV deployments.[26]

The President and Kissinger were unwilling to challenge the military's stance on MIRV. They not only agreed with the Pentagon assessment that MIRVs were necessary to make a counterforce option a reality, they also considered them an integral part of the administration's SALT strategy. Without MIRVs, the United States would have no bargaining chips, except Safeguard, at the SALT table.[27]

SALT I: Round Two, April–June 1970

Before the second round of SALT began in April 1970, Nixon decided to offer the Soviets two options drafted by Kissinger. The first called for the limitation of ICBMs and SLBMs at the U.S. total of 1,710, with an additional limit of 250 on modern, large ballistic missiles, like the Soviet SS-9 ICBM. Bombers would be frozen at their current levels (527 for the United States and 195 for the Soviet Union). This option also would freeze at then current levels medium- and intermediate-range ballistic missiles as well as submarine-launched cruise missiles (jet-powered drones capable of delivering nuclear or conventional warheads). This option also called for a MIRV ban with on-site inspection and a limit of one ABM deployment around each national capital. Both Nixon and Kissinger realized that this option would probably not be accepted by the Soviets because of its on-site inspection requirement. But the offer, even if rejected, Kissinger pointed out, would allow the administration to assume "the positive public posture of having favored comprehensive limitations."[28]

If, as expected, the Soviets rejected this option, the United States would present the second option. It called for major, phased reductions in offensive systems and an annual decrease of 100 launchers until the combined total of ICBMs and SLBMs for each side reached 1,000 (after January 1, 1978). While this option did not ban MIRVs, it did propose that ABMs be banned altogether or at least be limited to one ABM deployment around the national capitals.[29]

The Soviets, predictably, rejected the first option. While they accepted the ABM proposal, they rejected the MIRV ban offer because of its collateral requirement of on-site inspection. But they also rejected it because it would have prevented MIRV testing as well as production and deployment. Had the Soviets accepted the American MIRV ban proposal, the Soviet Union would have been denied a MIRV capability, something the United States was about to acquire.[30]

The Soviets then proceeded to reject the second American offer. They had little interest in the proposal because it would have limited Soviet strategic programs, including a one-half reduction in the size of the Soviet SS-9 ICBM force, but would have placed no limits on American MIRVs. In addition, by reducing ICBMs, it would have forced the Soviets to build more SLBMs, an area where they were at a distinct disadvantage in technology and geography (before reaching their launch stations, Soviet submarines have to traverse narrow straits that could be closed easily during wartime). Moreover, reductions in Soviet ICBM strength would have increased the strategic importance of American forward-based systems.[31]

After turning down the American offers, the Soviets tabled their own proposals. They were based on the supposition that limitations must precede reductions. They called for restrictions on strategic offensive weapons, which they defined as all weapons capable of striking targets within the territory of the other side. The Soviets wanted to establish an unspecified aggregate for ICBMs, SLBMs, and strategic bombers, with each side free to replace units of one type by those of another. The Soviet proposals included a MIRV ban, but one limited only to the production, not the testing, of MIRV systems. Desiring to prevent the deployment of a nationwide American ABM, the Soviets offered to limit ABM launchers and certain associated radars. However, they rejected on-site inspection, stating that national means of verification would suffice to monitor the limitations.[32]

Kissinger considered the Soviet plan "a scheme so egregiously one-sided that even the most rabid proponents of limitations could not accept it." To Kissinger, it was "a neat device by which the Soviets would continue their own development and flight testing of MIRVs while freezing our deployments until they could catch up." As a result of the negative reaction of the Soviets, and at the insistence of the Pentagon, Kissinger and Nixon decided to drop the MIRV ban idea entirely from the SALT I agenda. They would instead concentrate on obtaining purely quantitative restrictions on strategic weapons deployments in SALT I and would leave qualitative restrictions, like MIRV, to SALT II.[33]

The failure to include a MIRV ban in SALT I was a critical turning point in the evolution of the strategic arms race. While the United States had a MIRV monopoly, it saw no value in a MIRV ban. But once the Soviets achieved a MIRV capability in 1973, that advantage would disappear. When combined with the greater number of ICBMs the Soviets possessed, Soviet MIRVs would increase the vulnerability of America's land-based ICBMs. Continued MIRV deployments would also complicate the strategic arms control talks. "Too late it was realized," Smith wrote in his memoirs, "that no stable agreement to limit offensive strategic weapons could exist without stringent MIRV control." Even Kissinger came to realize as much. In December 1974 he stated: "I would say in retrospect that I wish I had thought through the implications of a MIRVed world more thoroughly in 1969 and in 1970 than I did." Smith put it more directly: "a MIRVless world would have been much safer."[34]

The failure to conclude a MIRV ban agreement, however, was not solely the responsibility of the United States. The Soviet side put little effort behind their MIRV ban proposal, probably reflecting the unwillingness of the Soviet military to surrender this promising augmentation of their strategic nuclear arsenal. At any rate, when the United States began to deploy the first MIRVed Minuteman III ICBMs in June 1970, from the Soviet viewpoint a ban on MIRVs was out of the question. With the United States, in effect, unwilling to halt its only ongoing strategic program, the Soviets for the time being lost any interest they had in restricting offensive forces. In June, with the talks stalemated, the American delegation returned to Washington.[35]

SALT I: Round Three, November–December 1970

The third round of SALT I opened in Helsinki in early November 1970 and was over by December 18. It was as stormy as it was short. The session started with the Soviets formally rejecting a new American proposal which completely did away with the MIRV ban as well as phased reductions in ICBMs. In their place, the new proposal called for a mutual limit of 1,900 ICBMs, SLBMs, and bombers, with a sublimit of 250 SS-9s and a total ban on land-mobile ICBMs, as well as a ceiling of 100 ABMs for the defense of either national capital, or no ABMs at all. The plan also offered not to count Soviet intermediate-range ballistic missiles as a part of their strategic total as long as the Soviets agreed not to include American forward-based systems in the American strategic aggregate. It also dropped the earlier demand for on-site inspection, stating that the restrictions could be monitored effectively by each country using its own means of verification. The Soviets rejected the American proposal primarily because they insisted that forward-based systems must be included in any offensive limitation agreement, a condition the United States again refused to accept.[36]

In order to break the stalemate, President Nixon agreed to make another concession, the so-called Helsinki formula. The United States would agree to assess ways to limit forward-based systems, but only after both sides had ratified an agreement limiting central strategic systems. The Soviets flatly rejected this concession. Not counting forward-based systems in any treaty limiting strategic weapons, chief Soviet negotiator Vladimir Semenov said, was like the "American farmer who when shown a camel remarked that there was no such animal." On December 1 the Soviets suggested that the discussions on offensive limitations be set aside because of the FBS impasse; they would wait until a treaty restricting ABM was concluded. The Soviets then presented a formal proposal for a treaty covering defensive systems and limiting ABMs to an agreed number of interceptors at each nation's capital.[37]

Not surprisingly, Nixon rejected the new Soviet offer. The administration was unprepared to give up its ABM bargaining chip while no restraints existed on the Soviet ICBM program. It then offered the Soviets a new proposal, the so-called four-to-one option, which would have allowed the United States to build four ABM sites at ICBM fields while limiting the Soviets to one ABM site around Moscow. The Soviets considered the new American offer "manifestly inequitable," and not only because of the discrepancy in numbers of ABM sites. To the Soviets, four ABM sites covering American ICBM fields looked like a long step toward a nationwide ABM system and an effective American first-strike capability.[38]

The Back Channel

With SALT stalemated, the administration in January 1971 opened a separate and secret "back channel" to the Soviets, with Kissinger and Soviet Ambassador Anatoly Dobrynin as the conduits. The back channel was initiated without the knowledge of the American SALT delegation in order to prevent leaks of sensitive information, over which Nixon was becoming increasingly anxious, as well as to hasten the decision-making process which, Kissinger was forced to admit, the Verification Panel greatly retarded. In addition, Kissinger wrote in his memoirs, "Nixon was determined that the credit for SALT should go to him and not to [Gerard] Smith." To get SALT moving, Kissinger, through the back channel, offered the Soviet leadership a proposal whereby the United States would agree to negotiate an ABM agreement if the Soviets agreed to halt the construction of new ICBM silos while negotiations to limit offensive weapons continued. As an added incentive, Kissinger offered to spur talks on a Berlin accord and held out the prospect of expanded Soviet-American trade, both of which the Soviets badly wanted.[39]

The "carrots" Kissinger offered the Soviets apparently were irresistible.

The arms control logjam began to break up shortly after the back channel went into operation. On February 11, 1971 both superpowers signed an agreement to prohibit the emplacement of nuclear weapons on the seabed or ocean floor. The treaty was approved by the Senate in February 1972 by an 83 to 0 vote and entered into force in May 1972. Even though the Seabed Treaty barred nuclear weapon deployments in an area where they were never likely to occur, the completion of its negotiation was an indication that the superpowers were determined to resume the task of curbing the nuclear arms race. As such, the completion of the treaty served as a catalyst for the progress that occurred in the fourth round of SALT, held in Vienna from March 16 to May 28.[40]

SALT I: Round Four, March–May 1971

The culmination of the fourth round came on May 20 when both sides announced that they had achieved a breakthrough in the ABM-ICBM stalemate. The breakthrough agreement proposed that both countries would concentrate on negotiating an ABM treaty while simultaneously pursuing an offensive limitation agreement. Moreover, the Soviets agreed to drop forward-based systems from the talks in exchange for an American modification of its proposal for numerical equality in delivery systems. The new American position would allow the Soviets to preserve an edge in numbers of ICBMs, an advantage the Soviets insisted they needed to compensate for American superiority in numbers of warheads. In effect, both sides agreed to conclude an ABM agreement first, accompanied by only an interim agreement on offensive systems, thereby leaving an overall settlement on offensive systems for SALT II.[41]

Gerard Smith was stunned by the announcement of the SALT breakthrough. He was not only angered by Kissinger's willingness to negotiate behind his back, he was chagrined by the fact that the administration had given up its quest for an agreement limiting all central strategic systems—ICBMs, SLBMs, ABMs, and bombers—in favor of one limited essentially to ABMs with only short-term restraints on ICBMs. He was particularly upset by the absence of any limitations on SLBM deployments. Since the United States had no nuclear submarines under construction while the Soviets had an ongoing submarine program, Smith feared that the Soviet advantage in numbers of submarines would continue to grow. Kissinger soon realized the magnitude of his error and spent the better part of a year trying to get the Soviets to accept an SLBM freeze as a part of the SALT agenda.[42]

SALT I: Round Five, July–September 1971

Before the fifth round of SALT began in Helsinki on July 8, 1971, the administration instructed its delegation to inform their Soviet counterparts that the United States had dropped its four-to-one option and was no longer interested in an ABM site near Washington. Congressional hostility to a Washington ABM site proved too strong for the administration to overcome. The United States was now willing to settle for three ABM sites in its ICBM fields, and Smith was permitted to reduce that number to two at his discretion.[43]

The Soviets, however, had no use for the new American ABM proposal. Their acceptance of it would have forced them either to dismantle their Galosh site around Moscow, in order to build ABMs around their ICBM sites, or would have compelled them to keep only the Moscow site while allowing the United States to build three ABM sites around its ICBMs. The Soviets insisted that any ABM agreement must be based on the principle of equality, but should not permit a foundation that threatens to support a nationwide ABM system. Accordingly, they proposed an agreement limiting ABMs to both national capitals. The Americans, however, rejected the Soviet proposal. The Pentagon wanted a minimum of two Safeguard sites, and the administration refused to accept the principle of ABM parity until the Soviets were willing to reach an agreement on offensive systems.[44]

Smith personally thought it would have been better had the administration offered the Soviets a complete ban on ABMs. A total ban, he believed, would have strengthened the assured destruction capabilities of both sides by removing a major obstacle that their retaliatory forces would have to overcome if significant numbers of ABMs were deployed. Apparently, the Soviets came to share Smith's appreciation of this advantage. In July, Semenov told him that the Soviet government would accept a total ban on ABMs if the United States proposed it. "That," Smith observed, "should have been good enough for us— had we been seriously interested." But the administration was not interested. Nixon argued that a complete ABM ban would require more in the way of offensive limitations than the Soviets were likely to accept. Instead of attempting to determine the Soviet reaction to a total ABM ban, the President on August 12 rejected the total ban idea once and for all.[45]

Although the two sides remained divided on the issue of ABM numbers as the fifth session closed, progress had been made on a number of collateral ABM issues. Both sides agreed to accept restrictions on radar deployments that could have an ABM application. In addition, the United States proposed, and the Soviets accepted in a later session, a ban on the deployment of "futuristic" ABM systems—weapons using lasers and other types of beams against missiles. "We felt," Smith recalled, "that not to shut off future type ABM systems would be a clear invitation to weaponeers to try to fill this treaty loophole."[46]

Another area of agreement in this Helsinki session concerned the prevention of an accidental nuclear war. Both sides agreed to take steps to facilitate rapid communication in the event of an accidental launch, primarily by switching the Hot Line communication from cable to satellite transmission.[47]

Yet no progress could be made on the offensive limitation issue during this round of SALT because the Americans wanted a formal, detailed agreement of five years' duration. The draft of the American treaty included provisions for verification, a Standing Consultative Commission to monitor the treaty, a withdrawal option, and a commitment for ongoing negotiations leading to a more comprehensive limitation of offensive systems. The Soviets, on the other hand, wanted a general and informal understanding of short duration. Moreover, the Soviets insisted on exempting SLBMs from the limitations. They believed they needed more submarines to compensate for American forward-based systems, the qualitative superiority of American submarines, and the nine Polaris submarines Britain and France were building. The Soviet stance on SLBMs was at complete variance with the position taken by the Joint Chiefs of Staff. The latter realized that there was no way the United States could soon catch the Soviets in numbers of SLBMs, since American shipyards were already busy installing MIRVs on existing submarines. With the deployment of the first Trident submarines years off, the Joint Chiefs believed that only a SLBM freeze would restrain the Soviet SLBM buildup. With the offensive issue stalemated, the fifth round of SALT finally came to an end on September 24, 1971.[48]

SALT I: Round Six, November 1971–February 1972

The President's instructions to the American delegation for the sixth round of SALT, which began in Vienna on November 15, 1971, were the same as those presented in Helsinki. In Gerard Smith's words, they were to "concentrate on the interim freeze and 'insist' that the Soviet delegation do the same." In addition, the delegation again was instructed to make a strong effort to include SLBM submarines in the freeze. In an astonishing about-face for the Soviets, probably reflecting their desire to conclude a SALT agreement when Nixon visited Moscow in May 1972, they agreed to work diligently on ABM and offensive limitations simultaneously. But they refused to budge on the SLBM issue. And they continued to reject the administration's two-for-one ABM proposal—that is, the suggestion of one ABM site around the national capital or two in ICBM fields.[49]

Faced with continued Soviet stubbornness on ABMs and SLBMs, the administration decided to apply some military leverage. On December 2, 1971 Kissinger wrote Laird recommending that the new defense budget should include funds for four ABM sites, even though he knew that SALT would probably permit no more than two. In addition, Kissinger told Laird to expand the nation's submarine production "in a way that was highly visible to the Soviets." Laird responded with the first increase in spending for strategic weapons in several years, some sixteen percent over the previous year. The largest part of the increase went for the Trident submarine, whose deployment date was pushed up from 1981 to 1978 (however, because of cost-overruns, shoddy workmanship, and overburdened shipyards, the first Trident, the *Ohio*, was not commissioned until November 1981). In addition, funds were budgeted for the B-1 bomber, an ABM defense for Washington, as well as cruise missiles. But the Senate refused to appropriate funds for more than two ABM sites, thereby depriving Kissinger of the enhanced ABM bargaining chip he wanted.[50]

Both Nixon and Kissinger believed that the military buildup was necessary not only to prod the Soviets in SALT, but also to reinforce American diplomacy around the world. The administration's defense buildup coincided with the outbreak of war in 1971 between India and Pakistan over the status of East Pakistan, now Bangladesh. The Soviets supported the Indians; the United States sided with the Pakistanis. An augmentation of American military power, the President believed, would demonstrate to the Soviets that they could not use SALT to paralyze American responses to Soviet "adventurism."[51]

At the same time, Nixon's military buildup was partly a reaction to rising conservative criticism of SALT and détente. A military buildup would keep the hawks in line while Nixon attempted to play the role of dove. Clearly, he had no intention of appeasing the hawks by abandoning SALT. Too much of his prestige was at stake for such drastic action. Moreover, 1972 was an election year. A trip to China in February, the first for any American president, capped by the signing of a SALT agreement in Moscow in May, would practically assure Nixon's reelection.

SALT I: Round Seven, March–May 1972

While the seventh and final round of SALT I began in Helsinki on March 28, 1972, Henry Kissinger prepared for another round of back channel diplomacy. On April 20 he flew to Moscow for personal talks with the Soviet leadership. During the talks, Brezhnev offered Kissinger three proposals. One called for a two-site ABM limit for each nation. One site could be located at the national capital; the other in an ICBM field. In another proposal, the Soviets finally accepted an SLBM freeze. Apparently, however, the Soviets agreed to the SLBM freeze only after Kissinger agreed to a Soviet-American grain deal and to a provision in the treaty that would give the Soviet Union superiority in numbers of SLBMs, 950 to 710 for the United States. And, finally, the Soviets accepted a provision for a five-year moratorium on then current ICBM levels, which, in effect, would give the United States time to develop new ICBMs that could then be deployed at the end of the moratorium.[52]

Although Kissinger accepted these proposals, he rejected Brezhnev's call for a mutual no-first-use-of-nuclear-weapons pledge. He believed it would have undermined the NATO alliance, upset China, and amounted to an "abdication" of America's responsibility to defend the free world. In June 1973, in response to American rejection of a no-first-use proposal, the Soviets accepted what Kissinger described as "a bland set of principles that had been systematically stripped of all implications harmful to our interests." In other words, the threat of using nuclear weapons still remained a major component of American strategic doctrine.[53]

While Kissinger and Nixon were elated by the Moscow breakthrough, Secretary of State William Rogers and Gerard Smith were not. Both feared the negative "psychological effect" on Congress and the American people that would result from allowing the Soviets superiority in numbers of SLBMs as well as ICBMs. Smith was also afraid that the virtual freeze on American SLBM production would make Congress less likely to fund the Trident program.[54]

More threatening to an impending SALT agreement than the hostility of either Smith or Rogers, however, was the intrusion of the Vietnam War. Only a few weeks before the summit was scheduled to begin, Nixon, according to Kissinger, wanted to use the threat of canceling the summit to bring Soviet pressure on the North Vietnamese to end the war. Kissinger persuaded him to go ahead with the summit while mining the ports of North Vietnam. Smith's first reaction was, "Here goes the ball game." But surprisingly, the Soviets responded as though nothing had happened. Smith was "confident that the Soviets really wanted SALT agreements."[55]

The Moscow Summit, May 1972

On May 22, 1972 Nixon and Kissinger arrived in Moscow to begin what turned out to be a short but very intensive round of negotiations with the Soviet leadership. Pointedly, Smith and the American SALT delegation were not invited to Moscow. This was to be Nixon's own show.[56]

After four days of hard bargaining, both sides signed two major SALT agreements—one limiting ABMs, the other restricting ICBM and SLBM deployments.

The ABM Treaty was the most significant of the two agreements. The treaty permitted each side to build only two ABM sites, one around the national capital, the other around an ICBM field. It also limited each ABM site to a hundred interceptor missiles and prohibited the construction of ABM-related radars outside the two permissible sites. In effect, the treaty, by ruling out the possibility of nationwide ABM defenses for each side, preserved the retaliatory capability of both sides. In addition, while the treaty permitted the modernization and replacement of ABM systems, it also prohibited the development and deployment of "futuristic" ABM weapons.[57]

The ABM Treaty, both sides agreed, would be monitored by national means of verification, thereby bypassing the old obstacle of on-site inspection. And what was even more significant in the history of arms control, the treaty barred each side from interfering with the verification procedures of the other. To facilitate verification, the treaty provided for a Standing Consultative Commission, whose functions were to establish procedures for implementing the treaty, for considering suspected violations of the treaty, and for discussing additional measures to limit strategic weapons. The Standing Consultative Commission would meet at least twice a year, as well as at any time a party requested its convocation.[58]

The ABM Treaty, while of unlimited duration, could be reviewed every five years. Each party could withdraw from the treaty, with six months notice, if it determined that "extraordinary events related to the subject matter of this Treaty have jeopardized its supreme interests." In a unilateral statement, the United States indicated that its supreme interests could be affected by the failure to negotiate a more comprehensive offensive limitation agreement within five years.[59]

The Interim Offensive Arms Agreement was less comprehensive than the ABM Treaty and of definite duration—five years. For that period, beginning July 1, 1972, it prohibited both sides from constructing additional fixed, land-based ICBM launchers. The agreement, in effect, limited the United States to 1,054 ICBMs and the Soviet Union to 1,618. The agreement also prohibited both sides from converting "light" ICBMs (those with low throw-weight capabilities) deployed prior to 1964 into land-based launchers for "heavy" ICBMs deployed after 1964. In effect, this provision barred the Soviets from replacing their light SS-11 ICBMs with heavy SS-9s. They were limited, as a result, to no more than 313 heavy missiles, the number operational or under construction on May 26, 1972. The Interim Agreement placed no limitations on MIRVs, bombers, or land-mobile ICBMs, which both sides agreed to discuss in subsequent negotiations.[60]

Limitations on SLBMs and ballistic missile submarines were spelled out in a protocol attached to the Interim Agreement. It limited the United States to a total of 710 SLBMs and 44 ballistic missile submarines and the Soviet Union to 950 SLBMs and 62 submarines. However, to build up to these limits, both sides were required to dismantle an equivalent number of SLBMs, submarines, or ICBMs deployed prior to 1964. The Soviets accepted in a unilateral statement the right of the United States and its NATO allies to deploy among them as many as 50 ballistic missile submarines with a combined total of 800 SLBMs. However, they asserted that forward-based submarines would have to be included in subsequent SALT sessions, a condition the United States did not accept.[61]

The Interim Agreement permitted both sides to modernize and replace existing strategic offensive missiles. Yet the basic issue—how to define a heavy missile—was left unresolved. Both sides agreed that new, land-based ICBMs could not be larger than ten to fifteen percent of the volume of the missiles they replaced, and that there would be no increases in the size of missile silos. But the Soviets did *not* accept a unilateral American statement that the United States would consider any missile heavier than the Soviet SS-11 a heavy missile. In testimony before Congress, however, Kissinger gave the impression that the Soviets *did* accept the American definition of a heavy missile. Not surprisingly, after the Soviets converted their SS-11s to SS-19s, a conversion that did not require an expansion of silo size nor an increase in missile volume above the agreed fifteen percent, Senator Henry Jackson would accuse them, unjustifiably, of duplicity.[62]

During the Moscow summit, both sides supplemented the SALT agreements with an agreement on "Basic Principles of Relations" between the two countries. The new relationship they were trying to build, the document stated, would be based on "the common determination that in the nuclear age there is no alternative to conducting their mutual relations on the basis of peaceful coexistence." Both sides agreed that differences in ideology and social systems would not be permitted to prevent "the bilateral development of normal relations based on the principles of sovereignty, equality, noninterference in internal affairs and mutual advantage." Both sides pledged "their utmost to avoid military confrontations and to prevent the outbreak of nuclear war." In addition, they agreed to continue their efforts to limit other armaments and to expand commercial, economic, scientific, technological, and cultural ties.[63]

The Reaction to SALT I

The Moscow summit marked the high point of détente and the post-World War II effort to control the nuclear arms race. "Never before," Kissinger said, "have the two world's most powerful nations . . . placed their central armaments under formally agreed limitation and restraint."[64]

But Kissinger's elation was not universally shared. Liberals objected that the agreements did nothing to prevent qualitative improvements in strategic weaponry, such as the B-1 bomber and the Trident submarine. Nor did SALT I do anything to prevent the enormous expansion of warheads that resulted from the unrestricted MIRV programs of both nations. Conservatives, on the other hand, believed the agreements did too much. The Interim Agreement, they pointed out, awarded the Soviets a significant measure of ascendancy over the United States both in overall numbers of offensive missiles and in combined throw-weight of its ICBM launchers. In effect, conservatives believed that the SALT agreements reduced America to a second-rate status in the nuclear equation and thereby made her vulnerable to Soviet nuclear blackmail.[65]

In reply to the liberal argument, Kissinger said that SALT I was only the first round in the strategic arms negotiations; that the administration wanted to place limitations on other weapons as well as ICBMs, SLBMs, and ABMs. Yet he also insisted that the United States "must continue those strategic programs which are permitted by the agreement and those research and development efforts in areas that are covered by the agreement in case the follow-on agreement cannot be negotiated." Without ongoing weapon programs, he said, the Soviets would have no incentive to reach arms control agreements with the United States. Liberals, however, feared that weapon systems developed for bargaining purposes in SALT would inevitably be deployed, and would thereby make future arms control talks that much more difficult to conclude.[66]

Kissinger counterattacked the conservative argument by stating that America's strategic forces were "completely sufficient" to counter any attempted Soviet blackmail. "The Soviets have more missile launchers," he stated, "but when other relevant systems such as bombers are counted there are roughly the same number of launchers on each side. We have a big advantage in warheads. The Soviets have an advantage in megatonnage." And, he pointed out, warheads, not launchers, could kill people. To Kissinger, the bottom line was his belief that the Soviet nuclear threat would have been much greater without the agreements.[67]

Kissinger's reassurances had little impact on conservatives. What they feared most was what might happen after the Soviets had MIRVed their ICBMs. The Soviets would then have more throw-weight *and* more warheads. Belatedly admitting the nature of this threat, the administration accepted an amendment to the Interim Agreement drawn up by Senator Jackson providing that future strategic arms limitation agreements would be based on the principle of equal strategic forces. However, the administration also accepted Jackson's amendment to ensure that he and other hardliners would vote to accept the SALT agreements.[68]

Jackson's amendment was not the only bargain struck by Nixon and Kissinger to ensure acceptance of the agreements. To obtain Pentagon support, the President had to assure the Joint Chiefs of Staff that he would not only support the continued modernization of America's strategic forces, he would accelerate it. In reality, it was no major concession. Admitted Kissinger, "This we were determined to do anyhow. It was one reason why we had promoted the freeze in the first place." It was an admission that reinforced the view of liberals that the administration had engaged the Soviets in SALT only to gain time to build new offensive weapons, not to end the nuclear arms race.[69]

Yet liberals had no choice but to support the SALT agreements concluded by the administration. In spite of all their shortcomings, the agreements made a first step toward controlling the nuclear arms race. Without them, there would have been no restraints on the expansion of strategic nuclear arsenals. As a result, liberals joined with conservatives in the senate to ratify the ABM Treaty on August 3 by a vote of 88 to 2. The Interim Agreement, because it was not in treaty form, had to be approved by both houses of Congress. It passed in the Senate on September 14 by a vote of 88 to 2 and in the House by a margin of 329 to 7. On September 30 the Interim Agreement was signed by the President in an elaborate White House ceremony. Both the ABM Treaty and the Interim Agreement became effective in October 1972.[70]

10

Nixon, Ford, and the Decline of Détente, 1972–1977

The Purge: Fall 1972–Spring 1973

Shortly after the signing of the Interim Agreement on September 30, 1972, President Nixon discussed SALT with its chief antagonist, Senator Henry Jackson. Nixon reaffirmed his intention to make the Jackson Amendment to the Interim Agreement the basis of the American position in SALT II. He also apparently accepted Jackson's argument that the "soft-headed" leadership of the Arms Control and Disarmament Agency had to be changed. After Nixon's reelection in November, the President required all key executive branch officials to submit their resignations. In almost all agencies, the resignations were received, but not accepted. In the ACDA, however, the resignations of virtually all the top officials were accepted. Of the agency's seventeen top officials, only three remained in 1974. In addition, its total staff of 230 was reduced by 50 and its budget by one-third in 1973. Among those who left was the ACDA's chief, as well as head of the SALT delegation, Gerard Smith. But unlike many agency staff members, Smith's departure was voluntary; he had announced well in advance of the Nixon purge that he intended to leave government after the ratification of the SALT I agreements.[1]

In another response to Jackson's pressure, the White House on January 4, 1973 announced that the director of the ACDA and the head of the SALT delegation would no longer be the same person. U. Alexis Johnson, a career Foreign Service Officer having no experience with nor, some said, enthusiasm for arms control, was appointed the new chief of the SALT delegation. In April 1973 Fred Iklé, a Rand consultant who had written studies for Senator Jackson, was named head of the ACDA. Iklé, according to one National Security Council staffer, was "to the right of the JCS [Joint Chiefs of Staff] on SALT" and less willing to cooperate with Kissinger's brand of détente diplomacy. With Iklé's appointment to the ACDA and Johnson's elevation to the SALT delegation, Senator Jackson, in effect, was able to institutionalize his opposition to SALT.[2]

SALT II: Fall 1972–Fall 1973

SALT II began with an unproductive session that opened in November 1972 and ended the following month. The lack of initial progress resulted partly from differences within the administration on the negotiating strategy it intended to pursue. Verification Panel meetings conducted in February and March 1973 attempted to resolve these differences. The main issue of disagreement was the position to be taken on MIRV. The State Department favored a moratorium on MIRV testing. But the Joint Chiefs and the Defense Department strongly objected. Without MIRVs, the Pentagon argued, the Trident program would be killed by Congress.[3]

The only position the Pentagon did favor was one that would have provided equal numbers in every category of strategic weapon systems, regardless of any difference in the design of the American and Soviet systems, as well as equality in throw-weight. While Kissinger agreed with the Joint Chiefs that the Soviet throw-weight advantage would have to be reduced, he did not think it could be accomplished by SALT. If SALT set the throw-weight ceiling at the American level, the Soviets would have to make drastic cuts in their force structure or completely rebuild it in the American image. That, Kissinger thought, would be a "highly improbable outcome of SALT." On the other hand, if the throw-weight ceiling were set at the Soviet level, the United States would have to build larger missiles, or double the number of existing Minuteman missiles, while the Soviets froze their missile programs to enable the Americans to catch up—an equally improbable outcome to SALT.[4]

What really upset Kissinger was the view of the Joint Chiefs that limitations on MIRVs would be inconsequential. "If this is true," he asked, "can someone explain what the hell this negotiation is all about?" Without limitations, he argued, both sides would be free to MIRV their entire forces. Equal ceilings on numbers of launch vehicles, he explained, "would become nearly meaningless" as a result of increasing the number of deployed warheads while the number of targets remained relatively the same. Moreover, without a MIRV limitation, the Soviets would not only have more warheads but, because of their greater throw-weight capacity, more deliverable megatonnage. In Kissinger's estimation, limiting Soviet MIRVs was the paramount objective of America's SALT II strategy.[5]

Kissinger, however, could not sway the Joint Chiefs in their opposition to a limitation on MIRVs during the waning months of the Nixon presidency. And Nixon, increasingly preoccupied with his Watergate problems, lacked either the desire or the clout to impose a MIRV limitation on a constituency that would likely support him to the bitter end of the impeachment process. Faced with the paralysis of government that Watergate was bringing about, the instructions forwarded to the American delegation in May were a conglomeration of contending viewpoints. They were instructed to seek equal numbers at a ceiling of 2,350 delivery vehicles, some 250 below existing Soviet numbers and 150 above the American total. The American proposal also included a ban on the testing of MIRVs as well as a freeze on further deployments of land-based MIRVs. "This," Kissinger realized, "neatly shut the Soviets out of MIRVing their ICBM force, which comprised 85 percent of their total throw-weight, without significantly curtailing any program of our own." Needless to say, the American proposal was unacceptable to the Soviets.[6]

Stalemated on the main issue of SALT II—MIRVs—both sides agreed to a vague statement of Basic Principles when Brezhnev visited Washington in June 1973. The principles reaffirmed the proposition that subsequent accords must recognize the "equal security interests" of both nations. In what was a significant concession for both sides, they promised to include limitations on qualitative deployments, an indirect reference to MIRVs. Both sides also acknowledged that national means of verification would continue to be the means for verifying SALT agreements. Finally, in order to get the talks moving, both sides committed themselves to negotiate a more comprehensive and permanent offensive agreement by the end of 1974.[7]

The Washington summit produced another accord—the Agreement on Preventing Nuclear War. In it, both nations promised to refrain from the threat, or use, of force against the other party, against the allies of the other party, and against other countries in circumstances that might endanger international peace and security. If events appeared to be moving toward nuclear conflict, both nations agreed to "enter into urgent consultations with each other and make every effort to avert the risk."[8]

SALT II Stalemate: September 1973–June 1974

On September 25, 1973 SALT II resumed in Geneva. The Basic Principles concluded at the Washington summit, especially the goal of an agreement in 1974, provided a new sense of urgency to the talks. The sense of urgency on the American side was increased by the news that the Soviets had tested their first MIRVed missile in July. Nevertheless, the talks soon broke down. The Soviets offered a draft SALT II treaty that would have banned the deployment of all new strategic weapons for ten years. Naturally, Kissinger considered the Soviet proposal "one-sided" and "outrageous" because it would have prohibited the modernization of America's strategic forces *after* the Soviets had done much to modernize their own.[9]

The Geneva talks were also hampered by the outbreak of another Middle East war, a crisis that once again brought the superpowers to the brink of a nuclear conflagration. On October 6, 1973 Egyptian and Syrian military forces attacked Israeli positions in the Sinai Peninsula and in the Golan Heights. After several days of critical, defensive fighting, the Israelis were able to take the offensive, cross the Suez Canal, and encircle the Egyptian Third Army. The crisis reached the level of superpower confrontation when the Soviets threatened to intervene with troops to prevent another Arab humiliation. In response, President Nixon, at the urging of Kissinger, placed American strategic forces on an intermediate defense condition (DEFCON) level—the first alert at that level since the Cuban missile crisis. In the end, the Soviets backed off, and the Arab belligerents, as well as the Israelis, accepted a ceasefire and agreed to participate in Kissinger's unique brand of "shuttle diplomacy." But while a tenuous peace was restored to the Middle East, the aftertaste it left did nothing to improve the atmosphere of SALT. Both superpowers had threatened to go to war, even a nuclear war, only a little over a year after pledging that they would exercise restraint in the conduct of their diplomatic relations.[10]

With SALT again stalemated, with the efficacy of nuclear diplomacy seemingly demonstrated by the termination of hostilities in the Middle East, and with the Soviets about to deploy their first MIRVs, the Nixon administration attempted to gain additional military leverage. In October, after Nixon had approved funds for the development of a land-mobile ICBM, the MX, Kissinger directed the Defense Department to study the feasibility of a strategic cruise missile program. He wanted it not only to enhance the effectiveness of America's aging B-52 force, from which the cruise missile could be launched, but also as a bargaining chip in SALT. In December, Defense Secretary James Schlesinger, Laird's successor, announced the administration's new counterforce targeting doctrine as well as a plethora of new military programs to back it up. They included an improved guidance system for the Minuteman III ICBM; a new high-yield warhead, the Mark 12-A; and MaRVs, maneuverable reentry vehicles that could be guided to their targets, thereby giving the United States a "silo-killing" accuracy of a few hundred feet.[11]

Predictably, the Soviets reacted to the administration's new targeting doctrine and counterforce programs with new weapon programs of their own. In February 1974, timed to coincide with the reopening of the recessed SALT II negotiations, the Soviets tested four new missiles, three of them with multiple warheads. The SS-17 was tested with four warheads and the SS-19 with six. Although these missiles were not "silo-killers," the third missile tested, the SS-18, did have the requisite accuracy and payload capacity to destroy American ICBM silos. The SS-18 could launch as many as five warheads, each with a yield of between two and five megatons.[12]

In March 1974 Kissinger flew to Moscow in an attempt to achieve a "conceptual breakthrough" with the Soviets on strategic forces. He offered to extend the Interim Agreement for up to an additional three years if the Soviets agreed to a reciprocal inequality in numbers of MIRVed ICBMs. What he asked the Soviets to accept was a ceiling of 270 land-based MIRVed ICBMs while allowing the United States a limit of 550. In effect, the arrangement would have permitted the Soviets to keep an edge in total throw-weight, but it also would have required them to deploy less MIRVed ICBMs than the United States. Kissinger did not expect the Soviets to accept the numbers he proposed, but he hoped to use the proposal as a basis for thwarting the unrestrained deployment of Soviet MIRVs. The Soviets, however, were not interested in any agreement that would have left the United States with a substantial MIRV advantage.[13]

The Moscow Summit: June 25–July 3, 1974

The second Moscow summit which Nixon and Kissinger attended began on June 25, 1974. When it ended on July 3, the MIRV issue remained unresolved. A joint communiqué stated that both governments had abandoned the previous goal of reaching a permanent agreement by the end of 1974. Instead, it stated that they would direct their efforts toward concluding an agreement that would enter into force on the expiration of the Interim Agreement—that is, in 1977.[14]

Both sides, however, still believed that it was necessary to maintain the facade of détente, even though concrete progress on limiting strategic weapons was proving impossible to achieve. Accordingly, two nuclear arms control agreements were signed at Moscow. The first limited the number of ABM sites that each country could maintain to one, instead of the two permitted by the 1972 ABM Treaty. The second agreement, the Threshold Test Ban Treaty (TTBT), prohibited underground nuclear tests above a level of 150 kilotons. What was significant about the TTBT was the stipulation that both sides would not only exchange geological data to enhance verification, they would also refrain from doing anything that would interfere with the operation of the other's national means of verification, the method by which the treaty would be verified. An accompanying agreement, the Peaceful Nuclear Explosions Treaty (the PNE Treaty), was signed two years later. It established a 150-kiloton threshold for nonmilitary underground nuclear explosions. What was significant about this agreement was its provisions for some on-site inspection and for the monitoring of underground nuclear explosions. Both treaties were designed to have a five-year duration during which time neither party could withdraw from the PNE Treaty while the TTBT was in force.[15]

While the agreements looked good on paper, in practice they did little to restrain the nuclear arms race. The new ABM agreement simply formalized what was already recognized: ABM was unworkable. The ceiling of 150 kilotons on underground explosions in the TTBT and PNE Treaty was so high as to be meaningless in curbing nuclear weapon tests. However, because the high hopes for détente had waned, the nuclear test limitation treaties were not ratified. And, although both parties have since abided by their provisions, they remain unratified as of this writing.

The inability of Kissinger to produce a conceptual breakthrough at Moscow was undoubtedly affected by the administration's increasing preoccupation with the Watergate scandal. Kissinger privately expressed his fear that the Soviets were reluctant to conclude a new SALT agreement with a President who appeared to be on the way out of office. After Nixon resigned on August 9, following the initiation of impeachment proceedings by the House of Representatives, the Soviets indicated their eagerness to work with his successor, Gerald Ford.[16]

The Vladivostok Accord: Fall 1974

One of Ford's first acts as President was to invite Kissinger to remain as secretary of state. The new President, like most of his predecessors, had little experience with diplomacy or military policy and, as a result, was more than willing to rely on Kissinger to conduct the administration's foreign policy. Kissinger's attempt to direct the administration's SALT strategy, however, did not go unchallenged. Defense Secretary Schlesinger insisted that the United States "must face down" the Soviet desire to deploy a large number of MIRVs or, failing that, "buckle down for a five-year, all-out arms race." In the absence of an agreement reducing strategic forces or controlling modernization, he recommended an accord that would recognize an overall equality of strategic forces or one that would establish a MIRV ceiling so low that neither side could acquire a first-strike capability. Kissinger, on the other hand, wanted no part of a MIRV race, and he did not believe that the Soviets favored massive force reductions. Using CIA studies for support, he asserted that without a MIRV limitation agreement, the Soviets were prepared to deploy at least 3,000 delivery vehicles and anywhere between 1,500 and 2,000 MIRVs over the next five years. He also argued that Congress would not support an all-out arms race and, therefore, that the United States should pursue the goal of high ceilings for the present, and reductions and modernization controls later.[17]

Aided by Ford's passivity and by the support of the Joint Chiefs of Staff, whose modernization programs were not endangered by Kissinger's objectives, the secretary of state was able to devise a proposal much to his liking by the middle of October. Surprisingly, the Soviets displayed interest in the new proposal when Kissinger brought it with him to Moscow a week later. Even more unexpectedly, the American proposal became the basis of an agreement signed by Ford and Brezhnev at Vladivostok a month later.

The Vladivostok Accord stated that the future conduct of SALT would be based on these principles: (1) an overall ceiling of 2,400 delivery vehicles for both sides; (2) a ceiling of 1,320 MIRVs; (3) the inclusion of bomber-launched and land-mobile missiles in the overall total; (4) the limitation of Soviet heavy missiles at 313 deployments, with no new silo constructions permitted; (5) after the conclusion of an agreement, further negotiations beginning no later than 1980–81 on the issues of additional limitations and possible strategic force reductions to take effect after 1985; and (6) negotiations for an agreement based on these principles would resume in Geneva in January 1975. In signing the accord, the Soviets dropped their demand for compensation for American forward-based systems and British and French strategic forces. In return, the United States abandoned its effort to establish equality of throw-weights.[18]

Kissinger hailed the Vladivostok Accord as a "breakthrough" that put a "cap" on the strategic arms race. But Schlesinger was not as enthusiastic about the agreement. While calling it "a major step forward," he also said that if the overall ceiling on strategic forces were not reduced in subsequent talks, and the threat to America's ICBMs grew as the Soviets MIRVed their missiles, the

United States would "lean in the direction" of building more than the then currently planned ten Tridents.[19]

The critics outside the administration were much harder on the agreement than Schlesinger. They pointed out that, while the accord embodied the Jackson Amendment's principle of equality in numbers of strategic delivery systems, it made no provision for equality of throw-weight or for equality of warhead numbers. Conservatives feared that the Soviets would eventually convert their three-to-one advantage in throw-weight, which was left unaffected by the agreement, into a first-strike capability. Senator Barry Goldwater could only conclude that the agreement was "just another ploy by the Russians to try to fool some of our détente-happy people."[20]

Liberal criticism was equally sharp, though for different reasons. Liberals pointed out that the treaty imposed no restrictions on missile or bomber replacements or on almost any other aspect of the qualitative arms race. Nothing in the agreement, for example, ruled out the deployment of land-mobile missiles which, liberals argued, would make verification of arms agreements difficult if not impossible. But the key flaw in the accord, liberals asserted, was the height of the ceilings. The United States would actually have to build more strategic weapons to reach them. That, they believed, was not the way to curb the nuclear arms race.[21]

Trouble over Soviet-American Trade

By the time the Ford administration began to put the Vladivostok Accord into treaty form, an atmosphere conducive for effective arms negotiations had dissipated appreciably. By the end of 1974 détente had become a dirty word for many liberals as well as for the overwhelming majority of conservatives.

One of the major issues on which détente floundered was Soviet-American trade. To the Soviets, trade was almost as important as SALT. "Without trade," Brezhnev insisted, "no normal relations between two countries are possible." Kissinger, for his part, believed that increased Soviet-American trade would create a degree of mutual interdependence between the two countries that could contribute to the growth of shared interests, mutual restraint, and better relations.[22]

Spurred by the 1972 Moscow summit, the United States and the Soviet Union took a number of steps to expand their trade relationship. In July 1972 the two governments concluded an agreement providing $750 million in American credits for Soviet grain purchases. In the months that followed, the Soviets scored a financial coup by quietly buying about one-fourth of the American wheat crop at the relatively low price of $1.63 a bushel, causing American domestic wheat prices to soar to $3.00 a bushel by May 1973. In October 1972 a Lend-Lease Agreement was signed, whereby the Soviet Union agreed to pay the United States $772 million for its World War II debts in exchange for Export-Import Bank credits and most-favored-nation status (which, in effect, meant that Soviet imports would be taxed at a rate no higher than the tax rate on the imports of other countries).[23]

As early as September 1972 the Nixon administration realized that gaining Senate ratification of a new Soviet trade treaty would not be an easy task. In August the Soviet government had clamped a tax on Soviet citizens wishing to emigrate from the Soviet Union. The tax, which was as high as $30,000 per person, was designed to discourage the emigration of Soviet Jews to Israel. Senator Jackson jumped at the chance to embarrass the Nixon administration, while enhancing his own presidential ambitions. Declaring that "the time has come to place our highest human values ahead of the trade dollar," Jackson in October 1972 tacked an amendment onto the administration's trade bill prohibiting most-favored-nation status to any "nonmarket economy country" that limited the right of emigration, a very thinly disguised reference to the Soviet Union. A similar amendment, introduced in the House of Representatives by Ohio Congressman Charles A. Vanik, passed by a 319 to 80 vote in December 1973.[24]

The administration reacted with indignation to the congressional tampering with its trade bill. Kissinger warned that the Jackson-Vanik Amendment would not only set back the fight for freer emigration from the Soviet Union, it might also jeopardize the whole process of détente. He preferred to hold the Soviets to private assurances that they would not harass Soviet Jews who wished to leave, rather than attempt to pin down the Soviets to public assurances that would embarrass them and cause them to consider the Jackson-Vanik Amendment an unwarranted intrusion into their domestic affairs. Kissinger's warning proved to be prophetic. After Ford on January 3, 1975 signed a trade bill that was contingent on the increased emigration of Soviet Jews, the Soviet government informed the administration that it would not implement the 1972 trade agreement, nor would it pay its Lend-Lease debt.[25]

The Debate over Détente

The collapse of the Soviet-American trade agreement intensified the debate over the nature and purpose of détente. Its conservative critics argued that détente was a one-way street, with the Soviet Union giving little in return for the benefits it received. Exchange agreements concluded with the Soviet Union, conservatives charged, allowed the Soviets access to American technology without any meaningful Soviet reciprocation. Moreover, conservatives wondered why the American taxpayer was forced to pay the bill, in the form of trade credits, for bailing out the failing Soviet economy.[26]

What caused conservatives even greater consternation was their belief that, by subsidizing the Soviet economy, the administration was helping the Soviet Union augment the size and quality of its military establishment. While American defense expenditures were declining substantially (at an annual inflation-adjusted rate of 4.5 percent between 1970 and 1975), Soviet military spending increased by an annual inflation-adjusted rate of three percent. In that interval, the Soviets had not only achieved superiority in numbers of launch vehicles, but also had constructed an ocean-going fleet second only to that of the United States. "Under Kissinger and Ford," Ronald Reagan charged during his campaign for the 1976 Republican presidential nomination, "this nation has become Number Two in a world where it is dangerous—if not fatal—to be second best."[27]

On the other side of the political spectrum, the liberal critics of détente were particularly aggrieved by the way the Soviets continued to violate human rights in Eastern Europe and by the nonchalant manner in which they believed the administration had reacted. A prominent Soviet dissident, the writer Alexander Solzhenitsyn, was expelled from the Soviet Union in 1974 for "anti-Soviet" activities. At Kissinger's behest, President Ford refused to permit Solzhenitsyn to visit the White House in July 1975. Kissinger said that Solzhenitsyn's hostility to the Soviet leadership and détente would make the "symbolic effect" of a meeting with the President "disadvantageous" from "the foreign policy aspect." Conservative critics of détente pounced on the way Kissinger handled Solzhenitsyn as another example of what they considered to be the administration's appeasement of the Soviet Union.[28]

What particularly galled détente's critics—liberals as well as conservatives—was the way the Soviets seemed to aggravate world tensions in violation of the Twelve Basic Principles concluded at the 1972 Moscow summit, and yet still reap the benefits of détente with the United States. Soviet involvement in the Arab-Israeli War in 1973 was followed in the same year by massive arms shipments to North Vietnam in violation of the Paris Peace Agreement, which ostensibly ended the war in Vietnam. Soviet arms two years later enabled the North Vietnamese to crush the pro-American government in Saigon.[29]

The extent to which the Soviets violated the Basic Principles in the Middle East is, however, a matter of debate. Political scientist Alexander George argued that there is "evidence that Soviet leaders did operate with considerable restraint for a while in the Middle East and that they did make some effort to cooperate in crisis prevention." The administration ignored these efforts, George believed, because Kissinger was attempting to exclude the Soviets from the Middle East peace process. Kissinger's policy, argued another political scientist, George W. Breslauer, convinced the Soviets that he "was not interested in U.S.-Soviet Third World collaboration." When combined with the passage of the Jackson-Vanik Amendment, which virtually eliminated the prospect of expanded Soviet-American trade, Breslauer believed, Kissinger's Middle East policy did much to undermine the possibility of restraint in the Soviet Union's Third World policy.[30]

For their part, the Soviets never intended to abandon their effort to promote revolutionary activity in the Third World, despite American hopes that they might do so in return for better relations with the United States. Soviet ideologists consistently maintained that "the process of détente does not and never meant the freezing of the social-political status quo in the world." Nor could détente be used to prevent the Soviet Union from giving "sympathy, compassion and support" to those whom it chose to represent as "fighters for national independence." In the Soviets' eyes, détente was a necessary accommodation by the West to a "correlation of forces" that the Soviets viewed as increasingly favorable to themselves. The most important "forces" at work, they believed, consisted of two developments: the acquisition by the Soviet Union of nuclear parity and the increasing difficulty of the United States, as the Vietnam War demonstrated, to maintain its position in the Third World. With the new Soviet assertiveness in mind, Western critics of détente wondered out loud whether any agreement with the Soviets could be worth more than the paper it was written on.[31]

Kissinger reacted to the criticism of détente with a blend of grief and combativeness. He rejected the view that détente was just another form of appeasement. The United States, he said, would never tolerate Soviet expansionism. In fact, he asserted, the administration had repeatedly acted firmly to resist Soviet "adventurism"—in Cuba, when the Soviets tried to establish a submarine base in 1970; in the Middle East, to prevent Soviet military intervention on behalf of Egypt and Syria; and in Angola, where the Soviet Union and its Cuban client helped to establish a Marxist regime in 1975. The Angolan intervention by the Soviets was successful, Kissinger charged, because the Congress denied the administration both the carrot and the stick by which it expected to restrain Soviet behavior in the Third World. The Jackson-Vanik Amendment killed the prospect of expanded Soviet American trade. And another amendment, drafted by Iowa Senator Dick Clark, denied the administration funds to finance covert operations against the Marxist forces in Angola. As far as the Twelve Basic Principles were concerned, Kissinger insisted that they were not "a legal contract," but rather "an aspiration and a yardstick" by which the United States would assess Soviet behavior and react accordingly.[32]

Nor, Kissinger insisted, did détente impair the deterrent capabilities of the United States, as conservatives charged. Quite the contrary, he asserted, the Nixon and Ford administrations strengthened the nation's deterrent forces appreciably by spurring the development of the MX missile, the B-1 bomber, the cruise missile, the Trident submarine, and advanced warheads, while still engaging the Soviets in SALT. "No weapons system recommended by the Joint Chiefs of Staff and the Defense Department," he wrote, "was ever disapproved in the White House." It was the Democratic-controlled Congress, he argued, and not the defense policies of the Nixon-Ford administrations, that was primarily responsible for the decline in American military spending during the first half of the decade. As late as 1975, long after the Soviet military buildup had become glaringly obvious, the Congress still slashed $7 billion from the administration's defense budget.[33]

While attacking the defense cuts made by liberal Democrats, Kissinger also rejected the conservative argument that continued American strategic superiority was necessary to maintain the worldwide interests of the United States. "What in the name of God is strategic superiority?" he asked almost desperately. "What is the significance of it politically, militarily, operationally at these levels of numbers? What do you do with it?" He pointed out that "at current and foreseeable levels of nuclear arms it becomes increasingly dangerous to invoke them." The outcome of a nuclear exchange, he warned, could be as many as 100 million dead on each side. For that reason, he said, "in no crisis since 1962 have the strategic weapons of the two sides determined the outcome." And yet, he added, "the race goes on because of the difficulty of finding a way to get off the treadmill."[34]

To Kissinger, a stable strategic relationship was the best way to reduce the risks of a nuclear war and the disadvantages of an unrestrained nuclear arms race. A stable strategic relationship, in turn, could only result from the acceptance by both sides of nuclear parity. That, he repeatedly asserted, was what the SALT agreements represented. Although he admitted that SALT I had permitted the Soviets to acquire a substantial lead in ICBMs and SLBMs, he believed that the concessions were more than offset by American superiority in numbers of warheads and the quality of American delivery and guidance systems. More important, SALT I prevented the Soviets from deploying even more missiles than they otherwise could have deployed if they had not agreed to restrictions. If SALT II were to fail, he predicted, "tensions are likely to increase; a new higher baseline will emerge from which future negotiations would eventually have to begin. And in the end, neither side will have gained a strategic advantage. At the least, they will have wasted resources. At worst, they will have increased the risks of nuclear war."[35]

In his memoirs, Kissinger warned against the kind of simplistic military approach to the Soviet Union that conservative critics of détente seemed to favor. "American policy," he insisted, "must embrace both deterrence and coexistence, both containment and an effort to relax tensions." He rejected the view of conservatives that change in the domestic structure of the Soviet Union could be forced by the application of American power, whether economic, political, or military. "We would not accept it [pressure] from Moscow," he emphasized, "Moscow will not accept it from us. We will wind up again with the Cold War and fail to achieve peace or any human goal."[36]

Kissinger also discounted the criticism that the benefits of détente were one-sided in favor of the Soviet Union. Although admitting that Nixon had oversold détente, he nevertheless believed that the United States had gained much from the Cold War thaw, including the SALT agreements, a Berlin agreement, the Soviet Union as a market for American farmers, and Soviet diplomatic pressure on North Vietnam, which Kissinger believed helped bring the North Vietnamese to the negotiating table in 1972–1973. But to believe that détente must benefit only the United States, he stressed, was the height of political naiveté. The Soviets were not about to conclude agreements from which they did not expect to benefit any more than the United States would sign agreements which did not advance American interests. "Our test," he added, "was whether we were, on balance, better off with an accord than without."[37]

What particularly bothered Kissinger about the conservative critique of détente was its inherent assumption that the Soviet Union had more to gain from peaceful competition than the West. This assumption, he believed, demonstrated "an unwarranted historical pessimism, a serious lack of faith in the American people." Considering the weaknesses of the Soviet system, especially when compared to the strengths of the American system, Kissinger believed that, in time, the West could not but prevail. He wrote: "If Moscow is prevented by a firm Western policy from deflecting its internal tensions into international crises, it

is likely to find only disillusionment in the boast that history is on its side."
Time, he believed, was on the side of the West, if the people of the West were
wise enough to be patient.[38]

The Decline of SALT II: 1975–1976

Primarily because of the growing hostility to détente, the negotiation of a SALT
II treaty based on the Vladivostok Accord proved impossible before Kissinger
and Ford left office in January 1977. Brezhnev's trip to the United States to
sign a final agreement was postponed from June 1975 to September of that
year, then to November, then to "early 1976," and then it was indefinitely put
off.

The impasse was caused by three issues: verification of the MIRV ceiling,
cruise missiles, and the definition of what constituted a long-range bomber.
The Americans insisted that verification of MIRV deployments could only be
guaranteed by providing that all missiles of a type already tested with MIRVs
be counted in the 1,320 ceiling. The administration believed, incorrectly, that
the Soviets could not add MIRVs to their ICBMs without increasing the size
of the ICBM silos (in fact, the Soviets were able to convert their single-warhead
SS-11s to MIRVed SS-19s without increasing silo size), and that the number
of MIRVed Soviet ICBMs therefore could easily be monitored by counting the
number of modified silos. The Soviets initially rejected the American verification method because they wanted to deploy their SS-18s with single as well
as multiple warheads. It was not until the final session of the Helsinki Conference on August 2, 1975 that Brezhnev indicated that the Soviet Union would
accept the American proposal for MIRV verification.[39]

The impasse, however, was aggravated by two other problems—the Soviet
Backfire bomber and the American cruise missile. The nature of the dispute
centering on these two weapon systems was whether they were to be counted
as strategic vehicles and thus to be limited by the Vladivostok Accord. The
Soviets maintained that the Backfire, a swing-wing bomber capable of supersonic speed, was not a strategic weapon since its range was only 6,000 miles.
The Pentagon, however, insisted that the Backfire, flying from bases in extreme northeastern Siberia, could make a subsonic flight against targets ex-
tending from Southern California to the eastern tip of Lake Superior without
refueling, and thus should be considered a strategic weapon. At the same time,
the Soviets insisted that the American cruise missile, a pilotless, jet-powered
craft that could be launched from a bomber, a ship, a submarine, or from the
ground, should be considered a strategic weapon if its range extended beyond
360 miles. The Pentagon wanted only ballistic missiles, not cruise missiles, to
be included within the limitations of SALT.[40]

To break the impasse, Kissinger sided with the Soviets on the Backfire and
cruise missile issues. The Backfire, he said, was primarily a peripheral bomber
posing very little danger to the United States. Why, he asked, would the Soviets build a supersonic bomber for attacking the United States if it could reach

American targets only by flying at subsonic speeds? With regard to the cruise missile he stated that, if the Soviets accepted no restrictions on their deployment, the United States could potentially deploy 11,000 of them on existing bombers and an additional 10,000 on nuclear submarines. All would be capable of reaching targets in the Soviet Union. The result, he believed, would be to make meaningless the ceilings that the two sides were attempting to place on strategic weapons. The debate between the Pentagon and the Soviets on the Backfire and the cruise missile placed Kissinger in an embarrassing position. He had pushed a reluctant Pentagon into accelerating the cruise missile development program with the primary purpose of gaining a new bargaining chip to use in SALT. "How was I to know the military would come to love it?" he asked exasperated after the Pentagon refused to surrender it.[41]

What increased Kissinger's difficulties with the American military was the fact that 1976 was an election year. Ford became more and more concerned that support for the Soviet position on the Backfire and the cruise missile would only add ammunition to the conservative charge that his administration was "soft" on SALT. In fact, in early 1976 Senator Jackson, who was running for the Democratic presidential nomination, stated that even the Pentagon was being excessively generous on SALT. Jackson and other hardliners were angered when Ford fired James Schlesinger in November 1975 because he could not curb the defense secretary's hostility to SALT. Jackson charged that Ford had "silenced" Schlesinger since the defense secretary had "raised the tough questions" and was not "content with shallow rhetoric for an answer."[42]

The administration was also exposed to charges that it had been "soft" on alleged Soviet violations of the SALT I agreements. On December 2, 1975 Admiral Elmo R. Zumwalt, the retired Chief of Naval Operations, told the House Select Committee on Intelligence that Kissinger had not properly informed President Ford about "massive" violations of the accords by the Soviets. Kissinger angrily denied the charge, but he did little to sway the view of conservatives, and apparently a large percentage of the American people, that SALT was a bad deal for the United States. A March 1976 NBC public opinion poll indicated that the American people, by a margin of sixty-nine percent to twenty-two percent, did "not trust the Russians to live up to their agreements."[43]

Amid a strained environment even more polarized by SALT, Kissinger traveled to Moscow in January 1976 to see if he could work out an end to the stalemate. He was authorized by the President to offer the Soviets two proposals. The main feature of the first was a requirement that, with the exception of the 120 Soviet Backfires then currently deployed, all Backfires must count within the overall ceiling of 2,400 strategic weapons. If Moscow agreed, the United States would make the following concessions: aircraft used to launch cruise missiles would count against the ceiling of 1,320 launchers with multiple warheads as well as the ceiling of 2,400 strategic weapons; in addition, submarine-launched cruise missiles with a range beyond 375 miles would be banned. If this proposal were rejected by the Soviets, Kissinger would offer the second proposal. It called for a ten percent reduction of the 2,400 strategic weapons ceiling and permitted a limited number of Backfires above the new ceiling. While Brezhnev accepted the principle of reducing the overall ceiling on strategic weapons as well as a ban on submarine-launched cruise missiles with a range beyond 375 miles, he refused to accept any limitations on the Backfire. The administration made one last pre-election bid for an agreement by offering to ratify the Vladivostok Accord without any reference to Backfire or cruise missile limitation, except to say that they would be the subjects of future negotiations. The Soviets, however, considered the proposal a "step backward," and turned it down. With the presidential election approaching, the Soviet stance, in effect, meant the end of meaningful SALT negotiations for the remainder of the Ford administration.[44]

Horizontal Nuclear Weapons Proliferation

While the superpowers were preoccupied with the vertical proliferation of nuclear weapons (the ongoing development and deployment of nuclear weapons by the nuclear weapon states), the threat of uncontrolled horizontal proliferation (the acquistion of nuclear weapons by nonweapon states) increased. In May 1974 India became the sixth nation to explode a nuclear device. Although the Indians insisted that their test was conducted solely for peaceful purposes, the world was shocked by the realization that nuclear explosives developed for peaceful purposes could easily have military application.[45]

The Indian nuclear test made a mockery of the international effort to control the horizontal proliferation of nuclear weapons. Technically, the Indian government broke no law, treaty, or agreement in exploding a nuclear device. India was not a signatory of the Nuclear Nonproliferation Treaty (NPT). And although India was a member of the International Atomic Energy Agency (IAEA), membership in that organization does not obligate a country to accept IAEA safeguards, and India did not fully do so. Moreover, the IAEA statute did not bar peaceful nuclear explosions, which the Indians claimed was the nature of their nuclear test. Nor did the Indians violate any bilateral agreements with the United States or Canada. Both countries provided India with nuclear technology and materials without first insisting that India ratify the NPT and agree to abide by all the nuclear restrictions of the IAEA. The Indian nuclear device contained plutonium that was obtained from a nuclear reactor provided by Canada in 1956. The heavy water (deuterium) which the reactor employed was provided by the United States. In effect, India demonstrated that it was far easier for a nonparty nation of the NPT to acquire the means to build nuclear weapons than for the nations that had agreed to abide by the provisions of that treaty.[46]

Henry Kissinger soon appreciated the danger of the Indian test. If other states followed the Indian example of "peaceful" nuclear testing to acquire the means to produce nuclear weapons, the risks of a nuclear accident, unauthorized use of a nuclear weapon, or even a preemptive nuclear strike would increase dramatically. Indeed, any multiplication of nuclear powers could make the maintenance of a stable deterrence relationship between the United States and the Soviet Union virtually impossible. With a multiplicity of nuclear states, a nuclear strike could come from any quarter. In spite of the ominous implications of the Indian test, Kissinger reacted cautiously. He did not want a public confrontation with the Indians for fear that he would drive them more deeply into their newly concluded alliance with the Soviets. Privately, however, he sought to obtain from the Indians a commitment that plutonium produced from nuclear fuel supplied by the United States would not be used for any type of nuclear explosion. When the Indians refused to comply, the administration delayed a delivery of enriched uranium fuel for India's Tarapur reactor, which was built with American technical and financial assistance.[47]

To correct the impression that the United States was doing too little to discourage the expansion of the nuclear arms race to the nonweapon states of the world, the administration announced a number of unilateral steps to strengthen its own domestic controls. The United States would now require the application of IAEA safeguards on exports of American nuclear material and technology; ban American assistance to those contemplating the detonation of any nuclear device, whether for civilian or military purposes; apply more stringent safeguards for the physical security of nuclear equipment and materials; and establish the Nuclear Regulatory Commission (which, along with the newly created Department of Energy, replaced the Atomic Energy Commission). On July 2, 1974 the administration suspended long-term enrichment contracts. Two years later, on October 28, 1976, it took stronger action by announcing that the United States would suspend for three years the export of reprocessing and other nuclear technologies that could contribute to nuclear weapons proliferation. The administration, in effect, stated that the danger of proliferation took precedence over commercial considerations. But to many nations, the administration's actions also meant that the United States no longer could be counted on to be an unrestricted supplier of nuclear fuel and technology for the global nuclear energy industry that it helped to create and promised to maintain.[48]

Although progress was made in tightening domestic nuclear controls, it was more difficult for the Ford administration to garner support for stronger action on the international level. In August 1974 the United States arranged a conference of the major states supplying nuclear technology, equipment, and materials (Britain, France, Australia, Canada, West Germany, and the Soviet Union). Its purpose was to consider a series of uniform guidelines and principles proposed by Kissinger to prevent the inadvertent transfer of weapons-manufacturing potential and to discourage free-market competitiveness in the bidding of nuclear technology sales. To that end, Kissinger proposed the adoption of a code that would make nuclear importers subject to tighter international safeguards, inspections, and other controls administered by the IAEA.[49]

The administration's proposals for stronger international action, however, encountered the strong opposition of France and West Germany, both of which were reluctant to curtail their own lucrative nuclear-export industries. It was not until January 1976—after West Germany agreed to sell Brazil a complete nuclear fuel cycle, including uranium enrichment, fuel fabrication, and plutonium reprocessing facilities, as well as several nuclear reactors—that the supplier states, excluding France, agreed to the guidelines outlined by Kissinger.[50]

Many Third World nations felt that Kissinger's guidelines did not go far enough in addressing the proliferation problem. At the first Review Conference of the Nonproliferation Treaty in 1975, the Group of 77, an organization of Third World nations led by Mexico, Nigeria, Rumania, and Yugoslavia, placed the blame for the proliferation of nuclear technology squarely on the nuclear powers. It was the nuclear powers that had disseminated nuclear technology and materials; it was the nuclear powers that had continued to expand their own nuclear arsenals in direct opposition to Article VI of the NPT. Among the suggestions made by the Group of 77 to reduce the threat of nuclear proliferation were the following: (1) an end to underground nuclear testing; (2) a substantial reduction in existing nuclear arsenals; (3) a pledge not to use or threaten to use nuclear weapons against nonnuclear parties to the NPT; (4) substantial aid to developing countries in the peaceful uses of nuclear energy; (5) creation of a special international framework for conducting peaceful nuclear explosions; and (6) international agreement to respect all nuclear-free zones.[51]

One delegate to the Review Conference, Alfonso Garcia Robles of Mexico, hoped to link international security with progress toward an underground test ban and nuclear arms reduction. First, he proposed that the nuclear powers agree to suspend all underground testing as soon as the number of full parties to the NPT reached one hundred (there were ninety-six in 1975). The moratorium on underground testing would become permanent as soon as all nuclear weapon states (including France, China, and India) ratified the NPT. Second, Garcia Robles proposed that the superpowers reduce the Vladivostok Accord ceilings on strategic delivery vehicles by fifty percent as soon as the number of full parties to the NPT reached one hundred, and then further reduce ceilings by ten percent each time more states became parties to the treaty.[52]

Garcia Robles' proposals were ignored by the United States and the Soviet Union. The superpowers resented Third World meddling with their strategic armaments; they considered SALT their exclusive domain. Nonweapon states, for their part, were embittered by the insensitivity of the weapon states to Third World concerns and suggestions. Only arrogance, Group of 77 leaders believed, could motivate the weapon states to continue to insist that the nonweapon states refrain from acquiring nuclear weapons, while the weapon states went on expanding their own nuclear arsenals. The insensitivity of the superpowers was highlighted on the very eve of the Review Conference, when the Soviet Union conducted an underground nuclear test. At least the United States waited until the middle of the conference to test a nuclear device.[53]

Confronted by the hostility of the weapon states, the first Review Conference ended in failure. While no nonweapon state indicated its intention to withdraw from the NPT, the conference did nothing to encourage the near-weapon states from forgoing their nuclear options. Many believed that the prospects for the spread of nuclear weapons throughout the world and, even more ominously, the acquisition of nuclear weapons by terrorists, would be greater than ever. To some, the disintegration of civilization under the threat of nuclear terrorism, if not its destruction in a nuclear war, was becoming a distinct possibility.

11

Carter and SALT II, 1977–1981

The Transformation of an Idealist

In his inaugural address, President Jimmy Carter announced that his ultimate goal was the "elimination of all nuclear weapons" from the face of the earth. Nine months later, he told the U.N. General Assembly that the United States was willing to cut the size of its nuclear arsenal by fifty percent if the Soviets agreed to do the same. As first steps in that direction during his first year in office, the President canceled production of the B-1 bomber and slowed down the development of the MX ICBM. In addition, he said he would take steps to limit the proliferation of nuclear weapons to nonweapon states.[1]

Yet in the final analysis, his effort to end the nuclear arms race stood in marked contrast to the idealistic goals he had set early in his presidency. During his four years in office, he ordered production of the components of enhanced radiation weapons ("the neutron bomb") and directed the continued development of the Trident, cruise, and MX missiles. He also obtained NATO acquiescence to the stationing in Western Europe of 108 Pershing II intermediate-range ballistic missiles (IRBMs) and 464 ground-launched Tomahawk cruise missiles (GLCMs). In addition, he expanded the sale of American nuclear technology and materials to countries suspected of pursuing a nuclear weapons-producing capability. Carter's greatest reversal, however, was his decision to withdraw from the Senate's consideration a SALT II Treaty whose negotiation his administration and the Soviets had successfully concluded in June 1979.

Vance and Brzezinski

Why did Carter apparently abandon the goal of ending the nuclear arms race?

Part of the explanation lies in the fact that Carter, like most of his predecessors, entered the White House knowing next to nothing about international relations. Unlike these presidents, however, Carter did not permit one individual to oversee the administration's foreign policy. Instead, throughout most of his presidency, he relied in varying degrees on two foreign policy specialists, Secretary of State Cyrus Vance and Presidential Assistant for National Security Affairs Zbigniew Brzezinski, whose philosophies of international relations and diplomatic styles were often in conflict. For a president who felt comfortable directing the nation's diplomacy, for example, a person like Franklin Roosevelt, conflicting advisers often served as catalysts for effective policy. But for a president as diplomatically inexperienced as Carter, the consequences of listening to the conflicting advice of Vance and Brzezinski were confusion and vacillation in the conduct of the administration's foreign policy.[2]

For at least the first six months of the administration, the views of Cyrus Vance were preeminent with Carter. A Wall Street lawyer when he was not serving in the government, Vance was the administration's leading exponent of a diplomatic approach to the Soviet Union. While he admitted that the Soviet Union was engaged in a policy of "unceasing probing for advantage in furthering its national interest," he believed that, if the United States acted with "patience and persistence" to check these probes, it would be possible to reach mutually advantageous agreements with the Soviet Union. Nuclear arms control, Vance believed, was one such area of possible agreement since both states had a common interest in avoiding a final holocaust. But while Vance intended to build on the SALT efforts of the Nixon and Ford administrations, he rejected the linkage strategy on which their SALT policy was based. He believed that issues as important to American security as nuclear arms control should not be linked to progress in other areas of the Soviet-American relationship having less dire consequences for the fate of the world.[3]

While Vance's views were preeminent early in the Carter administration, they were increasingly challenged by those of Brzezinski. Unlike Vance, the Polish-born, former professor of political science considered expansionism the primary motivation of Soviet foreign policy and American military power rather than diplomacy the most effective way to check it. Where Vance looked for areas of cooperation with the Soviet Union, Brzezinski concentrated on the competitive aspects of the Soviet-American relationship, particularly in the Third World, which he considered the key arena of East-West competition. Where Vance emphasized local factors as the causes of Third World instability, Brzezinski blamed the meddling of the Soviet Union. The primary way to deal with Third World unrest, Brzezinski believed, was to "respond forcefully" to Soviet intervention in the developing nations. For Brzezinski, this response included not only the use of, or the threat of using, American conventional power, but also the linkage of Soviet actions in the Third World with other issues, particularly economic relations and SALT. Unlike Vance, who believed that SALT was central to the effort to improve Soviet-American relations, Brzezinski saw the talks only as another means of restricting the expansion of Soviet military power.[4]

Brzezinski admitted that he had a difficult time converting Carter to his brand of power politics. The President, he wrote, wanted to be remembered primarily as a great peacemaker, with Woodrow Wilson as his model. But Brzezinski cautioned Carter: "You first have to be a Truman before you are a Wilson." He stressed that it was necessary first "to revive global respect for American power, after the debacles of Watergate and Vietnam, lest our pursuit of principle be confused with weakness." According to Brzezinski, "Carter agreed cerebrally, but emotionally he thirsted for the Wilsonian mantle—and this was sensed on the outside and generated the unfair but damaging charge of vacillation."[5]

Vance, on the other hand, not only believed the vacillation charge was justified, he strongly suggested in his memoirs that Brzezinski was primarily responsible for it and for the damage it produced—not only to the administration's foreign policy but also to Carter's reelection bid. He came to resent Brzezinski's Cold Warrior approach to the world and especially his attempt to be the administration's foreign policy spokesman. When it became clear, in the wake of the Iranian hostage crisis and the Soviet invasion of Afghanistan, that Brzezinski's views had become the President's, Vance resigned.[6]

The March 1977 Proposals

As if Vance did not have enough trouble with Brzezinski, he also had to contend with Senator Henry Jackson. Shortly after the inauguration, Jackson sent Carter a SALT memorandum, prepared by Jackson staffer Richard Perle, which opposed the Vladivostok Accord and called for deep cuts in Soviet ICBMs and IRBMs. The Vladivostok Accord set limits that were too high, the memorandum argued, and when Soviet missiles became fully MIRVed, by the 1980s, America's Minuteman ICBMs would become vulnerable to a Soviet first-strike. Jackson also felt that the Vladivostok Accord placed too many restrictions on America's cruise missile program and not enough on the Soviet Backfire bomber.[7]

Carter was at first responsive to the Jackson-Perle memorandum. "Jackson," Vance explained, "would be a major asset in a future ratification debate if he supported the treaty, and a formidable opponent if he opposed it." Jackson's strength was demonstrated by his opposition to the Senate confirmation of Paul Warnke as the administration's chief SALT negotiator and head of the Arms Control and Disarmament Administration. Warnke had incurred Jackson's wrath by publicly questioning the wisdom of weapons close to Jackson's heart—the B-1 bomber, the Trident submarine, and multiple warheads—and by calling the Jackson-Perle memorandum "a first-class polemic." Largely because of Jackson's opposition, the Senate confirmed Warnke by only a 58 to 40 margin. The administration could not help but realize that the favorable vote was short of the two-thirds margin required for ratification of a SALT II treaty.[8]

Over the objections of Warnke, as well as Vance, Carter decided to make Jackson's proposal for deep cuts the administration's initial SALT proposal. The decision was prompted not only by Carter's desire to gain Jackson's support for SALT but also by his own dissatisfaction with the minimal arms reductions contained in the Vladivostok Accord. He was supported by Brzezinski and Defense Secretary Harold Brown, both of whom felt that deep reductions as well as greater restraints on missile modernization programs were the primary methods of impeding what they perceived as the growing Soviet threat to America's land-based ICBMs.[9]

Vance, like Warnke, thought the prospect of increasing Minuteman vulnerability was exaggerated. Both believed that the United States had more than enough bombers and SLBMs to devastate the Soviet Union even if the Soviets were able to destroy America's ICBMs in a first-strike. Moreover, they argued, there was no way, short of major concessions, that the Soviets would accept deep cuts in the number of their heavy missiles which, Vance emphasized, were "the mainstay of their nuclear forces and their counterweight to the larger U.S. lead in the number of strategic warheads and to our technological superiority." Vance felt that "accepting the Vladivostok framework offered the best prospect for a rapid conclusion of a SALT II Treaty"; the lengthy negotiations required for a deep cut agreement should be reserved for SALT III.[10]

Prompted by Vance's objections, Carter decided to compromise. He instructed the secretary of state to offer the Soviets two proposals when he met with Gromyko on March 28. The first, called the comprehensive plan, was preferred by the President. It included a proposal for deep reductions in the Vladivostok ceilings, from 2,400 launchers to 1,800 and from 1,320 MIRVed ICBMs and SLBMs to between 1,200 and 1,000. There was to be an additional sublimit of 550 MIRVed ICBMs. The comprehensive plan also called for the number of Soviet heavy missiles to be reduced from the Vladivostok ceiling of 313 to 150. It also included a number of proposals for qualitative restrictions on strategic weapon deployments. They included a ban on the construction of new ICBMs; a ban on the modification of existing ICBMs; a ban on the development, testing, and deployment of mobile ICBMs; and a range limitation of 2,500 kilometers for all cruise missiles. The comprehensive plan also offered to exempt the Soviet Backfire from the strategic weapon limitations if the Soviets agreed to a number of measures that would limit the Backfire's range. The second plan that Vance took with him to Moscow was sometimes called the "deferral plan." It was simply the Vladivostok Accord without any references to the Backfire or the cruise missile.[11]

The Soviet response to both American plans could not have been colder. If accepted, they would have required the Soviets to eliminate as many as 600 launchers, while the United States would have been required to dismantle none. (The Soviets had some 2,400 strategic launchers in their arsenal in 1977, compared to some 1,700 for the United States.) The plan would also have required the Soviets to forgo the planned deployments of 400 to 500 MIRVed ICBMs, with no corresponding American renunciations. While the comprehensive plan would have required the United States to cancel its mobile MX missile, still early in its developmental stage, it would have required the Soviets to halt the deployments of their already developed SS-17, SS-18, and SS-19 missiles. In effect, the comprehensive plan would have reversed the Soviet strategic deployment program while leaving the American program virtually untouched. Brezhnev warned that if the United States persisted in seeking deep cuts in

Soviet forces, especially heavy missiles, the Soviet Union would have the right to suggest liquidation of American bases in Western Europe, submarines belonging to NATO, medium-range bombers, and other vehicles capable of carrying nuclear weapons. He reminded Vance that he had agreed to exclude these systems from SALT in exchange for Kissinger's agreement to leave intact the Soviet heavy ICBM force. As for the second American plan, the deferral plan, Brezhnev was as adamantly negative about it as he had been when he rejected a similar proposal by Kissinger a year earlier. The Soviet leader wanted the cruise missile to be included in the SALT limitations, but not the Backfire bomber, which he insisted again was not a strategic weapon.[12]

Brezhnev was clearly upset by the Carter proposals. During the talks, Gromyko's deputy, Georgy Kornienko, took Warnke aside and told him, "You shouldn't have disregarded the fact that Brezhnev had to spill political blood to get the Vladivostok accords." Perhaps for this reason, Brezhnev hailed the Vladivostok agreement as a monumental achievement and the only basis for further progress in SALT.[13]

Yet the Soviets did agree to a number of other administration arms control proposals. As a result, working groups were set up to pursue a comprehensive test ban; nuclear proliferation; prior notification of missile tests; the demilitarization of the Indian Ocean; curbs on civil defense programs as well as on chemical, conventional, radiological, and anti-satellite weapons; and arms transfers to Third World countries.[14]

Nevertheless, in spite of the willingness of both sides to work on these issues and their agreement to meet again in Geneva in May to discuss SALT, several negative consequences of the March meeting in Moscow were significant. The Soviet rejection of the initial American SALT proposals delayed by several months the conclusion of a SALT II treaty, long enough for its conservative opponents to marshal their resources and public opinion against it. By offering an ambitious plan the Soviets were sure to reject, the administration appeared to add substance to the hardline charge that any future agreement with the Soviets would be the result of an American retreat. The stillbirth of the comprehensive plan also produced the first serious split in the Carter administration, with Brzezinski and Vance blaming the other for the negative Soviet reaction. Perhaps of more lasting damage, the comprehensive plan and its rejection by the Soviets helped to create the impression that the administration did not know what approach to take in SALT.[15]

The Geneva Breakthrough and Carter's B-1 Decision, May–June 1977

By the time Vance and Gromyko met again in Geneva on May 18, the acrimonious atmosphere produced by the failure of their March meeting had largely dissipated. After three days of talks, Vance announced that an agreement had been reached on a new framework for the negotiations. It consisted of three "tiers," or parts, which combined some of the elements of the Vladivostok

Accord with those of the March comprehensive proposal. The first tier consisted of a treaty whose duration would extend until the end of 1985 and would be based on the launcher and MIRV ceilings of the Vladivostok Accord. The second tier included a three-year protocol to the treaty that would place limitations on particular weapon systems (such as cruise missiles and mobile ICBMs), missile modernization, and new types of missiles. The third tier consisted of a joint statement of principles establishing a framework for future negotiations leading to a SALT III agreement.[16]

With the negotiating framework in place, the Soviet and American delegations began work in Geneva on the details of the new package. Numerous difficulties had to be overcome before a final agreement could be concluded. And two years of tough and complicated negotiations were required before they were finally overcome.

One of the initial and major difficulties which confronted the negotiators was the size of the reductions in the Vladivostok ceilings desired by the United States. The Carter administration was now prepared to accept a smaller reduction in the ceiling on launchers than the one it had proposed in March. If accepted, it would bring down the ceiling on total launchers from 2,400 to 2,160. But the administration still wanted a MIRVed ICBM subceiling of 550. Although Gromyko had told Vance in their May meeting in Geneva that the Soviet Union would accept a new and reduced launcher ceiling of 2,250, the Soviet government would not agree to a MIRVed ICBM subceiling of only 550. This would have required a forty percent reduction in the Soviet MIRVed missile force while leaving America's MIRVed missiles untouched. Yet the Soviets did indicate that they might accept a MIRVed ICBM ceiling lower than the one established by the Vladivostok Accord, but only if the United States made a major concession, such as counting heavy bombers armed with cruise missiles against the Vladivostok subceiling of 1,320 MIRVed launchers.[17]

The President, however, was too interested in the cruise missile during the summer of 1977 to accept a significant limitation on its deployment. In June he decided to cancel development of the B-1 bomber. To Carter, the B-1 was an excessively expensive and redundant weapon system. He believed that it would be much less expensive to arm the B-52 with cruise missiles than it would be to build a fleet of the B-1s. Needless to say, Carter's alternate plan would be unfeasible if cruise missile deployments were restricted by SALT.[18]

Not surprisingly, Carter's decision to cancel the B-1 generated a storm of protest from hardline critics of the administration. Back in February, Senator Jackson had warned the President that there must not be any constraints on American bomber modernization or on cruise missile deployments. Now Jackson and his allies attacked Carter. Instead of placing cruise missiles on the swifter B-1, they argued, the President was going to place them on aging and sluggish B-52s. Some who agreed that the B-1 program was too expensive deplored Carter's decision to cancel the bomber without first using it as a bargaining chip in SALT. To make matters worse for Carter, instead of attracting a welcome reaction from Moscow, the B-1 decision was condemned. The Soviets complained that now, more than ever, the United States would be tied to the development and eventual deployment of the cruise missile—a weapon system which because of its accuracy and potential numbers greatly alarmed Soviet defense planners.[19]

The Washington Agreements, September 1977

In September Gromyko visited Washington, and another SALT breakthrough was made. The Soviet foreign minister expressed the willingness of his government to accept a subceiling of 820 for MIRVed ICBMs in exchange for an American agreement to allow the number of Soviet heavy ICBMs to remain at their then current level of 308. In a major concession by the United States, the administration accepted Gromyko's offer and, in effect, gave up, once and for all, the American effort to reduce the number of Soviet heavy ICBMs.[20]

The American concession on Soviet heavy missiles was made in return for a major Soviet concession on the verification problem. Gromyko stated that his government would accept the American plan for verifying compliance of the numerical limitations on MIRVed missile launchers, the so-called three-part counting rule. Under this scheme, all missiles of a type tested with MIRVs would be considered MIRVed missiles, regardless of whether or not they were in fact MIRVed. In addition, all silos of a type that housed MIRVed missiles would be counted as MIRVed launchers, regardless of the type of missiles they contained. And finally, certain launchers that resembled MIRVed launchers would be counted as MIRVed launchers. This part of the counting rule applied specifically to 120 "look-alike" silos containing single-warhead SS-11 and MIRVed SS-19 ICBMs which the Soviets deployed together in two missile fields near Derazhnya and Pervomaisk in the Ukraine. Because it was difficult for American photo-reconnaissance satellites to differentiate between the two types of missiles that were deployed in these fields, Moscow agreed to count all 120 silos as containing MIRVed launchers.[21]

The Washington talks also produced progress on other issues. Tentative agreements were reached on a range limit of 2,500 kilometers for air-launched cruise missiles (ALCMs), a range limit of 600 kilometers for deployed ground- and sea-launched cruise missiles (GLCMs and SLCMs), and a ban on the testing and deployment of mobile ICBMs for the duration of the three-year protocol. More significant, in late September both sides agreed to extend, for the duration of the SALT II negotiations, the life of the SALT I Interim Agreement beyond its original October 3 expiration date. Elated by the progress made since September, Carter predicted that the negotiations would be concluded "within a few weeks."[22]

The President, however, underestimated the reaction of the treaty's opponents. Senator Jackson and his allies were upset by press reports that the administration had abandoned its earlier effort to reduce the number of Soviet heavy ICBMs. They were also unhappy with the subceiling of 820 MIRVed ICBMs that the administration had accepted. It was a substantial increase over the original figure of 550 contained in the comprehensive plan. The administration,

Jackson's aide Richard Perle commented, had "given away the store." Jackson also condemned the decision of the administration to extend the Interim Agreement indefinitely beyond its October 3 expiration date without first consulting Congress. But the primary reason for the conservatives' opposition to SALT, at least according to Vance, stemmed from ideology: "The Right feared any negotiation with the Soviets on the theory that nothing the Communists would agree to could possibly be in our interest." By October, if not earlier, Vance realized that "no negotiable SALT agreement would satisfy them."[23]

Trouble with NATO

As if the administration's problems with its domestic critics were not enough, it also experienced increasing criticism of its handling of SALT from America's NATO allies. Some of the alliance's military planners began to fear that the Soviets were using SALT to limit the deployment in Western Europe of American nuclear weapons, particularly cruise missiles, which they hoped to deploy as counters to the deployment of Soviet SS-20 IRBMs and Backfire bombers in Eastern Europe. America's NATO partners were also aware that the Soviets were pressuring the administration to agree on a ban against the transfer of cruise missile technology to third parties, which, of course, included the NATO countries.[24]

The anguish that NATO military planners were experiencing from Carter's handling of SALT was only aggravated by the President's decision to defer, after initially approving, the production of enhanced radiation weapons, or "neutron bombs." Enhanced radiation weapons were designed to kill enemy troops with intensive neutron radiation rather than explosive blast, with the result that minimal damage would be suffered by surrounding structures. NATO leaders saw neutron weapons as an excellent way to deter a massive Soviet invasion of Western Europe, without having to threaten the devastation of the entire continent. Carter refused to deploy them not only because many Americans and Europeans feared that they would make nuclear war acceptable, but also because the allies, and particularly the West Germans, refused to share the onus of public reaction to their deployment. The allies turned down Carter's suggestion that, before the United States agreed to provide the weapons, the Europeans would formally request their deployment. The decision on the neutron bomb not only damaged Carter's public image, it also made the administration more determined than ever to go ahead with the deployment of the cruise missile, which was now seen as a major deterrent to a Soviet invasion of Western Europe.[25]

Partly to allay growing NATO concerns about SALT and the Soviet SS-20 deployments, and partly to restore the administration's standing with the allies in the wake of the neutron bomb decision, Carter agreed to sell the British the Trident I SLBM and to support a decision by the NATO military-planning group in December 1979 to deploy 108 American Pershing II intermediate-range ballistic missiles and 464 Tomahawk cruise missiles in Western Europe.

The move did much to aggravate Soviet-American tensions while doing nothing to inhibit the further proliferation of nuclear weapons.[26]

The Horn of Africa

SALT was also adversely affected by another crisis in Africa. In November 1977 the Soviets began to airlift arms and Cuban troops into Ethiopia to aid that country repel an invasion of its Ogaden province by the army of neighboring Somalia. In response, the Somalis, who were former clients of the Soviets, turned to the United States for help against the Ethiopians.[27]

The Somali request widened the split between Brzezinski and Vance. Brzezinski wanted Carter to dispatch an aircraft carrier to the Somali coast as a sign of American determination to check Soviet expansion into the Horn of Africa. He warned the President that a failure to take a forceful response could jeopardize not only Western interests in Africa and southwestern Asia but also the SALT II Treaty, particularly if the Soviet-Cuban offensive against Somalia coincided with the signing of the treaty.[28]

Vance, on the other hand, argued that the Soviet actions in Africa had to be dealt with in the "local context in which they had their roots" and not primarily as an aspect of the East-West conflict. Moreover, he argued that linking SALT to Soviet actions in Africa would be tantamount to "shooting ourselves in the foot" since the United States had as much to gain from strategic arms control as the Soviet Union. In addition, he believed, it would be a mistake "to threaten or bluff in a case where military involvement was not justified, or where Congress and the American people would not support it." Rather than a resort to force, Vance recommended a diplomatic solution to the crisis. He preferred to believe Soviet assurances that their forces would only be used to help the Ethiopians expel the Somali invaders, not to invade Somalia itself. With Harold Brown's support, Vance was able to convince Carter to seek a diplomatic solution and not to commit the aircraft carrier.[29]

Brzezinski felt that Carter's decision to reject a show of American power during the crisis was a mistake the President would live to regret. "Had we conveyed our determination sooner," Brzezinski recalled in his memoirs, "perhaps . . . we might have avoided the later chain of events which ended with the Soviet invasion of Afghanistan and the suspension of SALT."[30]

However, as events would prove, Brezinzski had lost only a battle; he was about to win his war with Vance for preeminent influence over Carter. In the wake of the Ethiopian-Somali crisis, the President would more and more follow a confrontational approach toward the Soviets. On March 2, 1978, while Vance was telling the Senate Foreign Relations Committee that "there is no linkage between the SALT negotiations and the situation in Ethiopia," Carter had decided that continued Soviet involvement in Africa "would make it more difficult to ratify a SALT agreement or comprehensive test ban agreement if concluded, and therefore the two are linked because of actions by the Soviets."

In an address at Wake Forest University on March 17 he said "that the Kremlin faced a stark choice between competition—even confrontation—and cooperation."[31]

Not surprisingly, the Soviets reacted harshly to the President's hardline approach. *Tass,* the press organ of the Soviet government, condemned the Wake Forest speech as a departure "from the earlier proclaimed course towards insuring the national security of the U.S. through negotiations, through limiting the arms race and deepening détente, to a course of threats and a buildup of tension."[32]

Vance, too, was not at all pleased by the change in the President's strategy. He wrote that the "lurching back and forth in public about linkage" between SALT and the Soviet and Cuban activities in Africa "was hurting the President politically. It was also undercutting our ability to conduct a consistent and coherent foreign policy." Vance, unlike Brzezinski, did not believe that the Soviets and Cubans could be pressured out of Africa. Therefore, he asserted, punitive actions, like delaying SALT, would be futile and would only hurt the United States. In the end, as Vance predicted, the Ethiopian crisis wound down without the necessity of an American military reaction. On March 14 the Somalis completed their withdrawal from the Ogaden and the Soviets and Cubans refrained from invading Somalia, just as they had assured Vance they would do at the beginning of the crisis.[33]

SALT II: Spring–Fall 1978

In spite of the troubles in Africa, Vance and Gromyko met in Moscow between April 20 and 22, 1978 and again in New York and Washington in late May to discuss SALT. During these spring meetings, progress was made on a number of issues. One was the treaty's force ceilings. Both sides agreed to reduce the aggregate launcher ceiling to 2,250 and the subceiling of MIRVed missile launchers to 1,200. In addition, the Soviets made one of their most significant SALT concessions by agreeing to freeze the number of warheads that could be deployed on existing types of ICBMs. In effect, the Soviets would not be able to deploy more than ten warheads on their SS-18 ICBM, though it was capable of carrying twenty to thirty. This so-called fractionation freeze would apply to the other Soviet ICBMs as well.[34]

Other issues proved to be less amenable to solution. One was modernization. Neither side could agree to a definition of what constituted a "new type" of missile, which both sides had considered banning in the treaty, as opposed to an existing missile that was modified. The Soviets wanted only new MIRVed ICBMs to be prohibited, a proposal that would have prevented deployment of the American MIRVed MX ICBM but not a Soviet single-warhead replacement for their SS-11. Vance turned down the Soviet offer.[35]

Nor were both sides able to complete their earlier agreement on range limitations for cruise missiles. Vance wanted the range limitations to make allowance for the zigzag flight path of the cruise missile by measuring total distance from the launch point to the target. Gromyko, on the other hand, insisted that the range of the cruise missile must be the total distance flown by the missile.[36]

The tensions in SALT were not allayed by the intrusion of other, peripheral Soviet-American confrontations during the spring and summer of 1978. In May the FBI arrested two Soviet employees of the United Nations for alleged espionage. The Soviets retaliated the following month by arresting two American businessmen in Moscow for alleged currency violations. On July 12, two days before Vance and Gromyko were to meet in Geneva to discuss SALT, two prominent Soviet dissidents, Anatoly Shcharansky and Alexander Ginsburg, were placed on trial for alleged anti-Soviet activities.[37]

Considering the depths to which Soviet-American relations had sunk by mid-1978, it was amazing that Vance and Gromyko met at all in July. What was even more surprising was their ability to make progress on SALT. They agreed that new SLBMs would not be banned in the treaty, except insofar as they would be covered by a freeze on multiple warheads. In effect, no SLBM could be armed with more than fourteen warheads, the maximum demonstrated capability of either side. This ceiling would enable the United States to deploy the Trident I SLBM and continue with the development of the Trident II SLBM.[38]

Vance and Gromyko resumed their discussions on SALT in New York on September 27. The secretary of state agreed to accept Moscow's definition of cruise missile range, with no allowance for the missile's zigzag flight path, provided that Gromyko agreed, as he did, that the treaty would limit only nuclear-armed ALCMs.[39]

Another session between Vance and Gromyko in Moscow the following month was less productive. The big stumbling block was telemetry, the electronic means by which a missile or warhead sends back to earth data about its performance during a test flight. The Soviets wanted to be free to scramble, or encrypt, their telemetry to inhibit electronic eavesdropping by the United States. The Americans, however, wanted to prohibit the encryption of Soviet telemetry to be able to determine if the Soviets were abiding by the SALT limitations. According to Vance, "what the Soviets wanted was an understanding that telemetry encryption was permitted *unless* it impeded verification." The issue was complicated by the fact that the United States did not want to explicitly state what telemetry was being encrypted by the Soviets for fear of compromising American intelligence-gathering techniques. The Soviets, for their part, refused to consider bans on encryption procedures that were not explicitly stated.[40]

SALT II: Late 1978

By the end of 1978, in spite of the progress made on SALT during the fall, there were increasing doubts within the Carter administration concerning its ability to get a treaty ratified. The negotiation of SALT II had dragged on too long, conservative opposition to the treaty was growing stronger every day, and Brezhnev's health began to deteriorate visibly.[41]

To speed up the conclusion of SALT as well as to circumvent conservative opposition to the treaty, Carter considered submitting it to the Congress as an agreement. A SALT agreement would have required a favorable vote from a simple majority of both houses of Congress. A treaty, by contrast, would have required the approval of two-thirds of the Senate. Carter, however, changed his mind after Vance, Warnke, and a number of key senators warned him that an attempt to make the treaty an agreement would antagonize almost the entire Senate, which was determined to maintain its constitutional prerogatives.[42]

The slow pace of the negotiations was not in any way hastened by the administration's decision on December 15 to normalize relations with China, effective January 1, 1979, and to receive Chinese Vice Premier Deng Xiaoping as a visitor near the end of that month. The Soviets feared that normalization of Sino-American relations was simply the prelude to an eventual alliance between the two countries that would be directed against the Soviet Union. Vance recalled that Gromyko was angered by the normalization announcement at the very time that the United States and the Soviet Union were about to conclude a SALT II Treaty.[43]

With the Chinese grabbing the spotlight, little progress was made by Vance and Gromyko during their December Geneva meeting. They failed to agree on the limits for missile modernization or on the average number of cruise missiles that could be deployed on bombers. But they did conclude a compromise on the encryption problem. It provided that encryption would be explicitly prohibited whenever it impeded verification, but only when it impeded verification. Yet the compromise was not specific enough for the President. Acting at the behest of CIA Director Stansfield Turner, as well as Brzezinski and Defense Secretary Brown, Carter wanted unacceptable encryption more clearly defined. Accordingly, he directed Vance to tell Gromyko that the United States would consider the type of encryption used in a Soviet missile test on July 29, 1978 a violation of the treaty. While not rejecting the American definition of illegal encryption, Gromyko was clearly angered by the reference to a specific Soviet test.[44]

The Last Phase of SALT II: Winter 1978–Spring 1979

Besides these disagreements, another problem arose with the outbreak of the Iranian revolution in January 1979. The overthrow of the shah by the Ayatollah Khomeini forced the United States to dismantle intelligence-gathering installations in Iran at a time when the domestic debate on the encryption issue was at its most sensitive point. Senator Jackson charged that the loss of these installations did "irreparable harm" to America's ability to monitor Soviet compliance with a SALT treaty. The administration countered by asserting that, while the installations in Iran were important, they were not indispensable, considering the other surveillance systems the United States employed to monitor Soviet missile activities, including earth-orbiting satellites, ships, planes, and ground installations elsewhere along the Soviet periphery. Moreover, Defense Secretary Brown assured the Senate that replacements for the lost Iranian installations could be established within a year; nevertheless, the political and public relations damage was substantial.[45]

While the Senate debated the verifiability of the SALT Treaty, the Soviet and American negotiators overcame the last obstacles to an agreement. Both sides agreed in March that the expiration of the protocol and compliance with the treaty's reduced ceiling of 2,250 launchers would occur on December 31, 1981. The Backfire problem was resolved when the United States agreed that it would accept, at the eventual summit, a letter from Brezhnev stating that the production rate of the bomber would not be increased beyond its then current level, thirty per year, and assuring that it would not be upgraded to the status of a heavy bomber. The cruise missile problem was settled when Vance and Soviet ambassador Anatoly Dobrynin agreed that no more than twenty cruise missiles could be deployed on a heavy bomber and no more than twenty-eight on other types of aircraft. The telemetry problem was resolved in correspondence between Carter and Brezhnev. In a letter to the Soviet leader, the President stated that a repetition of the extensive encryption that occurred in test flights of the Soviet SS-18 missile on July 29 and December 21, 1978 would be considered by the United States a violation of the treaty's ban on deliberate concealment measures. Brezhnev responded by stating that he considered the problem resolved.[46]

The modernization problem was resolved on April 7, when Dobrynin announced that his government had accepted the American proposal that any ICBM that differed from existing missiles by more than five percent in size or with a different number of stages, or a different propellent, would be classified as a new type of ICBM. The last obstacle to an agreement was removed on May 7 when Vance agreed to a Soviet demand for a formal statement that the United States would not deploy more than three warheads on its Minuteman III ICBM. Two days later, Vance announced that the negotiations had been concluded successfully. On May 11 the White House and the Kremlin announced that Carter and Brezhnev would meet in Vienna from June 15 to 18 to sign the SALT II agreements.[47]

MX

Shortly before the Vienna summit, Carter made another major defense decision. On June 8 he announced that he had approved production and deployment of 200 mobile MX ICBMs. In so doing, Carter reversed an earlier decision made in December 1977 in which he had rejected full-scale production of the MX. Three months after his June decision, he approved a basing plan that called for each of the MX missiles to be deployed on a mobile launcher vehicle that would move along a roadway to one of twenty-three concrete shelters built at intervals of one mile. All 200 missiles and their mobile launchers would require the construction of 4,600 widely spaced concrete shelters and 10,000 miles of roadway in Nevada and Utah at an estimated cost of $33 billion.[48]

Privately, Carter admitted that the MX decision sickened him. "It was a nauseating prospect, with the gross waste of money going into nuclear weapons of all kinds," he confided to his diary. When the MX was discussed in a National Security Council meeting on June 4, he accused Brzezinski of "jamming a decision down his throat." He complained that the argument for the MX was based on the inaccurate perception "that the Soviet Union was stronger than the United States." He blamed the National Security Council for "much of the perception of Soviet superiority."[49]

But Carter's initial inclination to reject the MX again was overcome by his principal advisers. Both Brown and Brzezinski argued that the mobile MX was necessary to counter the growing accuracy and numbers of Soviet land-based missiles that had made the immobile Minuteman force increasingly vulnerable to a Soviet first-strike. Vance also argued in favor of the MX. Without the missile, he believed, there would be little chance of getting the Senate to ratify the SALT II Treaty.[50]

The MX decision brought howls of protest not only from liberal arms control advocates but also from the officials of the states in which the administration planned to deploy the MX system. Liberals regarded MX as a first-strike weapon. With ten highly accurate warheads per missile, 200 MX missiles could destroy a large portion of the Soviet Union's land-based ICBMs, which constituted seventy percent of the Soviet deterrent forces. As such, MX, they argued, would destabilize the current strategic balance by encouraging the Soviets, in a tense diplomatic confrontation with the United States, to launch their ICBMs before they could be destroyed by the incoming American missiles. Moreover, the construction of 4,600 shelters in Nevada and Utah, the governors of those states told a House subcommittee, would destroy forever "a chosen way of life."[51]

Paul Warnke thought the multiple-shelter concept was sheer madness. It not only addressed a problem—Minuteman vulnerability—which he thought was hypothetical at best, it also multiplied the number of targets the Soviets would have to destroy in a first-strike, which would only encourage them to throw more warheads into the heartland of the United States. Furthermore, Warnke believed that the multiple-shelter concept was unverifiable. The Soviets would have to know which shelters contained missiles in order to verify American compliance with the SALT limitations; giving the Soviets that information would obviously destroy the purpose of the multiple shelters. Not unexpectedly, the Soviets expressed "grave reservations" about whether the multiple-shelter concept would be "valid and permissible," but they never formally rejected the concept—either because they did not want to derail SALT or because they wanted to deploy a similar system themselves, or for both reasons.[52]

Warnke, however, was unable to change Carter's mind. In fact, the President not only refused to rescind his MX decision, he also did not dissuade

Warnke from leaving the government in late 1978. Carter's passive reaction to Warnke's resignation was considered another attempt, albeit ultimately a futile one, to mollify conservative critics of SALT. Another was the President's decision to appoint, over the objections of Vance, General George Seignious as Warnke's successor. Vance was bothered by the impression that would be created by naming a military man to head the Arms Control and Disarmament Agency. The appointment of Seignious, which was strongly recommended to the President by Brzezinski, was another important indication of Vance's waning influence in the administration.[53]

The Vienna Summit and the SALT II Agreements, June 1979

On June 18, 1979 in a glittering ceremony in the Redoutensaal in Vienna's Hofburg Palace, where Beethoven conducted his Seventh Symphony before the monarchs of Europe at the Congress of Vienna in 1815, Carter and Brezhnev signed the SALT II agreements.

The agreements, by far the most detailed and complex in arms control history, consisted of three major parts. One was the treaty. It contained nineteen articles, forty-six agreed statements, and forty-two common understandings, all of which were scheduled to expire in 1985. The second part was the protocol, with four articles, four agreed statements, and five common understandings that would be in effect until December 31, 1981. And the third part of the SALT agreement was the Memorandum of Understanding. It contained a data base, which listed numbers and types of strategic offensive arms for both sides, and a Joint Statement of Principles and Basic Guidelines for Subsequent Negotiations on the Limitations of Strategic Arms.[54]

The SALT II agreements placed both qualitative and quantitative restrictions on strategic nuclear weapons. The principal quantitative restrictions were these: (1) an equal aggregate ceiling of 2,400 launchers for strategic nuclear delivery vehicles, including ICBMs, SLBMs, heavy bombers, and air-to-surface ballistic missiles with a range in excess of 600 kilometers (this ceiling would be lowered to 2,250 by January 1, 1981); (2) an equal aggregate limit of 1,320 MIRVed ICBMs and aircraft armed with cruise missiles whose range is greater than 600 kilometers; (3) a limit of 1,200 on all MIRVed ICBMs and SLBMs; and (4) a subceiling of 820 on MIRVed ICBMs.[55]

Among the qualitative restrictions were these: until December 1981, the duration of the protocol, a ban would be in effect against the deployment of mobile launchers, ground- and sea-launched cruise missiles with ranges greater than 600 kilometers, and the flight-testing and deployment of air-launched cruise missiles. The treaty limited the substitution of more technologically advanced missiles for less advanced missiles and prohibited the conversion of light ICBM launchers into heavy ICBM launchers. A number of agreed statements and common understandings specified a number of characteristics of new ICBMs and MIRVs that were prohibited, including testing parameters of new missiles, procedures for "releasing or dispensing" warheads from reentry vehicles, and even limits on reducing the weight of reentry vehicles tested before May 1, 1979.[56]

The SALT II agreements also contained a number of provisions on verification. Article II of the treaty, for example, contained definitions for ICBMs, SLBMs, and heavy bombers, and described the characteristics of strategic cruise missiles and MIRVs. It also defined heavy and light missiles with reference to specific systems on both sides, a first for SALT. Another innovation designed to enhance verification were FRODs, functionally related observable differences. An example of a FROD was the requirement that the Soviet Bison bomber, when used for refueling purposes, would be equipped with distinguishing characteristics that would enable the United States to differentiate it from Bisons carrying bombs. FRODS were also required for cruise missiles with ranges shorter than 600 kilometers, as well as for ICBMs and SLBMs armed with MIRVs.[57]

At Vienna, Carter told Brezhnev that he expected the lion's share of reductions to come in SALT III. He "wanted immediate implementation of SALT II with its strict limits, an additional five-percent annual reduction in these limits for the five years of its duration, a commitment to lower SALT III limits by at least fifty percent below those of SALT II, and the application of similar restraints on limited-range nuclear weapons in Europe." In addition, he decided, with the support of Vance, Brown, Brzezinski, and General David Jones, the chairman of the Joint Chiefs of Staff, to seek "a total freeze in production and deployment of all nuclear weapons."[58]

Carter also proposed a number of other nuclear arms control measures to Brezhnev at Vienna. He told the Soviet leader that the United States was "ready to move beyond SALT II, with improved monitoring of nuclear arsenals from both sides; no encoding whatever of test data; prenotification of massive bomber flights and missile launchings; improved monitoring stations; on-site inspection under certain circumstances; a reduction in the number of launchers, warheads, and total size of all warheads combined; and a moratorium on the construction of all new weapons." Carter also proposed the conclusion of a comprehensive test ban treaty as well as a ban on the testing of anti-satellite systems. He indicated that the United States was also "willing to agree to non-first use of *any* military force—both nuclear and conventional."[59]

In his memoirs, Carter recalled that Brezhnev agreed to discuss much of what he had proposed for SALT III. The Soviet leader stated that both sides should halt production of nuclear weapons and should reduce existing stockpiles. But Brezhnev insisted that other countries, particularly China and America's NATO allies, also must be involved in the negotiations. He also insisted that American forward-based systems must be included in the next round of the talks. He disappointed Carter by stating that all strategic factors must be considered before further reductions were made, and for that reason, he felt that simple, five percent, annual reductions would not be advisable.[60]

Both sides realized, of course, that before serious attention could be given to SALT III, SALT II would have to be ratified. For the Americans, the elation of concluding the SALT II negotiations, Vance recalled, "was tempered by a sober appreciation of the political struggles that lay ahead."[61]

The SALT II Ratification Debate

By the summer of 1979, the opponents of the SALT II Treaty were more formidable than ever. As before, the opposition was centered on the right and was led by powerful organizations like the Committee on the Present Danger; the Heritage Foundation, a conservative think tank financed originally by the Coors brewery fortune; the American Security Council, a military-oriented organization with a quarter of a million members; and another powerful pro-defense lobby, the Coalition for Peace through Strength. All of these organizations were united by a common distrust for the Soviet Union in general and arms control in particular.[62]

More specifically, conservative critics of the treaty argued that it was not equitable. The treaty permitted the Soviets to retain their heavy missiles while allowing the United States to build none. The Soviet heavy ICBM advantage, they pointed out, would enable the Soviets to perpetuate their throw-weight superiority and, when the Soviet "heavies" were MIRVed, in the early 1980s, would give the Soviets the capability to destroy ninety percent of America's ICBMs with only one-third of the Soviet ICBM force. Moreover, conservatives argued, the treaty failed to take into account some 375 Soviet Backfire bombers which, they insisted, had a strategic capability.[63]

In addition to being inequitable, conservatives asserted that the treaty was not verifiable. By failing to ban all encryption of telemetry, they argued, the treaty seriously impeded America's ability to verify whether the Soviets were testing more than one new missile. Nor could the treaty guarantee with any degree of certainty, they believed, cruise missile ranges or the nature of their payloads.[64]

The treaty, its conservative critics argued, would also weaken NATO. The ambiguity of the treaty's provisions on noncircumvention, they believed, could preclude the transfer of technology and weapons to America's NATO allies, thereby making it difficult for NATO to counter either the Warsaw Pact's superiority in conventional power or the buildup of Soviet SS-20s in Eastern Europe.[65]

In fact, the arguments of conservatives against SALT II were similar to those they had used against SALT I. The SALT process, they believed, could not guarantee détente, nor stop Soviet "adventurism" in Africa or anywhere else in the Third World, nor could it limit or reduce arms expenditures, considering the high levels of the ceilings as well as the ample leeway for modernization allowed both sides.[66]

Liberals, on the other hand, argued that the treaty was unacceptable because it did not sufficiently curb the nuclear arms race. They were particularly unhappy with the way SALT II allowed both sides to continue to modernize and replace their current weapon systems with bigger and more powerful ones, including the MX ICBM and the Trident SLBM for the United States, and the SS-18 ICBM and Backfire bomber for the Soviet Union. Even more ominously, they believed, the treaty only minimally restricted the numbers of warheads these systems could carry. By 1985, the date of the treaty's expiration, both sides could have an additional 5,000 nuclear warheads in their strategic arsenals. Liberals were also angry with the administration's attempt to buy conservative support for the treaty with new weapon programs.[67]

Returning to Washington on June 18, the President gave a spirited defense of the treaty before a joint session of Congress. In so doing, he tried to placate both liberal and conservative critics of the treaty. To satisfy liberals, he admitted that the treaty did not end the arms competition, but would make the competition "safer and more predictable, with clear rules and verifiable limits, where otherwise there would be no rules and no limits." In an attempt to satisfy conservatives, Carter said that for the first time a SALT agreement placed "equal ceilings on the strategic arsenals of both sides, ending a previous numerical imbalance in favor of the Soviet Union." Moreover, he pointed out, the treaty would restrict the numbers of warheads on missiles, their throw-weight, and the qualitative development of new missiles. He added that "without the SALT II limits, the Soviet Union could build so many warheads that any land-based system, fixed or mobile, could be jeopardized." He tried to assure conservatives that America had the means to verify Soviet compliance: "Were the Soviet Union to take the enormous risk of trying to violate this treaty . . . there is no doubt that we would discover it in time to respond fully and effectively."[68]

The Demise of SALT II

In the end, Carter was unable to translate the successful negotiation of SALT II into a ratified treaty. As so often before, SALT II was overwhelmed by external events. In early July American intelligence officials began to leak information that they had detected a brigade of Soviet combat troops in Cuba. Facing strong conservative opposition in his reelection campaign, Senator Frank Church, chairman of the Foreign Relations Committee, cast aside his dovish reputation by stating that the SALT Treaty could not be ratified until the Soviets withdrew their brigade. The Soviets responded to the furor, which Church had helped to create, by denying they had any combat troops in Cuba. They insisted that the only Soviet soldiers in Cuba were there to train Cuban troops, a function the Soviets had performed for seventeen years.[69]

Under pressure from Church and conservatives anxious to use the Soviet brigade as an excuse to scuttle the treaty, the administration's initial reaction was hard. Carter stated that the United States would "not be satisfied with the maintenance of the status quo," even though he knew there was no way that he could get the Soviets to withdraw their troops. It took the President nearly a month, until October 1, to back off the rhetorical limb on which he had climbed. He had to settle for a Soviet pledge not to expand or upgrade the brigade or to use it to threaten other countries. In effect, Carter did something he said he would not do: he accepted the status quo.[70]

The latest Cuban "crisis" proved costly to the SALT Treaty. It skewed the momentum to the right, in favor of the treaty's conservative critics. The administration, still twelve to fifteen votes short of the necessary two-thirds majority in the Senate, lost time, and perhaps a vital vote or two. Moreover, so many influential conservatives and moderates were urging long-term increases in defense spending that the administration felt compelled to give an advance preview of its 1981 Pentagon budget and a revised five-year defense program that increased military spending by three percent annually after inflation.[71]

By November the Cuban "crisis" had abated and the treaty seemed to be back on track. On November 9 the Senate Foreign Relations Committee, by a vote of nine to six, approved a resolution favoring ratification. The committee report concluded that the treaty was "better for the United States at this point than no treaty at all." Buttressed by the report and by Senate Majority Leader Robert Byrd's estimation that the treaty could be brought to a floor vote before the Christmas recess, the administration once again took heart.[72]

But in the meantime another event struck a blow at the treaty. On November 4, 1979 the American embassy in Tehran was overrun by militant students angered by the temporary admission into the United States of the deposed shah, who was dying of cancer. As a result, sixty-three Americans were held hostage in Iran for over a year. The Iranian hostage crisis not only turned public attention away from the SALT II Treaty, it also contributed to the further discrediting of the administration that had successfully negotiated the treaty. Carter's stock in the public opinion polls plummeted after a military expedition sent to free the hostages in April 1980 failed. Opposing the decision to dispatch the expedition, Cyrus Vance, the primary architect of the SALT II Treaty, tendered his resignation.[73]

But the Iranian crisis was not the treaty's fatal blow. On Christmas Day 1979, American intelligence learned that Soviet troops were pouring into Afghanistan, apparently to preserve a Marxist regime that had come to power in the wake of a bloody military coup in 1978. The coup had triggered a resistance movement primarily Islamic in character, a development that disturbed the Soviets, who feared that the unrest in Afghanistan could spread to Moslem areas within the Soviet Union. The Soviets obviously had concluded that the ratification prospects of the SALT Treaty were so poor that they had little to lose in their relationship with the United States by invading Afghanistan.[74]

The Soviet invasion of Afghanistan seemed to vindicate hardliners inside and outside the Carter administration. Wrote Brzezinski, "had we been tougher sooner, had we drawn the line more clearly, . . . maybe the Soviets would not have engaged in this act of miscalculation." Prompted by Brzezinski and others (including pollster Pat Cadell, who advised that playing tough with the Soviets would pay political dividends in an election year), Carter reacted vigorously to the Soviet invasion. Among other actions, he sharply cut back on the sale of grain and technology to the Soviet Union and announced a draft registration program, the withdrawal of the United States from the Moscow Olympics, and a new counterforce targeting doctrine (contained in Presidential Directive 59).

He also asked Congress to fund a rapid deployment force to give the United States the capability to intervene militarily in the Persian Gulf, released classified information about new radar-resistant "stealth" aircraft, as well as new laser and particle beam anti-missile and anti-satellite weapons, and asked the Senate to postpone indefinitely its consideration of the SALT II Treaty.[75]

The Demise of Other Arms Control Efforts

SALT II was not the only casualty of the Afghanistan invasion. Another was the negotiation of a comprehensive test ban treaty (CTB). Until 1978, progress on the CTB had been substantial. In March 1977 Brezhnev had accepted in principle the American insistence on the on-site inspections to monitor compliance with the CTB. In September, however, the Soviets changed their position, stating that in order to be effective a CTB treaty must encompass all nuclear states, including France and China. In November the Soviets gave up their insistence that peaceful nuclear explosions must be excluded from the ban. Another major breakthrough was achieved early in 1978 when the United States, Britain, and the Soviet Union agreed in principle to allow up to ten seismic monitoring stations to be placed on each country's territory to facilitate verification. And on May 20, 1978 Carter authorized his negotiators to seek a five-year, total ban on all nuclear testing. It was the general consensus among the negotiators that a treaty could be concluded that year.[76]

But as negotiators in the Kennedy administration had found out, when the superpowers also had come close to concluding a CTB agreement, the opposition to a comprehensive ban proved too strong to overcome. Opponents of the CTB in Congress, the Pentagon, and the Department of Energy argued that a five-year ban on testing would reduce the reliability of America's existing nuclear warheads. Wolfgang Panofsky, director of the Stanford Linear Accelerator in Stanford, California, vigorously rejected this argument. It had been amply demonstrated, Panofsky stated, that the reliability of nuclear warheads could be verified without testing. Perhaps because of the President's sensitivity to the charges that he was "soft on defense" and a "unilateral disarmer," he decided in September 1978 to reduce the duration of the ban from five to three years.[77]

In the end, the CTB negotiations were overwhelmed by the problems that overcame the SALT II Treaty. So, too, were the other arms control initiatives launched by the Carter administration: the negotiations to ban deployment of anti-satellite weapons, to demilitarize the Indian Ocean, and to curtail conventional arms transfers. In effect, the Soviet invasion of Afghanistan and Carter's reaction to it represented the end of détente.

Further Decline of the Nonproliferation Regime

Carter's effort to strengthen the barriers against the horizontal proliferation of nuclear weapons was marred by the same lack of consistency that characterized his approach to vertical proliferation. When the administration took office, it adopted the same premise on which the Ford administration had belatedly operated—that the Nonproliferation Treaty-International Atomic Energy Agency (NPT-IAEA) system could work for the "once-through fuel cycle," which employs light-water reactors and nonweapon-grade uranium, but it could not work for breeders and reprocessing facilities, which could produce weapon-grade plutonium. In other words, the Carter proliferation policy presumed that, if the world used only light-water reactors and low-enriched (nonweapon grade) fuel, and the number of breeders and reprocessing plants were restricted, it would be much easier to prevent the diversion of nuclear facilities and materials from civilian to military projects.[78]

The administration initiated a number of steps to actuate this assumption. It attempted to defer indefinitely the development of a commercial breeder program in the United States by continuing and expanding the Ford moratorium on the domestic reprocessing and recycling of plutonium. For this reason, Carter attempted, unsuccessfully, to get congressional approval to cancel the construction of the Clinch River breeder reactor. Simultaneously, the administration placed an embargo on the export of enrichment and reprocessing facilities as well as highly enriched uranium, all of which could be used to produce nuclear weapons. At the same time, however, the administration also took steps to increase American production of low-enriched uranium in order to assure domestic and foreign reactor operators that the United States was still a reliable provider of nuclear fuel. But in the Nonproliferation Act of 1978 Congress insisted that by 1980 all recipients of American nuclear fuel must subscribe fully, in all their nuclear facilities, to the IAEA safeguard system.[79]

And there were critics abroad as well. Carter's nonproliferation policies were bitterly attacked by the Europeans and the Japanese. The Europeans considered his moratorium on breeders a poorly disguised attempt to maintain the American lead in light-water reactor and uranium-enrichment technology and sales by delaying the development of breeder technology, the one field of nuclear technology in which the Europeans led. The Japanese, who were more deficient in domestic energy supplies than the Europeans, were even more upset by the Carter policy. They considered the breeder a godsend that would enable Japan to lessen its dependence on foreign energy sources. America's allies also questioned Carter's appraisal of the dangers associated with the breeder. They believed the safeguards monitored by the IAEA were adequate to prevent breeder-produced plutonium from being used for military purposes.[80]

What particularly infuriated America's nuclear partners and customers, however, was the apparent duplicity of the Carter approach to the nuclear proliferation problem. While the President condemned the transfer to nonweapon states of sensitive nuclear materials and technology by the Europeans and the Japanese, he approved the sale of American uranium to India, which continued to refuse to fully accept the safeguard system of the IAEA. The sale was made despite a House resolution opposing it as a violation of the spirit, if not the letter, of the Nonproliferation Act. A similar resolution in the Senate was narrowly defeated. The administration argued that to deny the Indians nuclear fuel, which the United States had earlier committed itself to provide until 1994, would encourage India and other states similarly dependent on the United States to look elsewhere for nuclear assistance, a thinly disguised allusion to America's vigorous nuclear competitors. Moreover, the Carter administration concluded that it could not afford a "political breakdown" in relations with India at a time when, in the wake of the Soviet invasion of Afghanistan and the Iranian hostage crisis, tensions in southern Asia were increasing. Like SALT II and the other arms control measures pursued by the Carter administration, the nonproliferation policy was sacrificed on the altar of Afghanistan.[81]

12

Reagan and the "Rearmament" of America, 1981–1983

The Reagan Revolution

The election of Ronald Reagan in 1980 brought to the White House an individual committed to a nuclear arms race with the Soviet Union and the first president since Truman who was openly skeptical about the value of nuclear arms control agreements. During Reagan's first administration, defense spending nearly doubled. Major nuclear weapon systems that were shelved by Carter, the B-1 bomber and the neutron bomb, were revived. The deployment of developed systems, such as the MX ICBM and the Trident submarine, was initiated. And the development of new systems, like the ballistic missile defense (BMD) and anti-satellite weapons, was enthusiastically accelerated. Moreover, Reagan during his first term was not only unable to conclude any major nuclear arms agreements, existing agreements, like the ABM Treaty and the unratified SALT II Treaty, were threatened with revocation by the United States. Not surprisingly, in America and in Europe, the Reagan nuclear buildup and deemphasis on arms control did much to heighten fears of a nuclear conflagration during the first years of his presidency.

The Reagan Team

Like the overwhelming majority of America's Cold War presidents, Reagan entered the White House with no leadership experience in either diplomatic or military affairs. Before entering the political arena, he had been an actor in movies and in television. His only direct military experience occurred during World War II, when he served in the armed forces making training and documentary films. His first and only elected political position was the governorship of California, a position he held from 1966 to 1974. Reagan demonstrated little interest in foreign affairs until he ran unsuccessfully for the Republican presidential nomination in 1976.[1]

The military and diplomatic experience of Reagan's top national security advisers was also limited. Reagan's secretary of defense, Caspar Weinberger, whose primary responsibility would be overseeing the formulation and implementation of the administration's military strategy, had no background in defense policy. Before entering the Pentagon, Weinberger had served as Reagan's finance director in Sacramento and later as President Nixon's budget director and secretary of health, education and welfare. While the military experience of Reagan's first secretary of state, General Alexander Haig, was considerable, his diplomatic experience was negligible. Haig's inability to get along with the White House staff led to his forced resignation on June 24, 1982 and his replacement by George Shultz. Compared to Haig, however, Shultz's experience with national security affairs was almost nonexistent. A professional economist, Shultz had served as Nixon's budget director, secretary of labor, and secretary of the treasury.[2]

Partly because of the paucity of national security experience at the top of the Reagan administration, the President at first relied heavily on the Committee on the Present Danger, the nucleus around which the opponents of the SALT II Treaty had rallied, to provide the intellectual support for, as well as the staff to implement, his national security policy. Thirty-two of the committee's 182 members received key positions in the administration by the end of 1981. They included Richard V. Allen, Reagan's first national security adviser; Paul Nitze, chief negotiator for intermediate-range nuclear forces (INF); General Edwin Rowny, the head of the SALT delegation; and Eugene V. Rostow, the committee's chairman, who became Reagan's first director of the Arms Control and Disarmament Agency. With the appointment of members from the Committee on the Present Danger to positions of responsibility, the President returned to power an extreme hardline attitude toward the Soviet Union not seen since the Truman administration.[3]

The Idealogue as President

The Committee on the Present Danger, however, only refined and gave a degree of respectability to a hard core of anti-Sovietism that Reagan had molded for himself long before, in the Cold War milieu of the late 1940s and the 1950s, when he was fighting efforts of American communists to take over his Screen Actors' Guild. In essence, Reagan's attitude toward the Soviet Union had changed little since then. The Soviet Union, he said in 1983, was "the focus of evil in the modern world." Its goal is "the eventual domination of all peoples of the earth," with the American people its primary target.[4]

Reagan's anti-communist rhetoric appealed not only to well-educated, upper- and middle-class conservatives who equated the American welfare state with creeping Soviet communism, according to historian Robert Dallek, it became for many lower-middle-class Christian fundamentalists a key ingredient of a "crusade to restore traditional assumptions about God, family, and country to a central place in American life." Reagan's brand of anti-Sovietism also appealed to those who were affiliated with defense industries, whose profits had declined appreciably as a result of the relaxation of Soviet-American tensions brought about by détente and particularly by SALT I. In addition, anti-Sovietism, combined with superpatriotism, seemed to be the perfect antidote for a host of problems that had befuddled the Carter administration and dampened the spirit of the American people: the Iranian hostage crisis, the OPEC threat to Western oil supplies, and the Soviet invasion of Afghanistan, to name only a few.[5]

Reagan, of course, was not unique among America's recent presidents in his exploitation of anti-Sovietism. Where he was different, however, was in his apparent belief that agreements with the Soviet Union were not worth much. Paraphrasing Lenin, he said that to the Soviets "promises are like pie crust, made to be broken." The agreements that were observed by the Soviets, he asserted, were those that had cost the United States the most. Détente under Nixon, Ford, and Carter, he charged, was a one-way street, with the Soviets receiving all the benefits. The centerpiece of détente, the SALT agreements, he argued, had enabled the Soviets to augment their strategic forces to a point where America was vulnerable to a Soviet first-strike, the so-called window of vulnerability. The economic concessions Nixon, Ford, and Carter were prepared to grant the Soviets, Reagan believed, would have propped up an inefficient economic system and thereby enabled the Soviets to continue their military buildup and maintain their oppressive political structure.[6]

Rather than attempting to promote détente, Reagan and his advisers were prepared to engage the Soviets in a policy of confrontation and counterpressure. As outlined in a national security directive that was completed in May 1982, the Reagan strategy toward the Soviet Union amounted to what a senior White House official called "a full-court press." According to William P. Clark, who succeeded James Allen as national security adviser in January 1982, the goal of the Reagan strategy was "to convince the leadership of the Soviet Union to turn their attention inward," rather than outward in aggressive expansionism. The strategy not only called for the military containment of the Soviet Union but also for its political and economic isolation. Instead of assisting the Soviets to maintain an inefficient economic system with trade and credits, the Reagan administration intended to exploit the Soviet Union's economic weaknesses in the hope of undermining its military strength and political structure.[7]

If negotiations with the Soviets were going to take place, Reagan intended them to produce agreements that redressed the alleged imbalances of the past that favored the Soviet Union—or nothing at all. With respect to nuclear weapons negotiations, this would require, he stressed, major reductions in the size of the Soviet Union's nuclear forces, particularly its ICBM arsenal. For this reason, he warmly accepted the suggestion that SALT be renamed START— Strategic Arms Reduction Talks. However, in Reagan's view, Moscow would agree to major cuts in its strategic forces only if the United States demonstrated its willingness to match the Soviet Union weapon for weapon. In other words,

START would begin in earnest, if ever, only after the United States had "rearmed" and closed the alleged window of vulnerability.[8]

Preparing for Protracted Nuclear Warfare

The philosophical foundation of Reagan's nuclear strategy, which was largely formulated by the Committee on the Present Danger, was based on an ardent refutation of the mutual assured destruction (MAD) doctrine which had governed America's deterrent strategy since the early 1960s. To the Committee on the Present Danger, MAD offered no credible deterrent to Soviet aggression primarily because it made no provision for the failure of deterrence short of suicide by retaliation. What the committee feared most was the possibility that a Soviet first-strike aimed solely at America's military installations would be so devastating that the United States would have too few strategic forces remaining to retaliate effectively against the Soviet homeland. To retaliate with insufficient forces would be suicidal, the committee insisted, because America's cities would then be exposed to retaliation from the enemy's second and third strikes. Faced with such a prospect, to accept the destruction of America's military installations or risk the destruction of her cities through an expansion of the war, the Committee on the Present Danger believed that the United States would suffer a paralysis of will that could lead to American capitulation to Soviet nuclear blackmail.[9]

Not only did Reagan's strategists believe the Soviet Union had acquired a first-strike capability, some also contended that the Soviets were prepared to use it. If this were true, the Soviets had never accepted MAD as the basis of their nuclear strategy. Instead, they believed that a nuclear war could be won, and thus the nation better prepared for nuclear war could emerge from it as a viable, even though greatly weakened, society. For this reason, Reagan's advisers pointed out, the Soviets had dispersed their industries and developed a civil defense program far superior to that of the United States.[10]

To counter the possibility that the Soviets might attack only America's military installations and hold American cities hostage to Soviet nuclear blackmail, Reagan's analysts insisted that the United States must enhance its counterforce capabilities. But some, if not most, of Reagan's advisers felt that America's nuclear forces should not be limited to a retaliatory role. Colin Gray, a consultant to the Reagan ACDA, insisted that if necessary to preserve preeminent American interests, such as a non-communist Western Europe, the United States should be prepared to initiate, and win, a protracted nuclear conflict.[11]

The strategy Gray had in mind was outlined in a document, "Fiscal Year 1984–1988 Defense Guidance," which was leaked to *The New York Times* and published in that newspaper in 1982. It stated that American forces must be prepared to "render ineffective the total Soviet (and Soviet-allied) military and political power structure" and to destroy Soviet "nuclear and conventional military forces and industry critical to military power." It required that the United States be "capable of supporting controlled nuclear attacks over a protracted period" and of targeting during the "post-attack reconnaissance." The strategy also envisioned sabotage behind the Iron Curtain, military aid to China, and surgical nuclear "decapitation" of the Soviet Union's military and political leadership. The 1983 version of the same plan called for the military services to integrate their various strategies for firing long-, medium-, and short-range nuclear weapons so as not to "impose [an] arbitrary division between categories of nuclear weapons." [12]

The implementation of an American policy of global, coercive containment of the Soviet Union and its proxies, and the capability to negotiate with the Soviets from a position of overwhelming strength, would require, Weinberger told Congress, a military buildup of almost $1.7 trillion over a five-year period. Approximately ten percent of the money, or $180 billion, would be used for new strategic nuclear systems. Included in this category were 100 MX ICBMs; 100 B-1 bombers; 400 air-launched cruise missiles; 3,000 sea-launched cruise missiles; fifteen Trident submarines with 360 Trident I SLBMs; and command, control, communications, and intelligence systems (C^3I) that would be designed to survive a nuclear exchange. Funds were also appropriated for the development of the Stealth bomber; a new, single-warhead ICBM, named "Midgetman"; accelerated development of the Trident II counter-silo missiles; and anti-satellite weapons. The Reagan program also called for a seven-year, $4.2 billion civil defense program that would ostensibly allow the evacuation of most large American cities before a nuclear exchange. [13]

In addition to modernizing strategic forces, the administration embarked on a program to modernize tactical nuclear weapons and "dual-capable" delivery systems (weapons that can deliver nuclear or conventional munitions). By 1988, 400 nuclear-armed, sea-launched cruise missiles were to be deployed on some 600 attack submarines and surface ships. F-4 Phantom attack planes in the Marine Ground Task Force that were not certified for delivering nuclear weapons would be replaced by F-18 Hornets that were. Non-nuclear 105-millimeter howitzers would be replaced by the dual-capable 155-millimeter gun, almost doubling the Marines' nuclear artillery arsenal. Army and Marine 8-inch howitzers would be equipped to fire a neutron warhead, a weapon whose deployment, but not development, Carter had rejected in 1978. [14]

Of course, the ability to win a nuclear war also implied that civilization can survive a nuclear war and live in the aftermath. Colin Gray estimated that a nuclear war limited to military installations would cause twenty million American casualties. He believed that, while that number of casualties would be a disaster unprecedented in American history, it would certainly be preferable to the 200 million or so casualties that would result if the United States emphasized only a retaliatory strategy. Thomas K. White, Reagan's Deputy Secretary of Defense for Research and Engineering, Strategic and Nuclear Forces, told journalist Robert Sheer that the United States could recover from a limited nuclear war in just two to four years. He added, "If there are enough shovels to go around, everybody's going to make it." Explained Sheer, "the shovels were for digging holes in the ground, which would be covered somehow or another with a couple of doors and with three feet of dirt thrown on top, thereby providing adequate fallout shelters for the millions who had been evacuated from America's cities to the countryside." Said White, "It's the dirt that does it."[15]

Counterattack on Counterforce

While Reagan's strategic doctrine attracted criticism on several points, the critics zeroed in on the administration's counterforce strategy and its supposition that the United States could effectively wage a protracted, limited nuclear war. The critics found it difficult to conceive of how the destructive effects of a nuclear attack could be limited only to military installations. They pointed out that no matter how accurate missiles are, guidance or human errors are always possible. Nor could the effects of gravitational and magnetic forces on the accuracy of missiles fired over the North Pole be determined in advance with a high degree of confidence since, quite obviously, the Soviets never permitted the United States to test fire missiles over the pole into Soviet territory. As a result, American missiles targeted on Soviet military installations could conceivably strike Soviet cities.[16]

The problem of misguided missiles was only compounded by the fact that many American and Soviet military installations are located near heavily populated regions. The Defense Department admitted that there were sixty military installations in the city of Moscow alone that were targets for American nuclear warheads. According to a study conducted by the Congressional Office of Technology Assessment in 1979, a Soviet missile attack on the American ICBM complex alone would produce an estimated two- to twenty million American fatalities. Such enormous losses in human life—not to mention the damage to the social, economic, and ecological fabric of the nation—would make it difficult to distinguish between limited and unlimited nuclear war. Faced with such staggering destruction from a Soviet "limited" attack, the critics asked, would any American president confine his response to retaliating against only Soviet military installations? Concluded General David Jones, shortly before retiring as chairman of the Joint Chiefs of Staff, "I don't see much chance of a nuclear war being limited or protracted" without "a tremendous likelihood of it escalating."[17]

Many critics charged that the basic flaw of the Reagan strategy was the premise on which it was based—that America's deterrent forces were vulnerable to Soviet attack. The American strategic deterrent, the critics argued, was already credible without the massive nuclear buildup desired by the administration. To escape devastating American nuclear retaliation, the Soviets would have to destroy almost simultaneously every element of the American triad of strategic weapons. In 1983 this included 388 operational strategic bombers, 1,049 ICBMs, and 544 SLBMs. In addition, the Soviets would have to deal with hundreds of nuclear-capable fighter-bombers deployed in Europe and on aircraft carriers in the waters surrounding the Eurasian landmass. Henry Kendall, a nuclear physicist at Massachusetts Institute of Technology, calculated that a Soviet first-strike that disabled eighty-five to ninety-five percent of America's ICBMs and all of its bombers and submarines would still leave the United States with enough nuclear weapons to destroy the twenty-two largest Soviet cities and their thirty-two million people. Reflecting the pride of America's military in its deterrent forces, General Jones told Congress: "I would not swap our present military capability with that of the Soviet Union, nor would I want to trade the broader problems each country faces."[18]

What the Reagan buildup would produce, the critics argued, was another needless and expensive expansion of the American nuclear arsenal. Worse, they feared the more nuclear weapons produced, the more likely they would be used, whether by design or by accident. The possibilities, they argued, were not all farfetched. In an eighteen-month period that ended June 30, 1980, the North American Defense Command (NORAD) experienced 147 false alarms that were serious enough to require an assessment of whether they constituted a Soviet attack. In November 1979 fighters were scrambled after a NORAD computer falsely indicated that a Soviet attack was in progress. Twice in June 1980 a computer falsely warned that the Soviets had launched submarine- and land-based missiles at the United States. It was discovered that a computer chip costing forty-six cents was responsible for the false alarms. No one could say how many computer failures and accidents the Soviets have experienced. Ultimately, and ironically, America's security had come to depend as much on the efficiency of Soviet computers as on its own.[19]

The critics also attacked a second premise upon which the Reagan strategy was based—the assertion that the Soviets were inclined to wage nuclear war. It has been estimated that American retaliation after a Soviet first-strike would destroy at least sixty percent of the Soviet Union's industrial capacity from blast alone, disregarding the damage produced by initial radiation, fallout, and subsequent fires. Soviet fallout shelters, the critics asserted, would reduce the number of immediate fatalities, but obviously would fail to protect the livestock, crops, or the environment.[20]

For essentially the same reasons, the critics discounted the potential effectiveness of the Reagan "crisis relocation plan," a program to move urban dwellers out of the cities in the event of imminent nuclear war. They argued that the plan would require at least a one-week warning to evacuate American cities, thereby discounting the possibility of a Soviet surprise attack. Indeed, they asserted, the evacuation of American cities in a period of acute Soviet-American tension might provoke a Soviet preemptive attack. Nor did the Reagan plan have anywhere near the extensive provisions required for feeding, sheltering, and nursing those who did manage to escape. Ultimately, in 1985, the administration admitted that the crisis relocation plan had been abandoned, primarily because of the resistance of state and local governments.[21]

According to critics, the Reagan administration, by implying that a nuclear war was winnable, was making a nuclear war more likely to occur. And even if nuclear war did not result, the administration's nuclear buildup represented an enormous drain on the nation's resources, which, they believed, could be used more productively in the civilian sector. Critics argued that the Reagan

increases in defense spending, when combined with the tax cuts the administration pushed through Congress in 1981, helped to produce the largest annual
budget deficits in the nation's history. Some alleged that the President, by cutting domestic programs, including welfare, food stamps, health care, and housing, had attempted to finance his enormous military buildup at the expense of
the poor.[22]

The Reaction of Europe

Reagan's nuclear programs were also controversial in Europe. When Weinberger stated that the United States intended to deploy neutron weapons in
Western Europe and would "probably want to make use of" them in the event
of war, European hostility toward them again erupted. The Europeans feared
that use of neutron weapons would lower the nuclear "fire-break" between
conventional and nuclear weapons and make a nuclear war on their continent
more likely. The West German government stated flatly that it would not agree
to the deployment of neutron weapons on its soil. As a result, the administration felt compelled to announce that they would be stored in the United States.
The Soviets, who had stated that they would not begin production of neutron
weapons unless the Americans did, announced that they would respond in kind.[23]

But an even more controversial issue than the neutron bomb was the administration's decision to go ahead with the planned deployment of the 572 Pershing II IRBMs and Tomahawk cruise missiles in Western Europe, which were
approved by NATO's governments in 1979. Like the Carter administration,
Reagan believed that deployment of the missiles was necessary to counter the
deployment of Soviet SS-20s in Eastern Europe. But many West Europeans
feared that the deployment of the American missiles on the continent, combined
with the administration's protracted war strategy, would increase the risk of a
nuclear war.[24]

What made Europeans particularly uneasy was the prospect that the administration might try to limit a nuclear war to Central Europe to avoid the nuclear
destruction of the United States. That suspicion was strengthened on October
16, 1981 by Reagan's impromptu response to a reporter's question asking if
there could be a nuclear exchange limited to Europe. The President replied, "I
don't honestly know." He compounded European fears by saying, "I could see
where you could have the exchange of tactical weapons against troops in the
field without it bringing either one of the major powers to pushing the button."
The furor raised in Western Europe by the President's statements compelled
the administration to issue another statement denying the inference that the
United States was prepared to sacrifice Europe to prevent the destruction of
the American homeland. The matter, however, was raised again by the President on November 10, 1981 when he told a press conference: "I could see
where both sides could still be deterred from going into the exchange of strategic weapons if there had been battlefield weapons troop-to-troop exchanged
there [Europe]."[25]

Many West Europeans believed that the administration was more interested in deploying the Pershing IIs than it was in removing the Soviet SS-20s. "What we have decided, in the West European peace movement," British peace advocate E. P. Thompson wrote, "is that enough is enough. We are not even interested in 'balance'. There is sufficient nuclear weaponry placed now in Europe to blow up the continent twenty times over, and we don't much care if one side can do it twelve times and the other only eight times. We . . . have decided that the time has come when we can make concessions to the growth of militarism no longer."[26]

The Reagan nuclear buildup and rhetoric did much to revive the long-moribund, anti-nuclear movement in Europe. Massive demonstrations took place throughout Western Europe. In early October 1981, timed to coincide with Weinberger's visit to West Germany, an estimated 250,000 people amassed in Bonn to denounce the forthcoming deployment of American Pershing II missiles on German soil. That weekend nearly one-half million protestors in London and Rome joined their Bonn counterparts to demonstrate against the American missiles. In West Germany, a petition drive, the so-called Krefeld Appeal, collected 2.1 million signatures opposing deployment of the Pershing IIs and Tomahawks. Opposition to the deployment of the missiles within the Social Democratic Party prompted Chancellor Helmut Schmidt to threaten resignation if his party rejected the missiles. It also led to the formation of the so-called Green Party, which advocated a nuclear-free Europe from the Baltic to Portugal and the dissolution of East-West power blocs. A Soviet invasion of Europe, the party argued, could be deterred more safely by the creation of people's militias.[27]

Needless to say, the Soviets did all in their power to encourage the rising opposition to the deployment of the American missiles. The Pershing IIs could deliver warheads on targets deep within the Soviet Union four to eight minutes after they were launched. The Tomahawk cruise missiles, while much slower than the Pershings, were equally accurate and deadly. The result of an attack using the American intermediate-range missiles, the Soviets feared, would be the "decapitation" of the Soviet retaliatory capability. To preclude that possibility, Western critics of the missiles pointed out, the Soviets might be tempted to strike first at the Pershing and Tomahawk deployment sites before they were launched. Equally threatening to the critics was the possibility that the Soviets would institute a launch-on-warning system that would leave the decision to fire their missiles to computers. Moreover, the critics feared that the deployment of the Tomahawk cruise missiles would jeopardize arms control efforts because neither the number nor the nature of their warheads could be verified easily.[28]

The Soviets also considered the NATO missile deployments a violation of an understanding reached between Kennedy and Khrushchev in 1962 that helped to defuse the Cuban missile crisis. In 1963 the United States withdrew theater-nuclear missiles from Italy and Turkey as a *quid pro quo* for the withdrawal of Soviet missiles the preceding year from Cuba. The result of the understanding was to allow the Soviets to retain about 600 IRBMs in Eastern Europe, while the United States deployed only short-range Pershing I missiles in West Germany. The Reagan administration was now preparing to reintroduce intermediate-range missiles in Western Europe without offering to permit the Soviets to reintroduce their missiles in Cuba.[29]

The Nuclear Freeze Movement

The administration's nuclear buildup, combined with Reagan's hostile attitude and harsh rhetoric toward the Soviet Union, and the President's failure to make arms control talks with the Soviets the number one priority of his administration did much to fan fears of a nuclear war in America as well as in Europe. An NBC/Associated Press survey in mid-December 1981 found that seventy-six percent of the public believed that nuclear war was "likely" within a few years, an increase from fifty-seven percent in August. One defense expert after another issued the warning that nuclear war was inevitable if not imminent. Stated George Kennan: "Never in my thirty-five years of public service have I been so afraid of nuclear war." Said Admiral Hyman Rickover, creator of America's nuclear Navy: "I think we will probably destroy ourselves."[30]

The communications media gave full play to the nuclear war scare. On five evenings in 1982, CBS carried a series analyzing the nation's defense establishment using the backdrop of a possible nuclear war. ABC in November 1983 ran a movie about a fictional nuclear war, "The Day After," that drew the largest television audience in the history of that media. Immediately afterward, Secretary of State Shultz appeared to assure the nation that the administration was doing everything possible to prevent a nuclear holocaust of the kind depicted in the movie.[31]

Even the President seemed to have been affected by the nuclear war scare. On October 18, 1983 he told the *Jerusalem Post:* "I turn back to your ancient prophets in the Old Testament and the signs foretelling Armageddon, and I find myself wondering if, if we're the generation that's going to see that coming about." Reagan later said that his belief in the possibility of Armageddon would not prevent him from doing everything possible to avoid a nuclear holocaust.[32]

The public fear of nuclear war gave an enormous boost to the movement to halt the production of all nuclear weapons, the so-called nuclear freeze. The movement began in the summer of 1979, when the American Friends Service Committee arranged a meeting at the Princeton Club in New York attended by, among others, Richard Barnet of the Institute of Policy Studies and Randall Forsberg, the head of a small think tank in Massachusetts. With SALT II going nowhere in the Senate, the meeting concluded that a totally new approach to the strategic arms race was needed. Instead of arguing about numbers of launchers and delivery systems, as the superpowers had done in SALT, the freeze advocates believed that the numbers should be left where they were until an agreement halting the production of all new systems could be negotiated. Only then could reductions effectively take place.[33]

Anti-nuclear groups soon adopted the freeze idea. On Veteran's Day 1981, the Union of Concerned Scientists along with numerous co-sponsoring groups held teach-ins on 151 campuses in forty-one states, putting the new anti-nuclear movement on the map in the process. In May 1982 a nationwide Ground Zero education week on nuclear war prompted thousands in every state to sign a nuclear freeze proposal calling for the immediate halt in the testing, production, and deployment of nuclear weapons, to be followed by major reductions in weapons stockpiles. In June 1982 between one-half and three-quarters of a million people jammed Central Park in New York supporting an end to the arms race. It was called the largest protest meeting in American history. An increasing number of Americans had apparently reached the conclusion that Reagan was preparing to fight a nuclear war, one that would involve strategic intercontinental exchanges, as well as nuclear strikes in Europe.[34]

Institutional religions also gave their support to the freeze movement. The Catholic bishops of the United States began writing a pastoral letter on nuclear weapons which when completed in 1983 condemned first-strike weapons, like the MX, and called for a halt to the testing and production of nuclear weapons. A few Catholic clergymen took stronger action. Archbishop Raymond Hunthausen of Seattle urged Catholics in his diocese to withhold part of their income taxes as a way of protesting nuclear weapons expenditures. Catholics were joined by Protestants and Jews in condemning the nuclear arms race and supporting the nuclear freeze. Even Billy Graham, who once viewed U.S. military policy uncritically, called for the destruction of nuclear weapons. Alexander Schindler, a leader in the Union of American Hebrew Congregations, termed the nuclear issue the "central moral issue of the 1980s and 1990s."[35]

The freeze idea soon proved to be extremely popular with the American people. A *New York Times*-CBS poll taken in May 1972 showed seventy-two percent of those polled supported a mutual freeze, while only twenty-one percent opposed it. While the poll indicated that support for the freeze cut across racial, religious, regional, income, and party affiliation, it was strongest among the younger, more affluent, and better educated. In the 1982 congressional election, freeze initiatives were passed by voters in eight out of nine states and in municipal and county elections throughout the country. In fact, the election sent a pro-freeze majority to the House of Representatives.[36]

Recognizing a popular issue when they saw it, politicians soon jumped on the freeze bandwagon. In March 1982 Senator Edward Kennedy joined Republican Senator Mark Hatfield in introducing a freeze resolution in the Senate. The Kennedy-Hatfield Amendment drew the backing of 122 representatives and 17 senators, as well as the endorsement of several state legislatures. The resolution urged the United States and the Soviet Union to pursue a complete halt to the nuclear arms race, to negotiate a mutual and verifiable freeze on testing, production, and further deployment of nuclear arms, and then to pursue mutual and verifiable reductions in strategic stockpiles.[37]

While the freeze movement was building momentum, the Congress, after giving overwhelming approval to the administration's military buildup in 1981, began to question the wisdom of the President's defense programs. On March

29, 1982 the Senate Armed Services Committee voted to block administration plans for the interim deployment of the MX missile in existing, fixed silos until a more suitable basing scheme could be devised. The committee decided that this approach would not make the new missile invulnerable to a Soviet attack. The committee's action was an indication of a new willingness on the part of Congress to take a hard look at the administration's military programs rather than rubber-stamp them as they had in the first year of Reagan's presidency.[38]

With its nuclear buildup threatened, the Reagan administration counterattacked the freeze proposal repeatedly and vigorously. "The truth is," the President told a convention of the National Association of Evangelicals in March 1983, "a freeze now would be a very dangerous fraud, for that is merely the illusion of peace. The reality is that we must find peace through strength." A freeze, he added, "would prevent the essential and long-overdue modernization of U.S. and allied defenses and would leave our aging forces increasingly vulnerable." Moreover, he warned, it "would remove any incentive for the Soviets to negotiate seriously. . . . Instead, they would achieve their objectives through the freeze." Reagan also cautioned that "an honest freeze would require extensive prior negotiation on the systems and numbers to be limited and on the measures to ensure effective verification and compliance," requirements, he believed, were not met by the Kennedy-Hatfield freeze.[39]

Not satisfied with rebutting the arguments of the freeze advocates, Reagan also attacked the advocates themselves. On October 4, 1982 he called freeze proponents a group of "honest and sincere people" who were being manipulated by "some who want the weakening of America." The FBI, however, could find nothing to substantiate the President's charge. Speaking for the American Civil Liberties Union, Morton Halperin reacted to Reagan's charge by saying, "It is disheartening to see an American President returning to the tactics of McCarthyism. Charges of secret manipulation by foreign agents poison the well of public debate."[40]

Faced with the fact that his attacks on the freeze were doing little to diminish its popularity, the President tried to kill the issue by co-opting it in emasculated form. He announced his support for an alternative freeze proposal contained in an amendment drafted by Senators Henry Jackson and John Warner (Republican, Virginia). The Jackson-Warner Amendment called for a freeze at the point where the nuclear forces of both sides were equal. In effect, the amendment would permit the Reagan arms buildup to continue until a time when the Soviet "advantage" would be overcome.[41]

The President's effort to emasculate the Kennedy-Hatfield freeze resolution was successful. On August 5, 1982 the House rejected a resolution for an immediate freeze and voted 204 to 202 to support a resolution sponsored by Representative William S. Broomfield (Republican, Michigan), which called for a freeze at "equal and substantially reduced levels," a position similar to the Jackson-Warner Amendment. The following year a resolution calling on the United States and the Soviet Union to pursue negotiations for a mutual and verifiable freeze as well as talks on intermediate-range weapons and strategic arms was passed in the full House by a hefty margin of 278 to 149. But it did so only after an amendment was adopted that would require suspension of a freeze agreement if it were not followed by mutual arms reductions within a specified period. Freeze resolutions were handily defeated by the Republican-controlled Senate during 1983.[42]

Reagan's ability to defuse the nuclear freeze issue was largely a result of the public's confusion on the issue. While many polls showed Americans favoring a nuclear freeze, other polls showed them favoring a tougher stance toward the Soviets "even if that means risking war." Concluded historian Robert Dallek, "Americans wanted both a strong defense capacity and a smaller likelihood of nuclear war, however contradictory these two goals seem."[43]

13

Reagan and Nuclear Arms Talks, 1981 to the Present

The INF Talks Begin

In spite of the deep-seated distaste for arms control agreements on the part of Reagan and his advisers, pressure from Congress, the allies, and the burgeoning anti-nuclear movement compelled the administration to engage in nuclear arms negotiations. If it did not at least create the impression of a serious interest in reaching nuclear arms agreements, the administration feared that its military programs would be jeopardized in the Congress and in the parliaments of the NATO nations. As a result, it reluctantly began preparations for strategic arms talks as well as for the second "track" of the NATO deployment program—negotiations on intermediate-range nuclear forces (INF) in Europe. Because pressure was at first greater from the Europeans than from the Congress, it was to the INF talks that the administration first turned its attention.

The Zero Option

During the summer of 1981, Assistant Secretary of Defense for International Security Affairs Richard Perle, whom Defense Secretary Weinberger authorized to conduct the Pentagon's battle in the INF deliberations, recommended that the administration should adopt the so-called zero option plan as its INF proposal. It called for the United States to cancel plans to deploy its Pershing IIs and Tomahawks if the Soviets agreed to dismantle all of their deployed intermediate-range SS-4s, SS-5s, and SS-20s, including those that were deployed in the Far East. Perle considered the zero option the only way to totally eliminate the Soviet IRBM threat to Western Europe as well as to steal the propaganda thunder from the Soviets.[1]

Perle's zero option proposal was vigorously opposed by the State Department. Secretary of State Alexander Haig considered it nonnegotiable. "It was absurd," he recalled in his memoirs, "to expect the Soviets to dismantle an existing force of 1,100 warheads . . . in exchange for a promise from the United States not to deploy a missile force we had not yet begun to build and that had aroused such violent controversy in Western Europe." Furthermore, Haig believed that some NATO intermediate-range missile deployments were necessary to avoid separating the nuclear defense of Western Europe from that of the United States. For these reasons, he supported a "zero-plus" plan, whereby both the planned NATO and Soviet IRBM deployments would be reduced to a lower, but equal, level. Reagan, however, liked the zero option, primarily because of its simplicity. He believed the Soviets should be told: "You do away with your [SS-20s] and there won't be any Pershing IIs."[2]

Accordingly, it was the zero option that Paul Nitze, the head of the American INF delegation, offered his Soviet counterpart, Yuli Kvitsinsky, when the talks opened in Geneva on November 30, 1981. Nitze was not surprised when the Soviets abruptly and adamantly rejected it. The American proposal not only would have required the Soviets to dismantle all of their SS-4s, SS-5s, and SS-20s, it would have left unaffected NATO theater-nuclear weapons deployed in Western Europe. These included 108 Pershing I medium-range missiles in West Germany, American dual-capable fighters in Europe and on aircraft carriers around the Soviet littoral, and the nuclear weapons of the British and French. Moreover, the Soviets argued that the addition of the Pershing IIs and Tomahawks would upset an existing balance that their SS-20s were simply maintaining.[3]

The Soviets also asserted that the Pershing IIs were much more of a menace to the Soviet Union than the SS-20s were to the United States. While the SS-20s were not targeted on Washington, the new American missiles could be aimed at Moscow. Moreover, the Soviets realized that the threat posed by NATO nuclear forces in Europe would become more potent with time if they were not restricted. The modernization programs of the British and French would give them a combined nuclear arsenal of 1,200 or more warheads, more than enough to destroy every major target in the Soviet Union. The Soviets apparently would not make large reductions in their INF without gaining strict limits on those of the United States and its NATO allies already stationed in Europe.[4]

Not to be outdone by the Americans, the Soviets countered with their own version of the zero option. It called for an immediate freeze in long-range INF deployments, followed by a two-thirds reduction in their numbers by 1990. The Soviet proposal was quickly turned down by the United States and its NATO allies, not only because it included British and French systems in the American totals, but also because, the administration believed, it would have maintained Soviet INF superiority. According to administration estimates, the Soviets had a total of 3,825 INF, not the 1,000 figure they claimed to have, while the United States had only 560. Furthermore, the Soviet proposal would not only have prevented the deployment of the Pershing IIs and Tomahawks, it also would have reduced greatly the number of American nuclear weapons committed to the defense of Europe, thereby accentuating the presumed Soviet superiority in conventional forces. At the same time, the Soviet proposal also would not have required the elimination of a single SS-20 in Europe nor prevented their deployment in Asia.[5]

The "Walk in the Woods"

In July 1982 during a walk with Kvitsinsky in the Jura Mountains outside Geneva, Paul Nitze unilaterally attempted to break the ice that had frozen the negotiations. The so-called Walk-in-the-Woods formula that resulted would have permitted a partial deployment of the new American missiles in Western Europe in exchange for significant reductions in the numbers of SS-20s in Eastern Europe and a freeze of SS-20 deployments in Asia. Under this formula, the United States would have been permitted to deploy seventy-five cruise missile launchers, each with four missiles, but no Pershing IIs, which because of their shorter flight times were considered the greater threat by the Soviets. For their part, the Soviets would have been required to reduce their 243 SS-20 launchers in Europe to 75. Because the SS-20 carried three warheads while the Tomahawks carried one warhead apiece, the formula would have given the United States seventy-five more warheads than the Soviet Union. The American warhead advantage was considered by Nitze to be just compensation for the faster flight time of the SS-20. On the other hand, the consequent Soviet monopoly in IRBMs would have been compensation for the ballistic missiles of the British and French. The freeze on Soviet Asian SS-20 deployments would ensure that SS-20s removed from Europe would not be targeted against China, Japan, and South Korea. In addition, the formula would have limited both sides to a ceiling of 150 long-range INF aircraft in Europe.[6]

According to *Time* diplomatic correspondent Strobe Talbott, the Walk-in-the-Woods formula was acceptable to key figures in the Reagan administration, including Secretary of State Shultz, who succeeded Haig in June 1982, and General John Vessey, the new chairman of the Joint Chiefs of Staff. But it was opposed by Weinberger and Perle who believed that, if West Germany were not obliged to deploy the Pershings, the other allies would abandon their commitment to accept cruise missiles. The President supported Weinberger and Perle. He could not see why the United States would have to surrender all of its ballistic missiles while allowing the Soviets to retain some of theirs. With the Pershing IIs back in the negotiations, the Soviets lost interest in the Walk-in-the-Woods formula and rejected it on September 29. By the time the administration was willing to offer any concessions to the Soviets in the talks, nearly another year would pass, Brezhnev would be dead, and the opportunity for an agreement before the initial deployment of the American missiles began would be gone.[7]

Stalemate

On December 21, 1982, shortly after Yuri Andropov succeeded Brezhnev, he presented a new Soviet plan designed to prevent the deployment of the new

American missiles. He offered to reduce the number of SS-20s that were aimed at Western Europe from 243 to 162, a figure which matched the combined total of British and French ballistic missiles. He also proposed to redeploy the remaining SS-20s at sites about 700 miles further east, on the far side of the Ural Mountains. The offer was rejected immediately by Washington, London, and Paris, not only because it would have limited British and French forces and precluded the deployment of the American missiles, but also because the allies feared that mobile SS-20s stationed beyond the Urals could very easily be moved back to positions in the western part of the Soviet Union. Nevertheless, despite the unacceptability of the Soviet proposal, Andropov did succeed in creating the impression that the Soviet Union was being more flexible than the United States. That impression increased European pressure on the Reagan administration to move away from its zero option.[8]

On March 30, 1983 the President finally gave in and modified his zero option. In its place, the administration offered an "interim agreement" that would have allowed equal numbers of American and Soviet warheads "on a global basis," at the "lowest possible levels." Although the levels were unspecified in the offer to the Soviets, the State Department had in mind an agreement whereby the United States would have been allowed to deploy 300 single-warhead Pershing IIs and Tomahawks in Europe while the Soviet Union would have been permitted to deploy in Europe and Asia a total of 100 triple-warhead SS-20s. As a result, each side would have had an equal number of deployed intermediate-range warheads. On April 2 the Soviets rejected the new American offer. They claimed that it violated a basic premise of their negotiating posture—that French and British forces must be included in the limitations. Moreover, they rejected the concept of global ceilings because it would have included Soviet missiles deployed in eastern Asia. "Why," Gromyko asked, "should we drag Asia into this?"[9]

With the first American missiles scheduled for deployment in December, the tempo of the negotiations quickened during the summer and fall of 1983. On August 26 Andropov met a long-standing American objection by proposing to "liquidate" all the SS-20s that he had previously offered to remove from Europe, thereby precluding the possibility that they could be returned during a combat situation. The offer was not acceptable to the administration, which by this point considered deployment of at least some Pershing IIs a matter of principle.[10]

However, on September 26 the President did announce a further modification in the American position. Although continuing to insist on a global equality in the number of American and Soviet intermediate-range warheads, Reagan stated that the United States would drop its demand for parity in European warhead deployments. In effect, the administration offered to keep some of its intermediate-range warheads in the United States. The President also stated that the scope of the talks might be broadened to include ceilings on bombers based in Europe. Two days later Andropov denounced not only Reagan's latest offer but the very "essence" of the administration's approach to arms control.[11]

In spite of the apparent finality of the statement, Andropov made another offer on October 26. He proposed to reduce Soviet SS-20 deployments in Europe to 140 and to halt further SS-20 deployments in the Far East in exchange for the cancellation of the Pershing II and Tomahawk deployments and a freeze on the level of British and French nuclear forces. But he also warned that, if the West went ahead with its deployments, the Soviet Union would walk out of the Geneva talks. The American side responded by rejecting the latest Soviet proposal, and then by modifying its own. The administration on November 14 proposed an equal global ceiling of 420 warheads—equal to the number of warheads on the 140 SS-20s that Andropov proposed as the Soviet European force level in his latest offer. But that offer had also been dependent on the willingness of the West to cancel all the NATO deployments, and it would have allowed the Soviets the right to deploy additional SS-20s in Asia. Since both conditions were excluded from the latest American proposal, it was rejected by the Soviets.[12]

When the British announced the arrival in Britain of the first new American cruise missiles on November 14, the Soviet delegation stayed at the negotiating table. Kvitsinsky then made a late-hour offer to Nitze. He stated that the Soviet Union would accept an American proposal to cancel its impending missile deployments in exchange for the reduction of the Soviet SS-20 force to 120 missiles in Europe, rather than the 140 in Andropov's last offer. Moreover, he stated that the Soviet Union was also willing to defer to the strategic arms talks the issue of compensation for future increases in British and French forces. But by this time, the administration wanted to avoid doing anything that would delay the imminent deployment of the first cruise missiles. Consequently, it instructed Nitze to reject the Soviet offer.[13]

Soon thereafter, on November 22, the West German Bundestag voted to reaffirm support for the NATO missile deployments in West Germany. The next day Kvitsinsky terminated the talks without agreeing to a date for their resumption. The administration expressed sorrow that the Soviets had abandoned the talks, but predicted that they would be back. Instead, the Soviets said they would not return to the talks unless the United States withdrew the missiles it had begun to deploy. They also announced a number of military countermeasures of their own. They included additional SS-20 deployments, the accelerated deployment of Soviet tactical weapons in Eastern Europe, and the stationing of more Soviet missile-launching submarines off the American coasts. "By trying to lessen our security," Kvitsinsky concluded, "the United States has lessened its own security."[14]

A Slow START

It was not until well over a year after taking office that the administration was able to offer the Soviets a START proposal—much longer than either Nixon, Ford, or Carter waited after entering the White House before submitting a SALT offer. As in the INF talks, the primary reason for the delay was the fact that the administration was more interested in deploying additional nuclear weapons than it was in dismantling them. According to ACDA Director Eugene Rostow, developing a "clear, credible and unchallengeable second-strike nuclear capability" could not be sacrificed to any negotiating effort. Apparently, it was only with considerable difficulty that Haig was able to convince administration hardliners that "the American people would never agree to a posture that supported only a major arms building but ruled out negotiation that might produce greater security at less risk."[15]

Convincing the hardliners that they had no choice but to begin START was one thing; getting them to support a negotiable proposal was quite another. In the interdepartmental meetings that were held to formulate the administration's START policy, the Pentagon argued that the main objective of START must be a major change in the imbalance in the throw-weight capabilities of American and Soviet missiles. On the other hand, the State Department, according to Haig, "tried to adhere as closely as possible to SALT II, while identifying and correcting its flaws." This amounted to a program of gradual reductions, with only indirect restraints on the throw-weight problem. This approach, Haig argued, "was negotiable while Defense's position probably was not."[16]

According to journalist Strobe Talbott, another reason for the delay in opening START was Reagan's inability to give meaningful direction to the American negotiating delegation. During deliberations on START as well as on INF, Talbott wrote, the President was "a detached, sometimes befuddled character. . . . There was ample evidence that he frequently did not understand basic aspects of the nuclear weapon issue and of policies promulgated in his name." As a result, he "found it extremely difficult to assert himself as either a moderator or a decision maker," according to Talbott. Thus many of his decisions "were compromises jerry-rigged from competing options favored by the different agencies."[17]

Haig's memoirs seem to support Talbott's assessment. In the administration's first START proposal, Haig recalled, "the President decided . . . to give each supplicant half the baby." The proposal contained the State Department's call for gradual reductions as well as the Pentagon's demand for major reductions in the throw-weight of Soviet ICBMs. The result, Haig believed, was a "flawed START position."[18]

The Eureka Proposal

The "flawed" START proposal Haig referred to in his memoirs was presented by Reagan in a speech he delivered at Eureka College on May 9, 1982. In the speech, the President announced that the first round of START would begin on June 29 in Geneva. He also stated that the major arms control objective of his administration would be to "reduce significantly the most destabilizing systems: ballistic missiles, the number of warheads they carry and their overall destructive potential."[19]

The reductions Reagan proposed in his Eureka speech would take place in two steps. In a first phase, designed to mollify Haig, all strategic ballistic missiles (ICBMs and SLBMs) for both sides would be reduced to an equal level of 850 missiles over a five- to ten-year period. The size of the proposed reductions was roughly one-half the then current American level of about 1,700 delivery vehicles, but almost two-thirds of the total Soviet force of roughly 2,350 launchers. Of the 850 launchers, no more than 210 could be "medium" missiles, like the MX and the Soviet SS-19, and no more than 110 could be heavy missiles, such as the Soviet SS-18. In addition, the total number of warheads on strategic missiles would be cut to about 5,000. Of that number, no more than 2,500 could be on ICBMs. Yet limits on bombers and cruise missiles were excluded from this first stage of START; they were the two areas in which the United States enjoyed superiority.[20]

In a second phase of the talks, which was included to mollify Weinberger, the administration tried to equalize missile throw-weight at a ceiling of about 4 million pounds. Since total U.S. throw-weight was then about 4.2 million pounds and the Soviet Union's was about 11.2 million pounds, the Soviets clearly would have to make a much larger reduction. In one of the few concessions that the administration was prepared to offer, it stated that it would be prepared to discuss bomber reductions and cruise missile limits in the second stage of the talks, which would be conducted at some time in the indefinite future.[21]

The Soviets, not surprisingly, rejected Reagan's Eureka proposal. It would have required Moscow to scrap about 1,500 strategic missiles compared to about 850 for the United States. The proposed reduction in land-based ICBM warheads to no more than 2,500 also would have hurt the Soviets more than the Americans. The United States was at that time already 350 warheads below the proposed ceiling. The Soviet Union, on the other hand, was some 3,400 warheads above it, the consequence of deploying some seventy percent of its nuclear deterrent on ICBMs, compared to about twenty-five percent for the United States. To get down to the ceiling of 2,500 ICBM warheads, the Soviets would have had to dismantle all 308 of their ten-warhead SS-18 heavy missiles and roughly 100 of their six-warhead SS-19s or their four-warhead SS-17s. This would have left them with about 400 MIRVed ICBMs, compared to their then current level of 818. The United States could have met the proposed, combined ICBM-SLBM ceiling of 5,000 warheads by dismantling 500 older single-warhead ICBMs and roughly 130 MIRVed Poseidon SLBM launchers. By not including bombers in the first phase of the reductions, the proposal was even more advantageous to the American side. American bombers carried about 3,000 nuclear weapons compared to less than 300 carried by those of the Soviet Union. Excluding cruise missiles in the proposal protected another area of American technological superiority. Finally, the arms control constraints called for in the President's proposal would not have had much effect on the administration's modernization program. The United States would have remained free to deploy the B-1, the MX, Trident II, as well as cruise and Pershing II missiles.[22]

After rejecting Reagan's plan, Brezhnev countered with a proposal for a nuclear freeze that would have taken effect as soon as the talks had begun. He said the freeze should apply both to the numbers of weapons and to the modernization of existing weapons which, of course, would have included the new weapons the administration was planning to deploy. Later, in mid-1982, the Soviets tabled a proposal that would have preserved the structure of SALT II, but also would have somewhat reduced launcher ceilings and subceilings. It was probably very similar to what the Soviets would have offered in SALT III, had the SALT II Treaty been ratified. The Soviet offer called for reductions to a level of 1,800 strategic nuclear delivery vehicles. It also called for unspecified limits on the total number of nuclear weapons, including cruise missiles and other bomber armament, and modest reductions in the SALT II MIRV sublimits.[23]

The Soviet proposal was unacceptable to the Reagan administration for several reasons. First, the freeze proposal would have left the Soviet ICBM advantages intact, while preventing the introduction of any new American nuclear weapon systems. It was also unacceptable because, like SALT, it used launchers as the principal bargaining units rather than missile warheads. In addition, like SALT, the Soviet proposal would have permitted trade-offs between bombers and ballistic missiles, instead of concentrating exclusively on missiles. Furthermore, it contained no provision for the limitation of ballistic missile throw-weight, thereby maintaining that Soviet advantage. On the other hand, the Soviet proposal would have cancelled an American advantage by prohibiting all long-range cruise missile deployments. However, the development, testing, and production of cruise missiles would not have been prohibited. This obviously was designed to allow the Soviets to catch up to the Americans in cruise missile technology. Finally, the entire Soviet proposal was contingent on the cancellation of American plans to deploy the Pershing IIs and Tomahawks in Europe, which the Soviets considered "strategic" because they could reach targets in the Soviet homeland. In effect, the Soviet START position was explicitly linked to their negotiating position in the INF talks, thereby ensuring that the deployment of the American intermediate-range missiles would jeopardize both negotiations.[24]

Other Nuclear Arms Control Actions

At the end of May 1982 the President announced an arms control decision that was somewhat surprising considering his past opposition to the SALT II Treaty and the other unratified agreements—the Nuclear Threshold Test Ban Treaty and the Peaceful Nuclear Explosion Treaty. Although he had called the SALT treaty "fatally flawed" and the other agreements unverifiable, more knowledgeable voices, including those of Haig and the Joint Chiefs of Staff, persuaded him that the United States would be worse off without the agreements since they placed major restrictions on Soviet nuclear activities as well as those of the United States. If, for example, the SALT II Treaty were cast aside, the Soviets would be able to add many more warheads to their existing missile forces than the United States would be capable of doing. As a result, the President announced on May 31 that the United States would abide by the agreements as long as the Soviets displayed similar restraint.[25]

But while Reagan gave the arms controllers a major concession, abiding by the unratified agreements, he also killed a major arms control effort. On July 20, 1982 it was announced that the President had decided to set aside the long effort to negotiate a comprehensive ban on nuclear testing. The ostensible reason for the decision was the inadequacy of existing verification measures. However, critics charged that the real reason behind the decision was the fact that the CTB negotiations would have prevented the testing of warheads the administration planned to deploy on its new nuclear weapon systems.[26]

The administration also upset arms controllers with actions the latter believed could contribute to the further proliferation of nuclear weapons. In spite of the President's announced intention to urge all supplier nations to apply comprehensive safeguards to all nuclear sales, the administration approved nuclear sales to India, Argentina, China, and South Africa. Yet these nations had refused to open all their nuclear facilities to international safeguard inspection, refused to ratify the Nuclear Nonproliferation Treaty, and either had exploded a nuclear device or were strongly suspected of developing the means to do so. In addition, the Chinese were believed to be helping the Pakistanis develop weapon-grade uranium. Reagan seemed to be doing all he could to implement his pre-presidential stance that the United States should not interfere with countries attempting to acquire the means to produce nuclear weapons. In January 1980 he said: "I just don't think it's any of our business."[27]

Saving the MX: The Scowcroft Commission

By the beginning of 1983, it was apparent that START was still going nowhere. Congress was growing impatient with the delay, and some members were becoming suspicious that the administration was deliberately scuttling the talks. Faced with the START stalemate, criticism of the administration's nuclear buildup, particularly the MX program, intensified. More and more congressmen were beginning to consider MX a potential first-strike weapon. Many also were beginning to question the administration's argument that MX was needed as a bargaining chip in START; the threat to deploy the missile certainly did not make the Soviets more amenable to compromise at the negotiating table. Quite to the contrary, the MX appeared to be a major obstacle toward progress in the talks.[28]

During the winter of 1982–1983, the Congress demonstrated its frustration with the administration by rejecting Reagan's "dense-pack" deployment plan for the MX. The plan called for building launch sites for 100 MX ICBMs and placing them 1,800 feet apart in a column fourteen miles long at Warren Air Force Base near Cheyenne, Wyoming. According to the administration, the close positioning of the MX launch sites would render some of the missiles invulnerable to Soviet attack. The explosive effects of the first Soviet warheads would destroy or disable subsequent Soviet warheads, thus ensuring that some of the MX missiles would survive. Many congressmen not only believed dense-pack was unworkable, they also argued that the plan was a blatant example of how much more determined the administration was to deploy additional nuclear weapons than it was to negotiate their reduction.[29]

To rebuild support for the MX, Reagan established a presidential commission on strategic forces in January 1983. It was chaired by retired Air Force Lieutenant General Brent Scowcroft, a former aide to Henry Kissinger and a national security adviser for President Ford. Not surprisingly, no opponents of the MX were selected to be members of the commission. Nevertheless, the commission's report, which was completed on April 6, 1983, only partially met the administration's objectives. While it called for the deployment of 100 MX missiles in existing Minuteman silos, it closed the alleged window of vulnerability by stating that, as a whole, America's retaliatory forces would survive a Soviet first-strike and retain the capability to devastate the Soviet Union. The report also stated that in the long term, by the 1990s, the United States would be better off with a force of small, mobile, single-warhead ICBMs, dubbed "Midgetman," than the MX, primarily because the Midgetman would neither threaten nor invite a first-strike, while the MX very well could. The Scowcroft Report also stated that the administration could and should work harder to achieve a strategic arms agreement, particularly one that could decrease the ratio of warheads to launchers on both sides, and thereby diminish the incentive and possibility for a Soviet first-strike. On April 19 the President, despite his misgivings with some parts of the report, gave it his endorsement.[30]

But the Scowcroft Report was roundly criticized by those who saw it primarily as a political rather than a military document. "The reason for building MX," said Jeremy Stone, chairman of the American Federation of Scientists, "was to counter the alleged vulnerability of the land-based Minuteman force by making a mobile missile. For the commission to recommend that MX be put in the same silos whose vulnerability provided its rationale is to stand reason on its head." Stone also pointed out that no amount of reinforcement would prevent the MX silos from being prime targets of a Soviet attack. The only way to avoid losing the MXs in the event of a war, Herbert Scoville of the Arms Control Association pointed out, would be to use them first. Realizing their potential as first-strike weapons, Scoville argued, the Soviets would have every incentive to strike at them first before they could be fired at the Soviet Union. The MX, in short, its critics argued, would weaken rather than strengthen the deterrent balance.[31]

The Congress, however, paid little heed to the arguments against the MX. On May 24, ironically only three weeks after the House adopted a nonbinding nuclear freeze resolution by a 239 to 186 vote, that body permitted funds to be used for MX flight testing and for housing the missile in reinforced Minuteman silos. The Senate passed an identical measure the next day. Two months later, on July 21, the House approved production funds for the first twenty-one MX missiles by a 220 to 207 vote. Similar appropriations were passed by the Senate. That summer both houses also approved funding for other administration-sponsored nuclear weapon programs: the B-1 bomber, anti-satellite weapons, and Pershing II missiles. Yet during the MX debate, several members from each chamber warned the White House that their continued support for the MX would be dependent on the administration's good faith in pursuing arms control efforts.[32]

Back to START

The administration attempted to maintain congressional support for its defense program by announcing modifications in its negotiating position at the beginning of the fourth round of the START talks. In a June 8 statement, Reagan said that the United States would retain the proposed ceiling of 5,000 on the number of missile warheads, but would raise its proposed limit of 850 deployed ballistic missiles to 1,200, the number favored by the Soviets, so that both sides would have the option and incentive to restructure their forces over the long run in the direction of smaller and less vulnerable single-warhead ICBMs. The administration also removed its previous requirement that the Soviet Union reduce its throw-weight advantage to the American level and, instead, required that the Soviet advantage be only "substantially" reduced. This concession was made after an interagency study indicated that improvements in missile accuracy by both sides made throw-weight a less critical factor. In response to the Soviet charge that the previous administration proposals were not at all comprehensive in nature, a revised administration draft treaty called for equal limits below the SALT II levels on the number of heavy bombers and the number of air-launched cruise missiles allowed on each bomber. Finally, the administration combined the original two-phased proposal into one, placing all categories of strategic weapons on the table to be negotiated in a single agreement.[33]

The Soviets responded almost immediately with a rejection of the President's latest proposal. *Tass* stated that the new American plan was designed to get Congress to approve deployment of the MX while forcing the Soviet Union to emasculate its own land-based missile force.[34]

In July, however, the Soviets modified, if only slightly, their own negotiating proposal. The new Soviet offer called for a phased reduction by 1990 of strategic delivery vehicles to a ceiling of 1,800 for each side. Within that figure were a number of sublimits. One would limit each side to a combined total of 1,200 MIRVed ICBMs, MIRVed SLBMs, and strategic bombers armed with cruise missiles. That number included a subceiling of 1,080 for MIRVed ICBMs and SLBMs and another subceiling of 680 MIRVed ICBMs. While the Soviet proposal would have sharply reduced the number of their own MIRVed ICBMs, it would still have left the Soviet Union with more than 7,000 ICBM warheads, compared to the 2,500 proposed by the United States, and over 11,000 ballistic missile warheads on land- and sea-based missiles, compared to the 5,000 ceiling the administration wanted. The administration obviously felt the Soviet plan did not go far enough. But in one respect, it went too far. It was again contingent on the cancellation of the American Pershing II and Tomahawk deployments in Western Europe.[35]

The Build-Down

With START going nowhere, a proposal emerged during the spring and summer of 1983 as an alternative not only to START but also to the nuclear freeze proposal—the so-called guaranteed, mutual build-down plan. As originally envisioned by its principal authors, Senators William Cohen and Sam Nunn, the build-down would have required both sides to eliminate two nuclear warheads for each new warhead they added to their nuclear arsenals. It also would have required both superpowers to reduce the number of nuclear weapons in their arsenals by an annual percentage even if they chose not to replace their nuclear forces.[36]

The build-down, its supporters argued, offered a number of advantages over past efforts to reduce the number of strategic nuclear weapons. Rather than emphasizing, as SALT did, reductions in numbers of launchers, the net effect of which was to encourage both superpowers to add more warheads to each launcher, the build-down idea emphasized reductions in numbers of warheads. Rather than the rigid categories and subcategories of ICBMs, SLBMs, and bombers proposed in SALT II, the build-down provided each side with the freedom to mix strategic weapon systems as long as warhead ceilings were not exceeded. One result, its proponents argued, would be to encourage both sides to substitute less lethal, single-warhead missiles for the more threatening, multiple-warhead missiles in their arsenals. Moreover, if the build-down were extended to include the intermediate-range missiles both the United States and the Soviet Union were deploying in Europe, a way of ending the Euro-missile crisis might be found. Another advantage the build-down idea enjoyed over the SALT II Treaty was its extensive support in the Senate. In October 1983 that body voted 84 to 13 in support of the build-down idea.[37]

More important, President Reagan agreed to marry the build-down to his START proposal in the next round of START, which began on October 5. However, the Cohen-Nunn build-down was somewhat modified in the START/build-down plan offered by the administration. Instead of the straight two-for-one warhead reductions proposed by the senators, the administration plan would have required the retiring of two warheads for each new warhead deployed on MIRVed ICBMs, three warhead retirements for every two new warheads deployed on MIRVed SLBMs, and a one-for-one replacement for each new warhead deployed on any single warhead missile. The administration's variation of the build-down also incorporated a scheme by Glenn Kent, a retired Air Force general and nuclear weapons expert, that would have penalized more heavily weapons with greater throw-weight, such as the Soviet heavy ICBMs. Because the plan included both throw-weight and warhead reductions, it was also known as the "double build-down."[38]

In addition, under the terms of the administration build-down, both sides would have had to reduce their arsenals of ballistic warheads by a minimum of five percent per year for ten years until they both reached a plateau of 5,000 warheads. The guaranteed reductions would prevent either side from attempting to avoid the build-down by ceasing all apparent modernization. In an attempt to satisfy one Soviet objection, the double build-down also would have placed limitations on bomber-delivered warheads, including bombs, short-range attack missiles, and air-launched cruise missiles.[39]

Nuclear freeze advocates were quick to condemn the administration's acceptance of the build-down. They feared that the negotiations for a build-down treaty would take years to complete, during which time the number of deployed nuclear weapons could increase enormously. Indeed, they strongly suspected that that was the President's primary motive in adopting the build-down; it would not only delay START, it would enable the administration to maintain congressional support long enough to complete its massive buildup of the American nuclear arsenal.[40]

The Soviets were equally critical of the build-down idea. An October 23, 1983 editorial in *Pravda* called the administration's START/build-down plan "trickery." The paper accused the administration of attempting to reduce Soviet ICBM numbers, which would be most heavily penalized under the Reagan build-down plan, in exchange for the scrapping of obsolescent American B-52 bombers. In effect, the Soviets charged, the United States through the build-down plan was attempting to dictate the structure of Soviet deterrent forces— that is, by forcing the Soviets to dismantle their MIRVed ICBMs and to replace them with single-warhead ICBMs and SLBMs. At the same time, Soviet acceptance of the build-down would have required them to abandon the structure and units of account contained in the SALT II Treaty and used in their own START proposal. To the Soviets, already suspicious, calculating, and notoriously reluctant to accept new arms control suggestions, especially when they are offered by Americans, the build-down proved totally unacceptable.[41]

Possibly the finality of the Soviet rejection of the build-down idea was as much a result of poor timing as it was a product of what the Soviets considered a flawed concept. Actually, the timing could not have been worse. It was offered not long after the Soviet air force on September 1 shot down a Korean airliner which the Soviets claimed was on a spying mission for the United States, a charge the administration vigorously denied. In the wake of the incident, Soviet-American relations had deteriorated to the shouting stage, with Reagan calling the Soviets barbaric murderers. In November, Andropov died as a result of kidney failure. Moreover, as the deployment date of the first American Tomahawks in Western Europe approached, and with little progress being made toward breaking the stalemate in the INF talks, the Soviets were not inclined to accept a major breakthrough on START.[42]

At the end of the negotiating round on December 8, 1983, the Soviets stated that developments in the INF talks required them to reexamine their START position. Accordingly, the talks adjourned with no announced date for their resumption. The following April, partly in reaction to the break down of the arms control talks, and partly in retaliation for the withdrawal of the United States from the Moscow Olympics in 1980, the Soviets announced that they would not participate in the Los Angeles games during the summer of 1984. With Soviet-American relations at a new low, the prospect for meaningful nuclear arms control in the near future was as bleak as could be.[43]

"Star Wars"

The Soviet suspicion that the President was more interested in building weapons than in reducing their numbers was reinforced by Reagan's decision in late November 1983 to request a five-year, $26 billion appropriation for research and development of a ballistic missile defense system (BMD). The President's BMD program, which was first announced in a March 23 speech by Reagan, was officially called the Strategic Defense Initiative (SDI), but the media soon dubbed it "Star Wars," the title of a popular science fiction movie.[44]

The BMD program the President had in mind was much more ambitious than the ABM proposals of the 1960s. The Safeguard system proposed by the Nixon administration in 1969 was intended to provide only a "point" defense of "hard" targets, such as missile silos and command bunkers. Reagan's plan took a different tack. According to Weinberger, "the defensive system the President is talking about is not designed to be partial. What we want to try to get is a system which will develop a defense that is thoroughly reliable and total." Reagan's BMD would be designed to protect "soft" targets, such as cities, as well as hard targets.[45]

Reagan acknowledged that achievement of his goal would be no mean accomplishment. An effective, nationwide defense would have to intercept and eliminate virtually all the 10,000 or so nuclear warheads that the Soviets were capable of committing to a major strategic attack. But he believed the possible benefits were well worth the effort. In Reagan's estimation, an American BMD could not only defend against an all-out Soviet attack, but could protect the country against an accidental launch.[46]

Quite possibly, the administration had other objectives for its Star Wars program. Retired Lieutenant General Daniel O. Graham, who wrote a study on BMD for the Heritage Foundation, a conservative think tank, stated that an American BMD program would "confront the U.S.S.R. with precisely the sort of armaments competition that the Soviet leadership most fears"—that is, one where the United States enjoyed overwhelming technological superiority. Said Graham, a BMD would "severely tax, perhaps to the point of disruption, the already strained Soviet technological and industrial resources." At the very least, argued Edward Teller, an American BMD would force the Soviets to increase their missile expenditures beyond what they could reasonably afford, to assure that their missiles got through. If it did that alone, Teller stated, "we would have accomplished something."[47]

While the administration initially considered several approaches to a BMD, all of them envisioned the creation of a multilayered defense system in which ground- and spaced-based lasers, beams of subatomic particles, X-rays, or homing rockets would attack Soviet ballistic missiles and their warheads before they could strike American targets. Later, in January 1985, Weinberger stated that an effective anti-missile system would also have to be backed by an enhanced bomber defense system to protect the United States completely.[48]

Reagan's Star Wars program was criticized on several grounds. For one, while the critics believed that the individual components of various defensive systems under discussion were technically feasible, putting them together in a fully workable system, they argued, would be impossible. The defense system would require, among other things, a surveillance mechanism to detect the launching of enemy salvos, a method to determine whether they were unfriendly, and a highly precise aiming system to destroy the targets. Most of the systems—including satellites, mirrors, and ground-based lasers—were also highly vulnerable to Soviet attack, which would accompany, if not precede, any first-strike missile attack on the United States. And even in the highly unlikely event that the United States produced a BMD that was ninety-five percent effective, the critics asserted, the remaining five percent of Soviet warheads that got through would kill at least one-half of the American people and destroy the United States as a viable society.[49]

Star Wars' critics believed the Soviets could easily take a number of countermeasures rendering the United States even more vulnerable than it was without the systems deployed to defend it. They could simply double or triple the number of their deployed ballistic missiles and overwhelm the American BMD. Or they could equip more of their submarines with cruise missiles and deploy them closer to the United States, where they could be launched and fly under the American defense screen. Equipping ballistic missiles with fast-burn boosters would also enable Soviet warheads to be released while they were still within the protective blanket of the earth's atmosphere, where beam weapons, such as X-ray lasers, could not penetrate. In addition, the Soviets could coat their warheads with highly reflective paint to diminish the usefulness of laser beams. In other words, for every ruble the Soviets might spend on such relatively inexpensive diversions, the United States would have to spend millions of dollars on devices that could differentiate between real warheads and duds.[50]

What bothered Star Wars' critics most, however, was not its high cost or, what they believed to be, its dubious effectiveness, vulnerability, and impact on alliance relations, but its great potential to destabilize the nuclear balance. Because of the technical problems and other uncertainties involved, a BMD would be much more effective against a small and disorganized retaliatory attack than it would be in defeating a massive and well-organized first-strike. In fact, critics feared, a full-scale BMD would make a first-strike appear a necessity in the event of a highly charged Soviet-American confrontation. In such a crisis, the Soviets would have to consider the distinct possibility that the Americans would strike first and destroy the overwhelming majority of Soviet strategic forces, mop up most of those that survived and were able to be launched using the American BMD, and then protect the United States from those Soviet warheads that got through the American defensive screen with the nationwide civil defense system, which the administration was proposing to build.[51]

Star Wars' critics also feared that this defense tactic could trigger a dangerous escalation of the nuclear arms race in a new theater—space—where it had been prohibited by existing Soviet-American treaties, including the Limited Test Ban Treaty of 1963, the Outer Space Treaty of 1967, and the ABM Treaty of 1972. Among other prohibitions, the ABM Treaty included a provision requiring the parties "not to develop, test or deploy ABM systems or components which are sea-based, air-based, space-based, or mobile land-based." The deployment of a BMD would not only abrogate the ABM Treaty, it would also result in the removal of all constraints on offensive missiles. Star Wars' critics found it impossible to believe that the Soviets would reduce their ballistic missile force while the United States was building a ballistic missile defense system. In other words, the critics charged, a nationwide BMD program would be incompatible with arms control.[52]

In what seemed to be an attempt to prepare the nation for an abrogation of the ABM Treaty, the administration accused the Soviets of repeatedly violating not only that treaty but also the various other arms control agreements concluded between the two nations. A presidential report to Congress in January 1984 listed, among other alleged Soviet violations, construction of a so-called phased array radar near Krasnoyarsk in the central Soviet Union. This Soviet action, the administration charged, constituted a violation of the ABM Treaty, which bars construction of radar stations and other equipment, except at national borders, that could enhance a country's defense against incoming missiles or serve as an early warning system. The Soviets denied the American accusations and responded with their own list of alleged American violations.[53]

ASAT

In addition to Star Wars, the administration accelerated the development of an anti-satellite (ASAT) program initiated in the mid-1970s. One system, still currently under development, consisted of an interceptor missile designed to destroy a Soviet satellite by colliding with it. The interceptor could be launched into space by an eight-foot, two-stage rocket carried to launch altitude by an F-15 fighter. The administration argued that the American ASAT was necessary to counter one already developed by the Soviet Union.[54]

Administration critics, however, considered the new ASAT program both unnecessary and provocative. The operational capability of the Soviet ASAT, they argued, was primitive, and at best, of questionable effectiveness since it consisted of an interceptor, launched by a missile, that employed a nonnuclear warhead filled with shrapnel to destroy its target. The Soviet ASAT, the critics asserted, potentially threatened low-flying American reconnaissance and navigation satellites. But the most important American satellites, such as those used for early warning and communications, are flown in a geosynchronous orbit at 36,000 kilometers above the earth, far beyond the demonstrated range of the Soviet ASAT. And the threat to satellites in lower orbits would soon be removed as various satellite survivability programs were implemented in the near future, including Navstar navigation satellites designed to circle the earth at an altitude of 20,000 kilometers. All in all, the critics believed the best way to head off further development of Soviet ASAT capability was an agreement banning all anti-satellite weapon deployments.[55]

The Soviets, who had dragged their feet on an ASAT ban during the Carter administration, applied considerable pressure on the Reagan administration to call off the anti-satellite arms race. On August 22, 1983 the Soviet Union presented a draft treaty in the United Nations which would have placed a ban on the use of force in outer space. It called for the dismantling of all existing ASAT systems and proposed a prohibition on the use of manned spacecraft, like the American Shuttle, for military purposes or to launch military satellites. The United States was the only nation to vote against the treaty in the United Nations.[56]

There were a number of reasons why the administration demonstrated little interest in the Soviet draft treaty. For one, it considered an ASAT ban unverifiable. The Soviet Union, the administration argued, had the capability to place more killer satellites into orbit than the United States had the means to detect. More important, however, the administration wanted to do nothing that would jeopardize its Star Wars program, of which weapon satellites were a key component. American anti-satellite weapons would be used not only to blind a Soviet first-strike by knocking out Soviet reconnaissance satellites, they could also be used to defend American satellites and other BMD components from attack by Soviet killer satellites. Moreover, the administration insisted that Moscow would agree to abandon its ASAT only if it were confronted by a successfully tested American counterpart. For the critics of the administration, Reagan's reluctance to negotiate a ban on ASATs and his refusal to accept a moratorium on testing them while negotiations proceeded, again only dramatized his lack of real interest in any form of arms control.[57]

About-Face?

Several developments during the spring and summer of 1984 caused the administration to assume a more flexible strategy toward nuclear weapons negotiations. One was the approaching presidential election. The Democratic nominee, Walter Mondale, denounced the inability of the administration to conclude any major arms control agreements with the Soviets. Mondale promised that if elected he would be ready to begin nuclear arms control talks with the Soviets six months after his inauguration. During that interval, he said, he would initiate a unilateral freeze on American nuclear weapon deployments as a first step toward meaningful arms talks.[58]

Mondale's pressure on Reagan was augmented by the Congress and the Soviets. On June 12, 1984 the Senate refused to appropriate additional money for ASAT development unless the President certified that he was "endeavoring in good faith to negotiate the strictest possible limitations on anti-satellite weapons." Eight days later, that body passed a nonbinding amendment, by a vote of 77 to 22, calling for the President to attempt to gain ratification of the two nuclear test ban treaties and to resume negotiations of a comprehensive test ban treaty. In September both houses of Congress voted to postpone until the following spring a decision to produce additional MX missiles. Taking advantage of the pressure applied on the administration by the Senate, the Soviets on June 29 proposed that the two sides hold talks in Geneva on September 19, only five weeks before the presidential election, on ways to ban the militarization of outer space. Until such a ban could be negotiated, they suggested a joint moratorium on all ASAT and BMD tests.[59]

Partly to steal the Democratic Party's thunder on the arms control issue, partly to counter the propaganda value of the Soviet ASAT-BMD initiative, and partly to ensure continued congressional funding for the administration's nuclear weapons programs, the President in a speech before the U.N. General Assembly on September 24 proposed the establishment of a new Soviet-American negotiating "framework." He suggested combining the various arms control talks under one "umbrella" which, he promised, would make it difficult for a stalemate in one of the talks to disrupt progress in other arms negotiations.[60]

Reagan's astute ability to defuse the anti-nuclear weapons movement by pressing the Soviets to return to the negotiating table was one contributing factor behind his landslide election victory in November. Faced with Reagan for another four years, the Soviets after the election expressed interest in the President's offer for "umbrella" talks. They obviously were intent on doing all they could, even to the point of agreeing to resume talks on strategic weapons and INF, to halt the threat of a nuclear arms race in space. As a result, on November 22 it was announced that Shultz would meet with Gromyko in Geneva on January 7 and 8, 1985 to negotiate an agenda on limiting nuclear arms.[61]

The New Geneva Talks Begin, 1985

In the Geneva meeting, both sides made major compromises in their earlier positions. Gromyko dropped his demand that the United States withdraw its Pershing IIs and Tomahawk cruise missiles from Western Europe as a condition for resuming the negotiations. He also made no mention of an earlier condition that called for a moratorium on ASAT testing. And Gromyko no longer maintained that talks on space weapons could not be combined with negotiations on strategic and intermediate-range weapons. On the American side, Shultz agreed to put Star Wars on the negotiating table, something the administration had previously resisted. In another American concession, the administration agreed to have all three negotiations (offensive and defensive strategic nuclear weapons as well as INF) handled by a single delegation. Named to head the combined American delegation was Max M. Kampelman, a Washington lawyer and former aide of Hubert Humphrey.[62]

The first two rounds of the new, combined talks, which began in March and ended in April, made no significant progress, however. A second round in June and July also was unproductive. Although there were sharp differences between the two sides on several issues, the main obstacle appeared to be Star Wars. The administration again proposed major reductions in strategic and intermediate-range nuclear weapons, refusing to discuss limitations on BMD until it could first determine which technologies worked. The Soviets, on the other hand, insisted that no reductions in offensive strategic and theater weapons were possible until both sides agreed to prohibit BMD research.[63]

In the meantime, the administration's effort to link the new arms talks to continued appropriations for both the MX program and Star Wars was successful. In March the Congress, unwilling to weaken the President's bargaining power in the negotiations, approved the production of an additional 21 MX missiles, although it later voted to cap the total number at 50, half the figure that Reagan wanted. It also approved continued funding for the administration's Star Wars program, although it gave the President $1 billion less than the $3.7 billion he had requested.[64]

The Soviets tried to counter Reagan's success with the Congress by introducing new nuclear arms proposals. In April a new Soviet leader, Mikhail S. Gorbachev, announced a six-month moratorium on the deployment of Soviet INF and called for a similar freeze by the United States. With only 100 Pershing IIs and Tomahawks deployed by that time, compared to over 400 Soviet SS-20s, the administration was in no mood to comply. Then on August 6, the Soviets unilaterally halted all nuclear test explosions until the end of the year (later extended repeatedly) in an attempt to initiate a resumption of the talks on a comprehensive test ban treaty. When the Soviets invited the United States to respond in kind, the administration rejected the invitation.[65]

It was not until the third round of the talks, which began on September 19, that visible progress was made in START. On September 30 the Soviets offered to cut nuclear weapons by fifty percent with the condition that space weapons research, testing, and deployment were banned. The proposal called for a combined ceiling of 6,000 on warheads and bombs. Of these, no more than 3,600 could be on land-based ICBMs. The Soviet proposal also offered to freeze the number of Soviet three-warhead SS-20s at 243 in return for a limit of 120 American single-warhead missiles, with the remaining number in French and British missiles, to bring both sides to an equal warhead total of 729. Apparently Moscow wanted the limits to apply to cruise missiles only, in an attempt to ban the Pershing IIs. But the administration felt it was an unequal proposal because it applied to nuclear weapons that could "reach the homeland of the other side" and thereby would exclude Soviet INF.[66]

The administration's counteroffer was formally presented to the Soviets on November 1. It called for a ceiling of between 1,250 and 1,450 ICBMs and SLBMs for each side. In addition, each side would be allowed a combined limit of 4,500 ICBM and SLBM warheads. Of this number, only 3,000 warheads could be placed on ICBMs. This was above the 2,500 figure originally proposed by the United States. The administration also offered to limit bombers to 350, instead of the 400 figure favored in the past, and proposed to reduce cruise missiles to 1,500, instead of the 4,000 proposed earlier. But it again insisted that the total throw-weight of both sides must be reduced by fifty percent. In addition, the administration called for a ban on mobile ICBMs in an attempt to force the Soviets to scrap their new SS-24 and SS-25 ICBMs. Yet this also would require the United States to give up the Midgetman mobile

ICBM. Finally, the administration proposed to limit long-range INF to 140 missiles for each side, excluding those of Britain and France. The Soviets, however, again refused to consider deep cuts in offensive forces until the United States agreed to ban BMD, which the administration demonstrated no willingness to do.[67]

A November Reagan-Gorbachev summit in Geneva produced no breakthrough. Reagan refused to budge on Star Wars, and consequently Gorbachev made no further concessions on missile reductions. The two leaders did, however, agree to meet again, in 1986 and 1987.[68]

The Nuclear Arms Talks in 1986

Realizing that the nuclear arms talks were again going nowhere, Gorbachev was free to make a grandiose proposal without fearing that the United States would accept it. On January 15, 1986 he proposed a broad timetable for the elimination of all nuclear weapons by the end of the century. Gorbachev's plan consisted of three stages that would culminate in the year 2000 with a "universal accord that such weapons should never again come into being." In the first stage, which would last five to eight years, the United States and the Soviet Union would reduce by one-half the nuclear weapons that could reach each other's territory, contingent on the willingness of the United States to limit research on space-based weapons. The remaining delivery vehicles would carry no more than 6,000 warheads. In the second stage, which would begin in 1990 and last no more than five to seven years, the other nuclear powers would begin to engage in nuclear disarmament while the Soviet Union and the United States dismantled their tactical nuclear weapons. In the final stage, beginning in 1995, the world's remaining nuclear weapons would be eliminated.[69]

Reagan responded that he was "grateful" for the Soviet proposal and then, ignoring it, offered a counterproposal of his own. In essence, the new offer was a restatement of the zero option of 1981, except that it called for the elimination of intermediate-range missiles in Europe and Asia over a three-year period. In the first year, intermediate-range missiles in Europe would be cut to 140 for each side, while the Soviets would be obliged to make proportional reductions in their SS-20 force based in Asia. In the second year, the remaining intermediate-range missiles would be cut in half and, in the third year, they would be totally eliminated. Gorbachev responded by calling Reagan's counterproposal inadequate.[70]

Reagan reacted by increasing the pressure on the Soviet Union. On May 27 he stated flatly that, in the light of the alleged violations of the SALT II Treaty by the Soviets, the United States would no longer abide by the unratified agreement. However, he hedged somewhat by stating that he might reconsider his decision if the Soviets reversed their pattern of violations, including the alleged deployment of two new ICBMs instead of the one they were allowed under the SALT II Treaty. Gorbachev responded by stating that Reagan's decision raised "the legitimate question of whether Washington really wants a new [summit] meeting." After America's allies reacted with hostility, and the

House of Representatives passed a nonbinding resolution repudiating his action, the President agreed to send delegates to discuss the treaty in a July conference. But the meeting produced no change in the administration's position on the treaty.[71]

While the Reagan administration was trying to scrap the SALT II Treaty, the Soviets again attempted to save the ABM Treaty. In late May they offered a proposal that would permit some research on a BMD, provided the United States agreed to maintain the ABM Treaty for another fifteen to twenty years. In exchange for upholding the treaty, the Soviets said they would be prepared to limit strategic forces on each side to 8,000 warheads or bombs and to 1,600 delivery systems. This concession would require more Soviet warhead reductions than the United States would have to make. The Soviets also stated that they would be willing to drop the total number of warheads and bombs to 6,000 if the United States agreed to a complete ban on anti-missile research.[72]

The Soviets also made a number of significant concessions in other areas of the nuclear arms talks in their May proposal. For one, they dropped their demand that all sea-launched cruise missiles with ranges of more than 370 miles be banned. Yet they apparently continued to seek such a ban on surface vessels. They also agreed to drop their earlier demand that American forward-based systems be included in a strategic arms agreement, although they also insisted that they must be frozen at their then current levels. In another major concession, the Soviets offered to exclude British and French intermediate-range missiles from an agreement that would remove all Soviet and American intermediate-range missiles from Europe, provided the British and French agreed to maintain their missiles at their then current numbers. The Soviets also accepted some of the American proposals to verify INF limits.[73]

The Soviet concessions apparently widened the division within the Reagan administration on nuclear arms policy. The State Department reacted favorably to the Soviet concessions as a basis for negotiations; the Pentagon and the Arms Control and Disarmament Agency opposed it. Weinberger again insisted that Star Wars had not been initiated "as something to be given away." Reagan once more sided with his defense secretary. On July 12 he said that Star Wars was not a "bargaining chip."[74]

Two weeks later, Reagan formally responded to Gorbachev's latest proposals. On July 25 he offered to delay the deployment of Star Wars for seven and a half years if the Soviets agreed to allow it after this interval. While the United States would not make any Star Wars deployments during this period, the President's proposal would allow both sides to proceed with research, development, and testing of space-based defensive technologies and components. Reagan also said that he was willing to share the benefits of such research with the Soviets. In early September, Gorbachev told Reagan that he would accept a fifteen-year continuation of the treaty, and might accept as little as a ten-year extension. The Soviets obviously realized that a seven-and-a-half-year delay on the death of the ABM Treaty was no advantage to them because it was unlikely that the United States would deploy any space-based weapons during that interval with or without the treaty.[75]

In the meantime, Congress again increased its pressure on the administration. In August the House passed a defense bill that not only appropriated \$33 billion less than the President wanted, but also required continued adherence to the numerical limits on launchers and warheads set in the SALT II Treaty. In addition, prompted at least to some extent by repeated extensions in the Soviet nuclear testing moratorium, the House approved, by an unexpectedly large margin of 234 to 155, a resolution calling for a one-year moratorium on all American underground nuclear weapon tests above a one-kiloton yield, the apparent threshold of verifiability. The resolution, however, was contingent on Soviet reciprocity and acceptance of on-site inspection. The Senate approved a resolution urging the President to begin immediate negotiations with the Soviets for a comprehensive test ban.[76]

The Iceland Summit and Its Aftermath

While the Congress was increasing its pressure on the administration, Gorbachev was intensifying the Soviet negotiation effort. On September 19, in a letter to Reagan, the Soviet leader suggested a preliminary summit meeting that would give impetus to the Geneva arms talks. Gorbachev's suggestion came at a time when both superpowers were at loggerheads over the arrest by the Soviets of an American journalist, Nicholas S. Daniloff, for alleged espionage just after the United States had seized a Soviet employee of the United Nations, Gennadi F. Zakharov, for spying. On September 30 both governments announced that an agreement had been reached to release both men and, as a result, Gorbachev and Reagan would meet in Reykjavík, Iceland, on October 11 and 12.[77]

On the eve of the talks, the President, saying he did not want his hands "tied" at Reykjavík, pursuaded the Democratic leadership of the House to withdraw two of their key resolutions on arms control. One resolution mandated continued American compliance with the SALT II Treaty. The other called for a one-year moratorium on nuclear testing. The House leadership dropped this resolution in exchange for a promise by Reagan that he would work for an agreement with the Soviets to restrict testing. In addition, the Democrats obtained Senate acquiescence to a one-year ban on anti-satellite weapon tests and White House agreement to send to the Senate for ratification the Threshold Test Ban Treaty and the Peaceful Nuclear Explosion Treaty.[78]

By most accounts, the Americans were surprised by what Gorbachev had in store for them in Reykjavík. Instead of a preliminary meeting that would set the stage for a full-scale summit later, the Soviet leader came prepared with detailed, nuclear arms proposals that included a number of new and significant Soviet concessions. The President reacted positively to the Soviet negotiating challenge and both leaders engaged in eleven hours of intense, face-to-face talks that produced a number of significant agreements.[79]

The first area of agreement concerned strategic nuclear weapons. Both sides agreed that over a five-year period ending in 1991 the number of warheads carried by ballistic missiles and ALCMs would be cut to 6,000 and the number of long-range missiles and bombers would be reduced to 1,600. (At the time of the Reykjavík summit, the Soviet Union had 2,500 missiles and bombers, and the Americans 2,200.) The Soviets also dropped their earlier demand for a common limit on missile warheads and bombs. They proposed instead that each bomber with multiple bombs and short-range attack missiles count as one weapon under the 6,000 ceiling, and each bomber with ALCMs as well as bombs and short-range missiles would count as two weapons under the total ceiling. The Soviets also agreed to exclude sea-launched cruise missiles from the 6,000 ceiling.[80]

However, neither side could agree about what would transpire in the second five years of the proposed ten-year agreement. Reagan suggested that both sides eliminate all offensive ballistic missiles, in effect leaving each with only cruise missiles and bombers to comprise their strategic deterrent forces. Gorbachev, however, favored the elimination of all long-range nuclear weapons, including bombers and cruise missiles.[81]

The second area of agreement concerned INF. Both sides accepted a global limit of 100 INF warheads, with the Soviet deployments restricted to Asia and America's constrained to the United States. Thus, both sides agreed to implement Reagan's original zero option; they would withdraw all their intermediate-range missiles from Europe. (By October 1986 the Soviets had deployed 513 warheads on 171 SS-20 missiles in Asia and 810 warheads on 270 SS-20s in Europe. The United States had deployed 108 Pershing II IRBMs and 160 Tomahawk GLCMs, for a total of 268 warheads.) Reagan and Gorbachev also agreed on verification steps and on freezing shorter-range missiles pending further negotiations.[82]

Nuclear testing was a third subject of agreement. Gorbachev accepted Reagan's proposal for a phased reduction of testing, starting with the verification of existing treaties and working toward an ultimate cessation of tests. The President refused to say what types of limits on testing he favored, but two types were under consideration within the administration. One would gradually lower the number of tests permitted each year, while another would gradually reduce the limit on the size of explosions below the threshold of 150 kilotons.[83]

However, as in the Geneva summit the preceding November, the Iceland meeting fell apart over Star Wars. Late in the last day of the summit, Gorbachev insisted that all the concessions he made were contingent on America's adherence to a strict interpretation of the ABM Treaty. He insisted that there be no testing and development of BMD, and that research be confined to the laboratory. Reagan, who had expressed his willingness to abide by the ABM Treaty for an additional ten years, instead of the seven-and-a-half-year period he had favored earlier, rejected Gorbachev's interpretation of the treaty's restrictions. The President favored a much looser interpretation that would allow extensive testing and development of BMD outside the laboratory. The Soviet interpretation, the President stated, was designed to kill SDI.[84]

The talks ended in mutual bitterness, with each side blaming the other for their collapse. Gorbachev accused the President of throwing away an historic opportunity for major reductions in nuclear weapons by refusing to abandon SDI. Said the Soviet leader, "SDI . . . does not allow us to find a way out of the threat that hangs over the heads of mankind."[85]

To counter the impression that the President was primarily responsible for the failure of the Reykjavík summit, the administration launched a massive public relations campaign emphasizing the agreements reached in Iceland while blaming the Soviets for the breakup of the talks. As White House communications director Patrick Buchanan put it: "The President made the most sweeping, far-reaching arms control proposal in history. Gorbachev said no. He made a nonnegotiable demand that the President give up SDI, and the President said no." Public opinion polls conducted shortly after Reykjavík indicated overwhelming support for the administration's interpretation of the summit. A *New York Times*/CBS poll, for example, found that forty-four percent of those questioned held Gorbachev responsible for the collapse of the talks, while only seventeen percent blamed the President.[86]

More than a few defense experts, however, were critical of Reagan's performance at Reykjavík. Senator Sam Nunn, the ranking Democrat on the Armed Services Committee, wondered why the President let pass an opportunity for major reductions in the number of Soviet nuclear weapons to preserve the Star Wars program, which the senator considered only a technological "possibility." Moreover, Nunn pointed out, "if the Soviets took us up" on Reagan's proposal to eliminate all ballistic missiles, "every general in the Army and the Air Force, and probably some admirals, too, would have a heart attack." Along the same line, retired Colonel John M. Collins, a defense specialist at the Library of Congress, stated that some ballistic missiles were necessary to counter what he believed to be the Warsaw Pact's superiority in conventional forces and to cause doubt in the minds of Soviet leaders about how the United States would respond to Soviet aggression in Western Europe and elsewhere. For this reason, France's foreign minister, Jean-Bernard Raimond, regarded the possibility of "the total disappearance of American missiles from Europe" as "terrible." Later, similar concerns by British Prime Minister Margaret Thatcher prompted Reagan to assure her that some ballistic missiles would be retained for the defense of Europe and to ensure that the Soviets would not cheat.[87]

And yet other analysts, like the administration, did not view the Reykjavík summit as the end of the nuclear arms talks but rather a major step toward an eventual agreement. The key to such a breakthrough would be the types of BMD research both sides would permit. Fortunately, a step in this direction was made within days of the Reykjavík summit, when the Soviets indicated that Gorbachev's position had been misinterpreted, and that he had not intended to ban all research outside the laboratory. The administration responded by stating that it would try to define explicitly what type of testing would be permitted.[88]

For their part, the Soviets would in all probability expect the United States to provide written assurances that it would adhere to any newly agreed-to limits on BMD research and not move to a more permissive interpretation of the ABM Treaty for the duration of its ten-year extension. The administration may be willing—or even compelled—to give such assurances. A number of problems SDI has encountered since the program was first initiated have proven more difficult than anticipated. They include discriminating real missiles from decoys, writing the computer software that would coordinate the entire system, launching the components into space (a problem aggravated by the setback to the space shuttle program caused by the *Challenger* disaster in January), protecting the components against attack once they were in orbit, and bringing costs down to an affordable level. These and other problems, combined with repeated congressional cuts in SDI appropriations, have forced the administration to scale back many of the research aspects of the program. Considering these limitations, the administration might regard as palatable a new agreement that permitted BMD research outside the laboratory but restricted it to dimensions that were acceptable to the Soviets.[89]

In the weeks following the Iceland summit, however, the administration gave no indication that it was about to compromise further on its SDI position. Quite to the contrary, it increased its pressure on the Soviets to change their position on BMD. On November 28 the United States exceeded the limitations of the SALT II Treaty when a B-1 bomber was equipped with more cruise missiles than that unratified agreement permitted. Moreover, in January 1987 the administration indicated that it was planning for the deployment of some SDI units as soon as the early 1990s, and reasserted its belief that such deployments would not violate the ABM Treaty.[90]

Even more ominous for the chances of concluding an eventual nuclear arms agreement, however, may be the effects of the Iran-Contra military aid scandal. In November the public learned that the administration had approved and facilitated the sale of American arms to Iran, and that some of the proceeds from the sale had been illegally diverted to administration-backed rebels, the "Contras," fighting the Sandinista government of Nicaragua. Some fear that the administration's preoccupation with the Iran-Contra affair may make it unwilling to deal with the Soviets on any nuclear arms agreement. Said one unidentified top aid to Shultz: "We can't be too eager to deal or we'll be taken for patsies, or accused of trying to deflect attention from Iran." Yet the same official also stated that "we can't be so cautious that we paralyze ourselves and miss a golden opportunity."[91]

Only time will tell whether the President—and the Soviets—will squander the "golden" opportunity for meaningful arms reduction that the Reykjavík summit presented. But one thing is certain: with less than two years remaining in Reagan's term, and with its last year sure to be consumed by the diversions of another presidential election campaign, time for concluding such an agreement during his administration is running out.

Conclusion

The Unending Nuclear Arms Race

Ronald Reagan was only one in a series of Cold War presidents who vowed to do all they could to terminate the nuclear arms race while simultaneously augmenting the quantity and quality of America's nuclear weapons. What accounts for this discrepancy between the efforts of these presidents to end the nuclear arms race and their actions to sustain it?

Power Politics

One of the basic reasons American presidents have been unable to end the nuclear arms race is directly related to America's adherence to a theory of international relations that has prevailed since the Renaissance. Called by various names—realism, *Realpolitik,* or power politics—this theory considers military power the ultimate method of acquiring or maintaining national interests, including territory, natural resources, markets, military bases, allies, colonies and clients, national prestige and, above all, national security. Even peace, itself, becomes a product of balancing contending powers.

The decision to develop atomic weapons in World War II was prompted initially by a desire to preserve American national interests that were threatened by the Axis powers. Americans feared that if they did not develop atomic weapons first, the Germans would. Moreover, most Americans at the time believed that the preservation of American security justified the use of atomic weapons against Japan. The moral problem of using a weapon as hideous as the atomic bomb primarily against civilian targets, like Hiroshima and Nagasaki, bothered American policymakers only marginally. Indeed, World War II itself helped to anesthetize American moral sensibilities against the horrors of killing civilians. More people were killed by the firebombing of Tokyo in one night than in either the destruction of Hiroshima or Nagasaki.

The United States traditionally has cloaked its realistic objectives under the mantle of idealism. Use of the atomic bombs, the American people were told, would save thousands of American and Japanese lives by precluding the planned invasion of Japan. But it is also true that no significant effort was made to seek a negotiated termination of the war that may have precluded both an invasion of the Japanese home islands as well as use of the atomic bomb. The fact is that use of the bomb against Japan also served another purpose not publicized at the time: countering another nation—the Soviet Union—that threatened American interests.

The Cold War

In the wake of the collapse of the German and Japanese empires, and even before their demise, American and Soviet interests collided in areas where both nations attempted to establish their preeminent influence—in Europe, the Middle East, and the Far East. However, while conflicting Soviet and American national interests were difficult to reconcile, what made them seem insurmountable was the depth of the ideological chasm that separated the two superpowers. To the overwhelming majority of Americans, Soviet communism was, and is, repugnant. Americans regarded the expansion of communism the primary Soviet objective, downplaying the legitimacy of the Soviet Union's national interests. To the Soviets, on the other hand, the American desire to spread democracy was simply a ruse to disguise an American quest for domination of the world's resources and, by this means, to destroy the Soviet Union.

During the war, and immediately thereafter, Stalin downplayed ideological differences with the United States, not only to buy time for Soviet scientists to counter the American atomic monopoly, but also to secure greatly needed American assistance both during and after the war. He asserted that the creation of a Soviet sphere of influence in Eastern Europe was not motivated by a desire to spread communism but rather by the security needs of the Soviet Union. He indicated that he wanted a buffer of friendly, but not necessarily communist, governments on the Soviet border that would shield the Soviet homeland from invasion.

Roosevelt, at least tacitly, was prepared to recognize the predominant influence of the Soviet Union in Eastern Europe, while giving lip service to the Wilsonian principle of self-determination. Truman attempted to continue the accommodationist approach to the Soviets pursued by his predecessor. Ultimately, however, hardliners both inside and outside the administration convinced him that Stalin's actions in Eastern Europe, the Middle East, and the Far East would make any attempt to accommodate Soviet national interests analogous to the appeasement of Germany and Japan that occurred before World War II. Accordingly, Americans who favored an alternative approach toward the Soviets, one emphasizing negotiations based on a mutual accommodation of contending interests, rather than confrontation, were quickly and repeatedly denounced as "appeasers." In the minds of hardliners, there could be no meaningful negotiations with the Soviets, for the more concessions the United States made to the Soviets, the more they would demand.

With the breakdown of Soviet-American diplomacy after World War II, differences between the two countries often became crises, and crises seemed to provide further justification to the United States for its exclusively confrontational approach to the Soviet Union. In response to the American containment policy, which included the Truman Doctrine, the Marshall Plan, and NATO, the Soviets initiated the Molotov Plan and the Warsaw Pact, and extinguished the last flickering embers of national self-determination in Eastern Europe. With the status quo frozen in Europe, the Third World quickly became the main arena for superpower competition. Soon the incessant Soviet-American rivalry in the developing countries of the world would become a major reason for the inability of the superpowers to end the Cold War and the nuclear arms race.

Nuclear Deterrence

For a variety of reasons, nuclear weaponry has been a major ingredient in America's effort to contain Soviet expansionism. In the wake of the rapid demobilization of America's conventional armed forces after World War II, nuclear weapons were considered by the nation's political and military leaders as the only feasible deterrent to the massive conventional forces of the Soviet Union. Compared to maintaining large conventional forces, which the American public was not inclined to support during peacetime, nuclear weapons were inexpensive to construct and deploy. More important, they were the ultimate instruments of destruction. And, until the Soviets developed a nuclear arsenal, they could not be countered in kind.

After the Soviets began to produce nuclear weapons and the means to deliver them on American targets, nuclear weapons became even more important to the United States. They were considered necessary to deter a Soviet nuclear attack on the American homeland as well as to counter a wide variety of lesser Soviet challenges. American nuclear threats were made during the offshore islands crisis with China in the 1950s, during the Berlin crises of 1948 and 1961, during the Cuban missile crisis in 1962, and during the Arab-Israeli War of 1973.

America's willingness to engage in nuclear diplomacy was to a large extent made possible by the overwhelming nuclear superiority the United States possessed until well into the 1970s. With the exception of ICBMs and ABMs, the United States has led the Soviet Union in developing and deploying every major strategic nuclear weapon system. American nuclear superiority was not only a matter of national security but also a matter of national prestige. To many, nuclear superiority was the clearest manifestation of America's supremacy.

Naturally, considering the Soviets were also practitioners of power politics, they felt compelled to play nuclear "catch up" with their more technologically advanced adversary. With time, the Soviets succeeded in matching virtually every major American nuclear weapon. Paradoxically, the ultimate and inevitable result of this so-called action-reaction cycle was an increase in American, as well as Soviet, insecurity rather than the diminution of their mutual vulnerability. The more nuclear weapons the Americans targeted on the Soviet Union, the more nuclear weapons the Soviets aimed at the United States.

Although both sides have given lip service to the prospect that the development of additional nuclear weapons would increase the likelihood that they would be used, whether by accident or miscalculation, the principles of power politics have become so deeply a part of the Cold War that neither the United States nor the Soviet Union has been capable of breaking its continued reliance upon them.

In the United States, new counterforce doctrines were developed in an attempt to explain the paradox of why more nuclear weapons had to be manufactured than could possibly be used. With the Soviets achieving rough nuclear parity in the late 1960s, American strategic analysts feared that a Soviet first-strike would leave the United States with too few strategic weapons to react effectively against the Soviet Union. To retaliate with insufficient forces would be suicidal, they argued, for America's cities would then be exposed to retaliation from the enemy's second and third strikes. To preclude that possibility, American counterforce proponents argued, the United States required not only more nuclear weapons, but also more accurate systems to ensure that Soviet ICBM silos could be destroyed.

Many have argued that a counterforce doctrine is in fact more destabilizing than the assured destruction strategy it was designed to replace, for it threatens Soviet retaliatory forces ostensibly designed to deter a first-strike from the United States. At least once—in the Cuban missile crisis—the Soviets attempted to meet America's growing counterforce capability with a step that almost produced the nuclear holocaust both sides have tried to avoid.

The Military-Industrial Complex

The birth of the atomic age coincided with a development that was unique in the history of the United States, what Dwight Eisenhower called a "conjunction of an immense military establishment and a large arms industry." This so-called military-industrial complex has contributed much to maintaining the momentum of the nuclear arms race.

The military component of the complex has wanted more nuclear weapons because they are considered necessary for deterrence and defense. But in arriving at that assessment, military planners usually have relied on worst-case scenarios of Soviet capabilities rather than more prudent estimates of what the Soviets are likely to do.

Nuclear weapons also have been important as bargaining chips in the intraservice competition for defense appropriations. In the forties and fifties, the armed services competed for the role of America's primary strategic deterrent force. Even though the Air Force won that contest, the Army and Navy were placated with nuclear weapon systems of their own. In the sixties and seventies, the Air Force lobbied vigorously for what would become the B-1 bomber and the MX ICBM, because it feared that without them the Navy's ballistic missile submarines could gain prominence and the Navy become America's preeminent strategic deterrent force. Ultimately, both the Navy and the Air Force received the strategic nuclear weapons they desired, and the United States built more nuclear weapons than it needed.

But the armed services have not been the only ones to benefit from a continuation of the nuclear arms race. Defense contractors, scientists, universities, labor unions, and politicians have all prospered from the nuclear weapons industry. The nuclear arms race has meant large contracts for companies doing business with the Defense Department and jobs for favored congressional districts. For many universities, grants for conducting research in nuclear warfare have amounted to generous, federally financed endowments.

No doubt, most scientists who have engaged in nuclear weapons work have concluded that the United States has had no choice but to stay ahead of the Soviets in nuclear weapons research and development. On the other hand, many scientists have believed that they have a moral responsibility to use their knowledge and prestige to halt the nuclear arms race. In fact, the first proposals for negotiated nuclear arms control agreements were made by scientists—Niels Bohr, Leo Szilard, and James Franck. But for those who have resisted the nuclear buildup from a position of authority, it has not been an easy struggle. The fate of J. Robert Oppenheimer, who was publicly disgraced and ousted from the inner circles of government primarily for opposing the H-bomb project, was the most prominent and extreme example of the penalties a scientist could incur by attempting to resist the development and deployment of nuclear weapons.

To arouse congressional support for increased defense spending, the military, defense contractors, and scientists have capitalized repeatedly on the public's fear of the Soviet Union with exaggerated estimates of Soviet capabilities. The result has been one "gap" scare after another—there was the bomber gap of the mid-1950s, the missile gap of the late fifties, the ABM gap of the late 1960s, the throw-weight gap of the seventies, and the space-weapons gap in the 1980s. And in its effort to gain congressional appropriations, the military-industrial complex has been assisted by more than a few politicians who have believed that nuclear buildups are not only necessary to counter the alleged Soviet leads but are also politically advantageous to themselves and to their party.

Most Americans have been susceptible to the scare tactics employed to manipulate them. To the general public, each augmentation of Soviet military power, whether real or not, has had to be met first by increased defense spending and only later by arms control negotiations. This explains the public support given to Kennedy's nuclear buildup during the alleged missile gap as well as the initial public support given to Reagan's "rearmament" programs. The news media, at least until the late 1960s, generally supported the quest of the military-industrial complex for larger defense appropriations by giving credence to its scare tactics. It was only in the late 1960s—as a result of Vietnam, the ABM debate, and the beginning of SALT—that the news media began to give nuclear weapons more critical coverage.

In reality, the general public, by practicing a form of psychic denial and not confronting what they feared most, has not wanted to think about the existing and possible effects of America's participation in the nuclear arms race. When they have had to, Americans have been easily overwhelmed by the complexity of the subject. As a result, they have been more than willing to defer to those in power. However, with but one significant exception—Dwight D. Eisenhower—America's Cold War presidents have lacked the military knowledge necessary to make and stick with a realistic assessment of the Soviet threat. One president after another has not only accepted worst-case scenarios as valid bases of American deterrence strategy, but also has used them to garner congressional support for nuclear weapon programs.

The Political System

From the beginning of the Manhattan Project until the late 1960s, when public and congressional support for America's containment strategy began to break down as a result of the Vietnam War, Congress rarely challenged the nuclear strategy that was fashioned by the executive branch. In fact, the first nuclear weapon programs were initiated without the knowledge of the full Congress. The legislative branch was virtually excluded from the decision to develop the atomic and hydrogen bombs. The secrecy surrounding both projects did much to preclude the possibility of congressional as well as public debate on the wisdom of developing these weapons. As a result of congressional and public acquiescence, or more accurately ignorance, the power to produce and, if necessary, to use nuclear weapons was assumed by the executive branch and, ironically and frighteningly, by presidents who for the most part had little personal expertise in military or diplomatic affairs, let alone nuclear strategy.

To be sure, all of America's Cold War presidents, to a greater or lesser extent, have felt compelled to seek advice to compensate for the shortcomings in their own military and diplomatic experience. But the advice they have accepted has come largely from a small, closely knit, narrowly based, elite group of policymakers drawn from the upper levels of the establishment. Consistency, not innovation, has been the hallmark of this so-called national security elite. And consistency has meant supporting the approved strategy for dealing with the communist challenge—one emphasizing military countermeasures and particularly the threat of using nuclear weapons.

The Failure of Nuclear Arms Control

And yet, while augmenting the power of America's nuclear arsenal, American presidents have felt compelled to attempt to end, or at least to control, the proliferation of nuclear weapons. Although the people of the United States have always supported American nuclear buildups, they also have favored steps to reduce the risks of nuclear war. Moreover, America's European allies, who would also share in the destruction of a nuclear holocaust, have given strong support to nuclear arms talks.

But the various nuclear arms control negotiations in which the superpowers have engaged since the late 1940s have had only limited success. They have curtailed the nuclear arms race in some important respects, including restrictions on nuclear weapon deployments, nuclear weapon tests, and the proliferation of nuclear weapons to nonweapon states, but they obviously have not ended the competition. Nuclear weapons tests continue, more nuclear weapons are being added to the arsenals of the superpowers, and more nonnuclear weapon states are developing or acquiring the means to produce nuclear weapons.

Given the ideological and national rivalry that has characterized Soviet-American relations, as well as the fear that has been generated by the ongoing threat of nuclear preemption or retaliation, the climate of trust that is a prerequisite of successful negotiations on arms control, or any other issues that have divided the two countries, has rarely and only superficially existed. In fact, both sides are so deeply engaged in power politics and ideological competition that they have attempted to derive as much of an advantage from nuclear arms talks as they have from their own nuclear weapon programs. Indeed, for both sides, the objective of arms control talks has been to place as many limits as possible on the other side's deployments and as few as possible on one's own.

In addition, America's negotiating posture has almost always been based on the assumption that the Soviets will not restrict their own deployment of nuclear weapons without the fear of new American deployments. In other words, this line of reasoning asserts that it is necessary to build new nuclear weapons in order to ultimately reduce them. Rather than attempting to persuade the Soviets to restrain their nuclear programs by engaging in a policy of American nuclear self-restraint, one administration after another has attempted to intimidate the Soviets with new nuclear weaponry. And yet it has proved extremely difficult to limit by treaty the deployment of weapon systems developed for bargaining purposes. The United States, for example, developed MIRVed missiles in the late 1960s to counter the development of a potentially large Soviet ABM system. Placing limitations on MIRV deployments proved impossible in SALT I and difficult in SALT II. Not surprisingly, the Soviets responded in kind. As a result of Soviet MIRV deployments, the Soviet first-strike scenario came much closer to becoming a realistic possibility.

The absence of trust in political solutions to the nuclear arms race also has been reflected in the American emphasis on the technical aspects of verifying arms control agreements. This was clearly demonstrated in the unsuccessful effort to conclude a comprehensive test ban agreement in the late 1950s and early 1960s. Despite the overwhelming political reasons for ending the nuclear arms race at that time—particularly the preservation of American nuclear superiority—technical issues, such as the number of annual inspections, determined the fate of those negotiations. More recently, the traditional American preference for technical solutions can be seen in the great emphasis the Reagan administration has given to the concept of ballistic missile defense, rather than negotiated agreements, as the favored method of ending the nuclear arms race. Since the verifiability of arms control agreements, and thereby their political acceptability, necessarily rests on technical capabilities, it is also true that the consideration of technical issues has made the pace of nuclear arms talks excruciatingly slow. As technological developments continue and become more sophisticated, nuclear arms agreements have become more and more difficult to conclude.

The asymmetrical relationship of the American and Soviet strategic arsenals has also complicated the task of defining a mutually acceptable balance of nuclear forces. Americans have refused to accept the Soviet argument that they need more ICBMs, SLBMs, and throw-weight to compensate for the superior accuracy and numbers of American warheads. The Soviets, on the other hand, have refused to acknowledge the threat to America's land-based missiles that would result if Soviet superiority in numbers of missiles and throw-weight was combined with substantial numbers of deployed Soviet MIRVs. Even after recognizing the validity of these asymmetrical threats, both sides have been reluctant to admit that they still possess effective retaliatory capacity no matter how potentially effective their opponent's first-strike forces could become—to do so, it seems, would take much of the force out of their arguments that additional nuclear weapons are needed.

In addition, the presence of American and Soviet intermediate-range nuclear forces on and around the Eurasian landmass and the existence of independent nuclear forces in Western Europe and in China have slowed the nuclear arms talks and created other hurdles. Because these forces were either independent of American control or nonstrategic in nature, the United States has always insisted that they should not count in the superpower, strategic-nuclear equation. The Soviet Union, on the other hand, has been steadfast in insisting that any nuclear weapon capable of striking Soviet territory could not be excluded from their negotiations.

The willingness of presidents to make nuclear arms control talks one more arena of Soviet-American military competition has made it increasingly difficult to maintain public and congressional support for arms control talks. Jimmy Carter, for example, failed to win strong support for the SALT II Treaty, not only because the treaty was too complex for most Americans to understand, but also because it was insufficiently restrictive to attract much needed liberal support.

Unlike the SALT II supporters, its opponents were extremely effective. They were not only well organized and financed, they also had a particularly shrewd strategy. Rather than attacking the value of arms control agreements directly, they simply stated that they wanted a better agreement than the one the Carter administration had negotiated. They then proceeded to tear apart and undermine support for the existing treaty without offering anything negotiable in its place.

The advantages the opponents of arms control have enjoyed have been compounded by the ease with which arms control agreements can be blocked in the American political system. The Joint Chiefs of Staff have possessed a virtual veto over arms control proposals. Without Pentagon support, it has been virtually impossible to gain the required affirmative vote from two-thirds of the Senate for the ratification of arms control treaties.

To mollify the Pentagon and other hardline opponents of arms control, some presidents have agreed to develop and deploy weapon systems they knew were militarily unnecessary. Kennedy had to promise to back a vigorous program of underground testing in order to get the Pentagon's support for the Limited Test Ban Treaty. Jimmy Carter approved the deployment of the MX ICBM, the Pershing II IRBM, and the Tomahawk cruise missile in a futile attempt to gain Pentagon and Senate support for the SALT II Treaty. By contrast, Ronald Reagan has used nuclear arms control talks to deflate overwhelming popular support for a nuclear freeze and to obtain continued congressional approval for his nuclear weapon buildup. The end results, however, have been the same—more deployed nuclear weapons.

Soviet Responsibility

While this book has concentrated on the role of the United States in the nuclear arms race, it has not attempted to absolve the Soviet Union from its share of the responsibility for maintaining the nuclear arms competition, In many respects, the factors that drive the nuclear arms race in the Soviet Union are a mirror image of the forces that propel nuclear weapon programs in the United States. As in America, Soviet civilian leaders have had to contend with pressure from their military leaders to build nuclear weapons. The secrecy that has surrounded American nuclear weapon programs is even more prevalent and strict in the Soviet Union. And Soviet leaders, particularly Khrushchev, have not been adverse to engaging in nuclear diplomacy or in participating in arms control negotiations from a position of excessive military strength.

Soviet domestic and foreign policies also have made it difficult to obtain the degree of trust from Americans that is a prerequisite of successful negotiations. Americans have found it easy to sympathize with the victims of Soviet oppression both in the Soviet Union and in its satellite states. Moreover, Soviet aggression—in Hungary, Czechoslovakia, Afghanistan, and elsewhere—has had the knack of undermining American congressional and public support for superpower arms control at crucial points in the negotiations. Perhaps it is asking too much of the Soviets to surrender such limited interests for the sake of concluding agreements to curb the nuclear arms race. But it is obviously asking too much of Americans to ignore Soviet aggressive actions. As a result, one administration after another has linked nuclear arms talks to Soviet behavior on other issues—with adverse effects to both the talks and to these unrelated issues.

Horizontal Nuclear Proliferation

The inability of the superpowers to effectively curb the growth of their own nuclear arsenals has set a bad example for nonweapon states contemplating the acquisition of nuclear weapons. Nor has the international effort to dissuade the spread of nuclear weapons through the dissemination of ostensibly peaceful nuclear technology and materials been completely effective. In addition to the six nuclear powers, numerous states have the capability to produce nuclear weapons, and several others are about to acquire it. It is extremely difficult, if not impossible, to prevent the military application of nuclear technology and materials designed for purely civilian application. And the United States has not diminished that possibility by providing nuclear aid to countries that have not fully participated in the NPT-IAEA safeguard system.

With the inability of the superpowers to end the nuclear arms race and the spread of nuclear weapons to the nonweapons states, it is possible to envision a world filled with nuclear weapons, increased opportunities for nuclear accidents and nuclear terrorism, and a deterrent relationship so complicated by the emergence of a multiplicity of nuclear weapon states that deterrence itself may become impossible and a superpower nuclear war inevitable. The continuation of the nuclear arms race, in other words, can only offer the ultimate prospect of a final world calamity—a nuclear Armageddon.

GLOSSARY OF ACRONYMS
AND TECHNICAL TERMS

ABM Anti-ballistic missile; a system designed to destroy or incapacitate strategic, offensive ballistic missiles.

ACDA Arms Control and Disarmament Agency.

AEC Atomic Energy Commission.

ALCM Air-launched cruise missile (see **cruise missile**).

ASAT Anti-satellite weapon; a weapon system designed to destroy or incapacitate enemy satellites.

ballistic missile A missile that does not rely on aerodynamic surfaces to produce lift and follows a ballistic trajectory when thrust is terminated.

BMD Ballistic missile defense; a system for defending against an attack by ballistic missiles.

breeder reactor A nuclear reactor that produces more fuel (primarily plutonium) than it consumes.

C^3I Command, control, communication, and intelligence systems.

counterforce Use of strategic nuclear forces to destroy or disable enemy military capabilities.

countervalue Targeting enemy population and economic centers.

cruise missile A guided missile using aerodynamic lift.

CTB Comprehensive Test Ban; an agreement that would prohibit all nuclear testing.

damage limitation A strategy designed to reduce damage to population centers and other nonmilitary sites during a nuclear exchange.

dual-capable weapons Systems capable of delivering either conventional or nuclear weapons.

encryption The encoding of communications or other data (e.g., telemetric data) for the purpose of concealing information.

enrichment The process of increasing the quantity of highly fissionable uranium-235 isotopes in uranium fuel or explosives.

233

EURATOM   European Atomic Energy Community.

FBS   Forward-based systems; American weapon systems based outside the United States capable of delivering nuclear warheads on Soviet targets.

FROD   Functionally related observable difference; a technique established by SALT II to distinguish weapons on the basis of their primary function.

GLCM   Ground-launched cruise missile (see **cruise missile**).

hardened site   A site constructed to withstand the blast and associated effects of a nuclear explosion.

horizontal proliferation   The acquisition of nuclear weapons by nonweapon states.

ICBM   Intercontinental ballistic missile; a fixed or mobile land-based missile capable of delivering a warhead to intercontinental ranges defined in SALT I and II as in excess of 5,500 kilometers.

IAEA   International Atomic Energy Agency.

INF   Intermediate-range nuclear forces; land-based missiles and aircraft with a range/combat radius between the battlefield range of tactical nuclear weapons and 5,500 kilometers.

IRBM   Intermediate-range ballistic missile; a ballistic missile with a range of between 1,500 and 3,000 nautical miles.

kiloton   Nuclear yield equal to that of 1,000 tons of TNT.

light-water reactor   The most common type of nuclear reactor, fueled with enriched uranium.

MAD   Mutual assured destruction; a concept of strategic stability under which rivals are deterred from launching a nuclear attack because each possesses the capability to devastate the other's homeland in a retaliatory attack.

MaRV   Maneuverable reentry vehicle; a multiple-warhead vehicle that can maneuver each warhead to its target.

megaton   Nuclear yield equal to that of one million tons of TNT.

MIRV   Multiple independently targeted reentry vehicle; multiple reentry vehicles carried by a ballistic missile, each of which can be directed to a separate target.

MRBM   Medium-range ballistic missile; a ballistic missile with a range of between 500 and 1,500 nautical miles.

MRV   Multiple reentry vehicle; a multiple-warhead vehicle in which the warheads cannot be guided to separate targets.

MX ICBM   An intercontinental ballistic missile capable of carrying ten nuclear warheads; also called "Peacekeeper."

NATO   North Atlantic Treaty Organization.

national technical means   Assets under national control, including photo-reconnaissance satellites and radars, used to monitor compliance with an arms agreement.

NPT   Nuclear Nonproliferation Treaty; an international agreement designed to check the horizontal and vertical proliferation of nuclear weapons.

nuclear fuel cycle   A process for developing, employing, and disposing of nuclear fuel.

PNE Peaceful Nuclear Explosion.

radiological weapon A weapon employing radioactive poison as the primary incapacitating agent.

reentry vehicle That portion(s) of a ballistic missile containing primarily the warhead(s) and guidance system.

reprocessing The process of separating uranium and plutonium from spent nuclear fuel.

SAC Strategic Air Command.

SALT Strategic Arms Limitation Talks.

SDI Strategic Defense Initiative; a ballistic missile defense program begun by President Reagan; also called "Star Wars."

SIOP Single Integrated Operational Plan; the U.S. strategic nuclear war plan.

SLBM Submarine-launched ballistic missile.

SLCM Submarine-launched cruise missile (see **cruise missile**).

Standing Consultative Commission A permanent U.S.-Soviet commission established to implement and maintain the SALT agreements.

START Strategic Arms Reduction Talks.

tactical nuclear weapon A short-range, low-yield nuclear weapon designed for battlefield use.

telemetry Data transmitted from missiles by electronic means.

throw-weight The weight of a missile that can be placed on a target trajectory.

TTBT Threshold Test Ban Treaty; an unratified agreement that limits U.S. and Soviet nuclear tests to 150 kilotons.

verification The process of determining whether parties to an agreement are in compliance with their obligations.

vertical proliferation The increase in numbers of nuclear weapons possessed by nuclear weapon states.

warhead The explosive part of a missile, projectile, torpedo, or rocket.

yield The energy released in an explosion, usually expressed in terms of TNT equivalent.

Notes

Preface

1. Robert S. Norris, Thomas B. Cochran, and William M. Arkin, "History of the Nuclear Stockpile," *Bulletin of the Atomic Scientists* 41 (August 1985), 106.

Chapter 1. Roosevelt and the Manhattan Project, 1939–1945

1. John Purcell, *Best Kept Secret: The Story of the Atomic Bomb* (New York: 1963), 48–49. Marion Yass, *Hiroshima* (New York: 1972), 18–19.

2. Arthur H. Compton, *Atomic Quest: A Personal Narrative* (New York: 1956), 18.

3. Walter Smith Schoenberger, *Decision of Destiny* (Athens, Ohio: 1969), 4. Henry De Wolf Smyth, *Atomic Energy for Military Purposes: The Official Report on the Development of the Atomic Bomb under the Auspices of the United States Government, 1940–1945* (Princeton: 1945), 24–25. Leo Szilard, "We Turned the Switch," *The Nation* 161 (December 22, 1945), 718.

4. Schoenberger, 5.

5. Purcell, 5.

6. Stanley A. Blumberg and Gwin Owens, *Energy and Conflict: The Life and Times of Edward Teller* (New York: 1976), 94. Spencer R. Weart and Gertrud Weiss Szilard, eds., *Leo Szilard: His Version of the Facts: Selected Recollections and Correspondence* (Cambridge, Mass.: 1978), 81–84, 94–97.

7. Weart and Szilard, 104–105. Roosevelt is quoted by Sachs, U.S. Congress, Senate Special Committee on Atomic Energy, *Hearings on Atomic Energy*, 79th Cong., 1st sess., 1945, part 1, 9.

8. Martin J. Sherwin, *A World Destroyed: The Atomic Bomb and the Grand Alliance* (New York: 1975), 28. Schoenberger, 8. Smyth, 43–44, 47.

9. Schoenberger, 9–10. Smyth, 47–48, 50.

10. Sherwin, 36. Yass, 25. Margaret Gowing, *Britain and Atomic Energy* (New York: 1964), 389–436.

11. Smyth, 53–54. Compton, 63. Sherwin, 37.

12. James MacGregor Burns, *Roosevelt: The Soldier of Freedom, 1940–1945* (New York: 1970), 251. Yass, 34. Anthony Cave Brown and Charles B. Mac-Donald, eds., *The Secret History of the Atomic Bomb* (New York: 1977), 350–352.

13. Yass, 38–39. Blumberg and Owens, 122. Richard G. Hewlett and Oscar E. Anderson, Jr., *A History of the United States Atomic Energy Commission, Vol. I: The New World, 1939–1946* (University Park, Pa.: 1962), 109.

14. Smyth, 85, 141, 153. Leslie R. Groves, *Now It Can Be Told: The Story of the Manhattan Project* (New York: 1962), 94. Schoenberger 22. Yass, 42–43. James W. Kunetka, *Oppenheimer: The Years of Risk* (Englewood Cliffs, N.J.: 1982), 19. U.S. Atomic Energy Commission, *In the Matter of J. Robert Oppenheimer—Transcript of Hearing before Personnel Security Board*, Washington, D.C., April 12, 1954 to May 6, 1954 (Washington, D.C.: 1954), 9–10. (Hereafter cited as *Oppenheimer Hearing*.)

15. Sherwin, 59. Weart and Szilard, 164–179.

16. Harry S. Truman, *Memoirs, Vol. 1: Year of Decisions* (Garden City, N.Y.: 1955), 10–11. Yass, 46.

17. Schoenberger, 26. Brown and MacDonald, 256–263. Joe [*sic*] Martin, *My First Fifty Years in Politics* (New York: 1960), 101.

18. Schoenberger, 23–24. Margaret Gowing, "Reflections on Atomic Energy History," *Bulletin of the Atomic Scientists,* 35 (March 1979), 51–54.

19. Sherwin, 94, 96–97. Alice Kimball Smith, *A Peril and a Hope* (Chicago: 1965), 5–11.

20. Smith, 8. Sherwin, 100–105, 108–114.

21. Gowing, *Britain and Atomic Energy,* 335. Sherwin, 71–73, 107. Churchill is quoted in William L. Neumann, *After Vichy: Churchill, Roosevelt, Stalin, and the Making of the Peace* (New York: 1967), 89.

22. Sherwin, 85. Robert Dallek, *Franklin D. Roosevelt and American Foreign Policy, 1932–1945* (New York: 1979), 417. U.S. Department of State, *Foreign Relations of the United States, Diplomatic Papers: The Conferences at Washington and Quebec, 1943* (Washington: 1970), 1–11, 188, 209–11, 221–2, 630–53, 894, 1096–7, 1117–1119.

23. Sherwin, 125. Smith, 13–14. Bush is quoted in Sherwin, 115.

24. Sherwin, 127, 286–288.

25. Ibid., 133.

26. Henry L. Stimson Diary (unpublished MS; Sterling Memorial Library, Yale University, New Haven, Conn.), February 15, 1945.

27. Hewlett and Anderson, I, 338.

28. Sherwin, 108–114. U.S. Department of State, *Foreign Relations of the United States, Diplomatic Papers, 1941, Vol. I: General: The Soviet Union* (Washington, D.C.:1956), 363, 365–366.

29. U.S. Department of State, *Foreign Relations of the United States, The Conference at Quebec, 1944* (Washington, D.C.: 1972), 492–493. (Hereafter cited as *FRUS, The Quebec Conference, 1944*.) Stimson Diary, September 9, 1943. Sherwin, 102–103. *Oppenheimer Hearing*, 163–80, 258–81.

30. Dallek, 471–472. Sherwin, 109. Hewlett and Anderson, I, 338.

31. *FRUS, The Quebec Conference, 1944,* 492–493.

32. Stimson Diary, March 5, 1945.

33. Ibid., March 15, 1945. Sherwin, 139.

Chapter 2.   Truman, Hiroshima, and Nagasaki, 1945

1. Leslie R. Groves, *Now It Can Be Told: The Story of the Manhattan Project* (New York: 1962), 221–223. Marion Yass, *Hiroshima* (New York: 1972), 49. Samuel A. Goudsmit, *ALSOS* (New York: 1947), 66–72. Anthony Cave Brown and Charles B. MacDonald, eds., *The Secret History of the Atomic Bomb* (New York: 1977), 210–232.

2. U.S. Atomic Energy Commission, *In the Matter of J. Robert Oppenheimer—Transcript of Hearing before Personnel Security Board*, Washington, D.C., April 12, 1954 to May 6, 1954 (Washington, D.C.: 1954), 32–33. Martin J. Sherwin, *A World Destroyed: The Atomic Bomb and the Grand Alliance* (New York: 1975), 145. Peter Weyden, *Day One: Before Hiroshima and After* (New York: 1984), 86–88, 185–187.

3. Walter Smith Schoenberger, *Decision of Destiny* (Athens, Ohio: 1969), 44–45. Henry L. Stimson and McGeorge Bundy, *On Active Service in Peace and War* (New York: 1947), 613.

4. U.S. Department of State, *Foreign Relations of the United States, the Conference at Quebec, 1944* (Washington, D.C.: 1972), 492–493. Robert Jungk, *Brighter Than a Thousand Suns: A Personal History of the Atomic Scientists* (New York: 1958), 175.

5. Grace Tully, *F.D.R. My Boss* (New York: 1949), 266. See also U.S. Department of Defense, *The Entry of the Soviet Union into the War Against Japan: Military Plans 1941–1945* (Washington, D.C.: 1955).

6. Fletcher Knebel and Charles W. Bailey II, *No High Ground* (New York: 1960), 78–94. Groves to Marshall, December 30, 1944, U.S. Department of State, *Foreign Relations of the United States, Diplomatic Papers: The Conference at Malta and Yalta, 1945* (Washington, D.C.: 1955), 383–384.

7. Harry S. Truman, *Memoirs, Vol. I: Year of Decisions* (Garden City, N.Y.: 1955), 10. Robert J. Donovan, *Conflict and Crisis: The Presidency of Harry S. Truman, 1945–1948* (New York: 1977), 10–14. Schoenberger, 300. Sherwin, *A World Destroyed*, 148, 194–195, 291–294.

8. Sherwin, 152, 154. Lisle Rose, *The Coming of the American Age, 1945–1946, Vol. I: Dubious Victory: The United States and the End of World War II* (Kent, Ohio: 1973), 91–110.

9. Schoenberger, 209. Truman, I, 260, 78, 86.

10. Schoenberger, 213. Edward R. Stettinius, Jr., *Roosevelt and the Russians, The Yalta Conference* (Garden City, N.Y.: 1949), 93–94. Minutes of White House meeting, June 18, 1945, U.S. Department of State, *Foreign Relations of the United States, Diplomatic Papers: The Conference of Berlin (The Potsdam Conference), 1945* (Washington, D.C.: 1960), I, 905–906. (Hereafter cited as *FRUS, The Potsdam Conference.*) U.S. Department of Defense, *The Entry of the Soviet Union into the War Against Japan: Military Plans, 1941–1945* (Washington, D.C.: 1955), 54.

11. Henry L. Stimson Diary (unpublished MS, Sterling Memorial Library, Yale University, New Haven, Conn.), April 25, 1945. Martin J. Sherwin, "The Atomic Bomb and the Origins of the Cold War: U.S. Atomic Energy Policy and Diplomacy, 1941–1945," *American Historical Review* 78 (October 1973), 964. Truman, I, 87. Stimson Diary, May 14, 1945.

12. Truman, I, 87. *FRUS, The Potsdam Conference*, I, 3–4, 41–52. Robert

E. Sherwood, *Roosevelt and Hopkins: An Intimate History* (New York: 1948), 902. Truman, I, 262–270.

13. Stimson Diary, May 15 and June 6, 1945.

14. Schoenberger, 124–125. Sherwin, *A World Destroyed*, 295–304. Arthur H. Compton, *Atomic Quest: A Personal Narrative* (New York: 1956), 236.

15. Compton, 238–239.

16. Sherwin, *A World Destroyed*, 297–299.

17. Schoenberger, 134.

18. Truman, I, 419. Stimson Diary, June 6, 1945.

19. Stimson Diary, June 6, 1945.

20. Alice Kimball Smith, *A Peril and a Hope* (Chicago: 1965), 28. Spencer R. Weart and Gertrud Weiss Szilard, eds., *Leo Szilard: His Version of the Facts: Selected Recollections and Correspondence* (Cambridge, Mass.: 1978), 205–208.

21. Smith, *A Peril and a Hope*, 28. James F. Byrnes, *All in One Lifetime* (New York: 1958), 284. Alice Kimball Smith, "Behind the Decision to Use the Atomic Bomb, Chicago, 1944–45," *Bulletin of the Atomic Scientists* 14 (October 1958), 296.

22. Smith, *A Peril and a Hope*, 30–34, 560–572.

23. Schoenberger, 141. Sherwin, *A World Destroyed*, 304–305.

24. Morton Grodzins and Eugene Rabinowitch, eds., *The Atomic Age: Scientists in National and World Affairs* (New York: 1963), 28–29. Weart and Szilard, 182–188.

25. Smith, *A Peril and a Hope*, 55–56. Edward Teller with Allen Brown, *The Legacy of Hiroshima* (Garden City, N.Y.: 1962), 13–14.

26. Schoenberger, 181. Barton J. Bernstein, "Shatterer of Worlds: Hiroshima and Nagasaki," *Bulletin of the Atomic Scientists* 31 (December 1975), 15. Gar Alperovitz, *Atomic Diplomacy: Hiroshima and Potsdam* (New York: 1965), 236–242.

27. *FRUS, The Potsdam Conference*, I, 903–910.

28. Henry L. Stimson, "The Decision to Use the Atomic Bomb," *Harper's* 94 (February 1947), 101–102, 105–107. Stimson Diary, May 10, 1945.

29. William D. Leahy, *I Was There* (New York: 1950), 259. U.S. Strategic Bombing Survey, *Japan's Stuggle to End the War* (Washington, D.C.: 1946), 12–13. U.S. Strategic Bombing Survey, *Summary Report (Pacific War)* (Washington, D.C.: 1946).

30. U.S. Department of Defense, *The Entry of the Soviet Union into the War Against Japan*, 62–68. *FRUS, The Potsdam Conference*, I, 905–906, 910. Ernest J. King and W. M. Whitehill, *Fleet Admiral King: A Naval Record* (New York: 1952), 606.

31. Joseph C. Grew, *Turbulent Era: A Diplomatic Record of Forty Years, 1904–1945* (Boston: 1952), II, 1446, 1456–1459.

32. Truman, I, 314–315. *FRUS, The Potsdam Conference*, I, 909.

33. Schoenberger, 192, 199–201. Grew, II, 1408–1421.

34. Schoenberger, 163–164.

35. Togo Shigenori, *The Cause of Japan* (New York: 1956), 275–276. *FRUS, The Potsdam Conference*, II, 1249.

36. Hopkins to Truman, May 30, 1945, *FRUS, The Potsdam Conference*, I, 160. Lewis L. Strauss, *Men and Decisions* (Garden City, N.Y.: 1962), 189. *FRUS, The Potsdam Conference*, II, 1589. Schoenberger, 170–171.

37. Ellis M. Zacharias, *Secret Missions: The Story of an Intelligence Officer* (New York: 1946), 321–384, 399–424.

38. Herbert Feis, *The Atomic Bomb and the End of World War II*, rev. ed. (Princeton, N.J.: 1960), 19. Stimson Diary, June 26, 1945. *Public Papers of the Presidents of the United States: Harry S. Truman, 1945* (Washington, D.C.: 1961), 95–96, 98.

39. Samuel Eliot Morison, *History of United States Naval Operations in World War II, Vol. XIV: Victory in the Pacific* (Boston: 1960), 340. Truman, I, 417.

40. Brown and MacDonald, 509–517. Oppenheimer is quoted in Len Giovannitti and Fred Freed, *The Decision to Drop the Bomb* (New York: 1965), 194–195.

41. Groves to Stimson, July 18, 1945, *FRUS, The Potsdam Conference*, II, 1361–1368. Oppenheimer is quoted in Giovannitti and Freed, 197. Kistiakowsky and Groves are quoted in Jim Garrison, *The Plutonium Culture: From Hiroshima to Harrisburg* (New York: 1980), 21.

42. *FRUS, The Potsdam Conference*, II, 1360–1361. Stimson Diary, July 18, 1945. Churchill is quoted in the Stimson Diary, July 22, 1945.

43. Harvey H. Bundy, "Remembered Words," *The Atlantic Monthly* 199 (March 1957), 57. Winston S. Churchill, *Triumph and Tragedy* (Boston: 1953), 638. Stimson Diary, July 23, 1945. *FRUS, The Potsdam Conference*, II, 1324.

44. Stimson Diary, July 19, 1945. *FRUS, The Potsdam Conference*, II, 1155–1157.

45. Schoenberger, 262. Truman, I, 416.

46. Groves, 141, 144–145. Weyden, 217–220. Joseph I. Lieberman, *The Scorpion and the Tarantula: The Stuggle to Control Atomic Weapons, 1945–1949* (Boston: 1970), 191–193. David Holloway, *The Soviet Union and the Nuclear Arms Race* (New Haven, Conn.: 1983), 15–28.

47. Stalin is quoted in Herbert F. York, *The Advisors: Oppenheimer, Teller, and the Super-bomb* (San Francisco: 1976), 31.

48. James Byrnes, *Speaking Frankly* (New York: 1947), 207–208.

49. Harrison to Stimson, July 23, 1945, *FRUS, The Potsdam Conference*, II, 1374. Stimson Diary, July 23 and July 24, 1945.

50. Stimson to Truman, July 2, 1945, *FRUS, The Potsdam Conference*, I, 892. Cordell Hull, *Memoirs* (New York: 1948), II, 1594.

51. Schoenberger, 204. Churchill, 641–644.

52. Groves, 305–306. Schoenberger, 182. Giovannitti and Freed, 248.

53. Truman, I, 419. Groves is quoted in Giovannitti and Freed, 246, 248.

54. Truman, I, 419. Schoenberger, 260.

55. Eisenhower is quoted in Alperovitz, 236–242. *Newsweek*, November 11, 1963, 107. Dwight D. Eisenhower, *Crusade in Europe* (Garden City, N.Y.: 1948), 443.

56. Robert J. C. Butow, *Japan's Decision to Surrender* (Stanford: 1954), 142–149. Kase Toshikazu, *Journey to the Missouri* (New Haven, Conn.: 1950), 207–211. Togo, 312–314. Butow, 148–149.

57. Giovannitti and Freed, 259. Caron is quoted in Garrison, 25.

58. Giovannitti and Freed, 262. Truman, I, 421.

59. Seijo is quoted in Giovannitti and Freed, 267–268.

60. Joanne Silberner, "Hiroshima and Nagasaki: Thirty-six Years Later, the Struggle Continuing," *Science News* 120 (October 31, 1981), 284–287.

61. William Hardy McNeill, *America, Britain, and Russia: Their Cooperation and Conflict, 1941–1946* (London: 1953), 637. John R. Deane, *The Strange Alliance* (New York: 1947), 276–277.

62. Giovannitti and Freed, 273. Togo, 317.

63. Giovannitti and Freed, 273, 281. Groves, 341–342.
64. Butow, 158–173, 244. Kase, 231–239. Togo, 320.
65. Truman, I, 428. Butow, 245.
66. Togo, 323, 327, 334. Butow, 193–208. Kase, 252–253. Schoenberger, 273–274.

Chapter 3. Truman and International Control of the Atom, 1945–1947

1. Henry L. Stimson Diary (unpublished MS, Sterling Memorial Library, Yale University, New Haven, Conn.), September 4, 1945.
2. Ibid., September 12, 1945.
3. Ibid.
4. Ibid. Barton J. Bernstein, "The Quest for Security: American Foreign Policy and International Control of Atomic Energy, 1942–1946," *The Journal of American History* 60 (March 1974), 1017.
5. Stimson Diary, September 21, 1945.
6. Forrestal is quoted in Walter Millis, ed., *The Forrestal Diaries* (New York: 1951), 95–96. Harry S. Truman, *Memoirs, Vol. I: Year of Decisions* (Garden City, N.Y.: 1955), 525–527. Barton J. Bernstein, "American Foreign Policy and the Origins of the Cold War," in Barton J. Bernstein, ed., *Politics and Policies of the Truman Administration* (Chicago: 1972), 45.
7. Wallace is quoted in Stimson Diary, September 21, 1945. Forrestal is quoted in Millis, 95.
8. Bush is quoted in Richard G. Hewlett and Oscar E. Anderson, Jr., *A History of the United States Atomic Energy Commission, Vol. I: The New World, 1939–1946* (University Park, Pa.: 1962), 421.
9. Ibid., I, 112. Daniel Yergin, *Shattered Peace: The Origins of the Cold War and the National Security State* (Boston: 1977), 133, 136. Gregg Herken, *The Winning Weapon: The Atomic Bomb and the Cold War, 1945–1950* (New York: 1980), 128, 130.
10. Groves is quoted in Yergin, 136. Ibid.
11. Byrnes is quoted in Herken, 53. Smith and Truman are quoted in Yergin, 137.
12. Bernstein, "The Quest," 1021–1022. U.S. Department of State, *Foreign Relations of the United States: Diplomatic Papers, 1945, Vol. II: General, Political and Economic Matters* (Washington, D.C.: 1967), 55–57, 60, 62. (Hereafter cited as *FRUS, 1945.*) Byrnes is quoted in Hewlett and Anderson, I, 419.
13. John Lewis Gaddis, *The United States and the Origins of the Cold War, 1941–1947* (New York: 1972), 253. Herken, 32.
14. Hewlett and Anderson, I, 423. Joseph I. Lieberman, *The Scorpion and the Tarantula: The Struggle to Control Atomic Weapons, 1945–1949* (Boston: 1970), 159. Hewlett and Anderson, I, 423. Lawrence S. Wittner, *Rebels Against War: The American Peace Movement, 1941–1960* (New York: 1969), 147.
15. Attlee is quoted in Francis Williams, *A Prime Minister Remembers* (London: 1961), 97–101.
16. Lieberman, 157. Truman's message to Congress, October 3, 1945, *Public Papers of the Presidents of the United States: Harry S. Truman, 1945* (Washington, D.C.: 1961), 366. (Hereafter cited as *TPP, 1945.*)

17. Truman, I, 534. *TPP: 1945*, 381–382. Truman is quoted in Herken, 39.

18. *TPP, 1945*, 437.

19. Lieberman, 163. Bush to Byrnes, November 5, 1945, *FRUS, 1945*, II, 69–73.

20. Bush to Brynes, November 5, 1945, *FRUS, 1945*, II, 69–73.

21. Bernstein, "The Quest," 1023.

22. *FRUS, 1945*, II, 75–76.

23. Molotov is quoted in Bernstein, "The Quest," 1026. Lieberman, 407.

24. Bernstein, "The Quest," 1026. *FRUS, 1945*, II, 578, 581.

25. Herken, 69.

26. *FRUS, 1945*, II, 92–96.

27. Hewlett and Anderson, I, 472. Herken, 71.

28. *FRUS, 1945*, II, 96–97. Arthur H. Vandenberg, Jr., *The Private Papers of Senator Vandenberg* (Boston: 1952), 228. Tom Connally, *My Name is Tom Connally* (New York: 1954), 290.

29. Truman, I, 547–548. Connally, 290; italics are Connally's. Lieberman, 190–191, 205–206. Minutes of U.S. Delegation, December 20, 1945, *FRUS, 1945*, II, 698. Herken, 81–82.

30. *FRUS, 1945*, II, 612–613, 698, 700–741, 744–747, 762–763. Bernstein, "The Quest," 1028–1029.

31. Yergin, 148. Leahy is quoted in Lisle A. Rose, *After Yalta* (New York: 1973), 158.

32. Truman, I, 551. Herken, 90.

33. Jonathan Knight, "American Statecraft and the 1946 Black Sea Straits Controversy," *Political Science Quarterly* 90 (Fall 1975), 451–475. Gary R. Hess, "The Iranian Crisis of 1945–1946 and the Cold War," *Political Science Quarterly* 89 (March 1974), 117–146. Yergin, 183. Truman, I, 552.

34. Robert Messer, *The End of an Alliance: James F. Byrnes, Roosevelt, Truman, and the Origins of the Cold War* (Chapel Hill, N.C.: 1982), 28–30.

35. Herken, 92.

36. Ibid., 138.

37. Kennan is quoted in Thomas H. Etzold and John Lewis Gaddis, eds., *Containment: Documents on American Foreign Policy and Strategy: 1945–1950* (New York: 1978), 50–63.

38. Churchill is quoted in Joseph P. Morray, *From Yalta to Disarmament: Cold War Debate* (Westport, Conn.: 1961), 43–50.

39. Stalin is quoted in Morray, 50. Walter LaFeber, *America, Russia and the Cold War*, 3rd ed. (New York: 1976), 31–32.

40. Herken, 115. Hewlett and Anderson, I, 428–530.

41. Dean Acheson, *Present at the Creation: My Years in the State Department* (New York: 1969), 124. Alice Kimball Smith, *A Peril and a Hope* (Chicago: 1965), 211–219. Herken, 119.

42. Herken, 124.

43. Smith, 373–388. Herken, 129–130.

44. Smith, 388–436. Herken, 131–135.

45. Wallace is quoted in John Morton Blum, *The Price of Vision: The Diary of Henry A. Wallace* (Boston: 1973), 569. Herken, 133.

46. Hewlett and Anderson, I, 532–533.

47. Lieberman, 235.

48. Ibid., 239. Bernstein, "The Quest," 1029–1030.

49. Lilienthal is quoted in Lieberman, 242. Ibid., 242–246.

50. Memorandum by Oppenheimer, February 2, 1946, U.S. Department of State, *Foreign Relations of the United States, Diplomatic Papers, 1946, Vol. I: General: The United Nations* (Washington, D.C.:1969), 749–754. (Hereafter cited as *FRUS, 1946.*)

51. U.S. Department of State, *A Report on the International Control of Atomic Energy* (Washington, D.C.: 1946). Hewlett and Anderson, I, 540–554. Lieberman, 243–246. D. F. Fleming, *The Cold War and Its Origins, 1917–1960* (Garden City, N.Y.: 1961), I, 364.

52. Bernstein, "The Quest," 1030–1031. P. M. S. Blackett, *Fear, War and the Bomb: Military and Political Consequences of Atomic Energy* (New York: 1949), 149–150.

53. Bernstein, "The Quest," 1031. Lieberman, 409.

54. Bernstein, "The Quest," 1033. Baruch to Truman, March 26, 1946, *FRUS, 1946,* I, 767–768.

55. David E. Lilienthal, *Journals of David E. Lilienthal, Vol. II: The Atomic Energy Years, 1945–1950* (New York: 1964), 30. Baruch is quoted in Yergin, 238. Oppenheimer is quoted in Lilienthal, II, 42–43.

56. Lieberman, 410. Bernstein, "The Quest," 1032–1033.

57. Hewlett and Anderson, I, 577–578. *FRUS, 1946,* I, 827–833.

58. Bernstein, "The Quest," 1034. Acheson, 155. Lilienthal, II, 42.

59. Nimitz to Baruch, *FRUS, 1946,* I, 853–854. Lieberman, 289.

60. Herken, 168.

61. Truman to Baruch, June 7, 1946, *FRUS, 1946,* I, 846–851. Herken, 169.

62. U.S. Department of State, *Documents on Disarmament, 1945–1959* (Washington, D.C.: 1960), I, 7–16. Bernstein, "The Quest," 1037. *Pravda* is quoted in Bernstein, "The Quest," 1039.

63. Bernstein, "The Quest," 1038. Hewlett and Anderson, I, 581.

64. *Documents on Disarmament,* I, 17–24. Lieberman, 309.

65. Herken, 174–178. Gromyko is quoted in ibid., 178.

66. Baruch to Byrnes, November 4, 1946, *FRUS, 1946,* I, 990–992. Bernstein, "The Quest," 1043–1044.

Chapter 4. Truman, The Cold War, and the Hydrogen Bomb, 1947–1952

1. Richard J. Barnet, *Intervention and Revolution: America's Confrontation with Insurgent Movements Around the World* (New York: 1968), 97–101. Lawrence S. Wittner, *American Intervention in Greece, 1943–1949* (New York: 1982), 1–69.

2. Ethridge to Marshall, February 17, 1947, U.S. Department of State, *Foreign Relations of the United States, Diplomatic Papers, 1947, Vol. V: The Near East and Africa* (Washington, D.C.: 1971), 23–45. (Hereafter cited as *FRUS, 1947.*) Kennan is quoted in Thomas H. Etzold and John Lewis Gaddis, eds., *Containment: Documents on American Foreign Policy and Strategy: 1945–1950* (New York: 1978), 86–89.

3. *Public Papers of the Presidents of the United States: Harry S. Truman, 1947* (Washington, D.C.: 1963), 178–179. Vandenberg is quoted in Walter LaFeber, *America, Russia and the Cold War,* 3rd ed. (New York: 1976), 45.

4. Thomas G. Paterson, "The Quest for Peace and Prosperity: International

Trade, Communism, and the Marshall Plan," in Barton J. Bernstein, ed., *Politics and Policies of the Truman Administration* (Chicago: 1972), 78–81, 90, 97–98. John Gimbell, *The Origins of the Marshall Plan* (Stanford, Calif.: 1976), 267–280.

5. D. F. Fleming, *The Cold War and its Origins, 1917–1960* (Garden City, N.Y.: 1961), I, 481. Adam B. Ulam, *Expansion and Coexistence: Soviet Foreign Policy, 1917–1973*, 2nd ed. (New York: 1978), 455. Paterson, "The Marshall Plan," 168.

6. Fleming, I, 504. Lloyd C. Gardner, "America and the German Problem, 1945–1949," in Bernstein, *The Truman Administration*, 113–148. Daniel Yergin, *Shattered Peace: The Origins of the Cold War and the National Security State* (Boston: 1977), 227. John Lewis Gaddis, *The United States and the Origins of the Cold War, 1941–1947* (New York: 1972), 326–327.

7. Yergin, 231, 298, 318, 331, 333–334, 372. Gaddis, 327–332. John H. Backer, *The Decision to Divide Germany: American Foreign Policy in Transition* (Durham, N.C.: 1978), 172–173.

8. Gardner, 142–144. Yergin, 368–369, 376–377. Fleming, I, 506.

9. Gardner, 144–145. Yergin, 378–387.

10. Yergin, 385, 395–396.

11. Gregg Herken, *The Winning Weapon: The Atomic Bomb and the Cold War, 1945–1950* (New York: 1980), 217. Etzold and Gaddis, 173–177.

12. U.S. Department of Defense, *History of the Office of the Secretary of Defense*, Alfred Goldberg, gen. ed.; Vol. I: Steven L. Rearden, *The Formative Years, 1947–1950* (Washington, D.C.: 1984), 13–14.

13. Samuel Huntington, *The Common Defense: Strategic Programs in National Politics* (New York: 1961), 369–381. President's Air Policy Commission, *Survival in the Air Age: A Report by the President's Air Policy Commission* (Washington, D.C.: 1948), 10–12, 133. Rearden, 313–316. Yergin, 268, 342, 360–361.

14. Herken, 271.

15. Eztold and Gaddis, 339–343.

16. David Alan Rosenberg, "U.S. Nuclear Stockpile, 1945–1950," *Bulletin of the Atomic Scientists* 38 (May 1982), 26. Herken, 196–198. Rearden, 439–440.

17. Rosenberg, "U.S. Nuclear Stockpile," 26, 28. Herken, 197, 239–240. Rearden, 440–441.

18. Rosenberg, "U.S. Nuclear Stockpile," 71–72. David E. Lilienthal, *Journals of David E. Lilienthal, Vol. II: The Atomic Energy Years, 1945–1950* (New York: 1964), 337, 270. Herken, 288–289.

19. Truman to the National Security Council, July 26, 1949, U.S. Department of State, *Foreign Relations of the United States, Diplomatic Papers, 1949, Vol. I: National Security Affairs, Foreign Economic Policy* (Washington, D.C.: 1976), 501–503. (Hereafter cited as *FRUS, 1949.*) Herken, 291. Rearden, 442–446.

20. Gallery is quoted in David Alan Rosenberg, "American Atomic Strategy and the Hydrogen Bomb Decision," *Journal of American History* 66 (June 1979), 70. *FRUS, 1949*, I, 505–507.

21. Herken, 292.

22. *Survival in the Air Age*, 14. Richard G. Hewlett and Francis Duncan, *A History of the United States Atomic Energy Commission, Vol. II: Atomic Shield, 1947–1952* (University Park, Pa.: 1969), 362–369. Herbert F. York, *The Advi-*

sors: Oppenheimer, Teller, and the Super-bomb (San Francisco: 1976), 34. David Holloway, *The Soviet Union and the Arms Race* (New Haven, Conn.: 1983), 20–28.

23. Lilienthal, II, 580, 577. Herken, 303–305.

24. York, 21. Edward Teller, "The Work of Many People," *Science* 121 (February 25, 1955), 268.

25. Edward Teller with Allen Brown, *The Legacy of Hiroshima* (Garden City, N.Y.: 1962), 50. Bradbury is quoted in U.S. Atomic Energy Commission, *In the Matter of J. Robert Oppenheimer—Transcript of Hearing before Personnel Security Board, April 12, 1954 to May 6, 1954* (Washington, D.C.: 1954), 487. (Hereafter cited as *Oppenheimer Hearing*). Bethe is quoted in Robert C. Williams and Philip L. Cantelon, eds., *The American Atom: A Documentary History of Nuclear Policies from the Discovery of Fission to the Present, 1939–1984* (Philadelphia: 1985), 134.

26. York, 63. Oppenheimer is quoted in Stanley A. Blumberg and Gwinn Owens, *Energy and Conflict: The Life and Times of Edward Teller* (New York: 1976), 207.

27. Hewlett and Duncan, 392–394. York, 63. McMahon to Truman, November 21, 1949, *FRUS, 1949*, I, 588–595.

28. Memorandum by the Joint Chiefs of Staff, November 23, 1949, *FRUS, 1949*, I, 595–596. Lilienthal, II, 580–581. Hewlett and Duncan, 393–399. Memorandum by Joint Chiefs of Staff, January 13, 1950, Department of State, *Foreign Relations of the United States, 1950, Vol. I: National Security Affairs, Foreign Economic Policy* (Washington, D.C.: 1977), 503–511. (Hereafter cited as *FRUS, 1950*.)

29. Williams and Cantelon, 117–120. Norman Moss, *Men Who Play God: The Story of the H-Bomb and How the World Came to Live with It* (New York: 1968), 21. York, 150–157. Herken, 307. *Oppenheimer Hearing*, 513.

30. McMahon to Truman, November 21, 1949, *FRUS, 1949*, I, 588–595. Oppenheimer is quoted in York, 55.

31. Statement appended to the Report of the General Advisory Committee, October 30, 1949, *FRUS, 1949*, I, 572–573. McGeorge Bundy, "The Missed Chance to Stop the H-Bomb," *New York Review of Books* 29 (May 13, 1982), 13–21.

32. Lilienthal, II, 591, 614. Johnson is quoted in York, 66. York, 67.

33. Lilienthal, II, 623–632. *FRUS, 1950*, I, 513–523.

34. Lilienthal, II, 632.

35. Bundy, 14. Lilienthal, II, 633. *FRUS, 1950*, I, 141–142. Rearden, 453–456.

36. Herken, 320.

37. Bundy, 13. Oppenheimer is quoted in York, 72–73.

38. Athan Theoharis, *Seeds of Repression: Harry S. Truman and the Origins of McCarthyism* (Chicago: 1971), 150–151.

39. Herken, 322. *FRUS, 1950*, I, 173, 524. John Major, *The Oppenheimer Hearing* (New York: 1971), 128. Blumberg and Owens, 228. York, 69. Ronald Radosh and Joyce Milton, *The Rosenberg File: A Search for the Truth* (New York: 1983), 5–19. Nancy Smith and Barton Bernstein, "Truman and the H-Bomb," *Bulletin of the Atomic Scientists* 84 (October 1984), 44–47.

40. Theoharis, *Seeds of Repression*, viii.

41. Truman is quoted in ibid., 153, and Athan Theoharis, "The Rhetoric of

Politics: Foreign Policy, Internal Security, and Domestic Politics in the Truman Era, 1945–1950," in Bernstein, *The Truman Administration*, 23.

42. Etzold and Gaddis, 385–442.

43. Bradley is quoted in William Manchester, *American Caesar: Douglas MacArthur, 1880–1964* (Boston: 1978), 627.

44. Truman is quoted in Herken, 333–334. Hewlett and Duncan, 522, 574.

45. Blumberg and Owens, 292. Roland Sawyer, "The H-Bomb Chronology," *Bulletin of the Atomic Scientists* 10 (September 1954), 290, 300. Teller, "The Work of Many People," 272–273.

46. York, 82. Blumberg and Owens, 295.

Chapter 5. Eisenhower and Massive Retaliation, 1953–1961

1. Robert A. Divine, *Eisenhower and the Cold War* (New York: 1981), 11. Edgar M. Bottome, *The Balance of Terror: A Guide to the Arms Race* (Boston: 1971), 72–73.

2. Richard H. Rovere, *Senator Joe McCarthy* (New York: 1959), 17–18, 32–33. Charles C. Alexander, *Holding the Line: The Eisenhower Era, 1952–1961* (Bloomington, Ind.: 1975), 55–57. Stephen E. Ambrose, *Eisenhower, Vol. II: The President* (New York: 1984), 55–66.

3. Walter LaFeber, *America, Russia and the Cold War*, 3rd ed. (New York: 1976), 182–183. Lawrence S. Wittner, *Cold War America: From Hiroshima to Watergate* (New York: 1974), 177. Alexander, 56–57.

4. Michael A. Guhin, *John Foster Dulles: A Statesman and His Times* (New York: 1972), 131. Dulles is quoted in Andrew H. Berding, *Dulles on Diplomacy* (Princeton, N.J.: 1965), 24.

5. Address by Dulles, April 23, 1956, *Department of State Bulletin* 34 (April 30, 1956), 708.

6. Guhin, 129–158. John Lewis Gaddis, *Strategies of Containment: A Critical Appraisal of Postwar American National Security Policy* (New York: 1982), 136–145, 161–163.

7. Dwight D. Eisenhower, *The White House Years: Mandate for Change, 1953–1956* (Garden City, N.Y.: 1963), 181.

8. U.S. Department of State, *Foreign Relations of the United States, 1952–1954, Vol. XV: Korea* (Washington, D.C.: 1984), part 1, 1068–1069. Divine, 19–20, 27, 29, 36–37. David Rees, *Korea: The Limited War* (New York: 1964), 402–420. Alexander L. George and Richard Smoke, *Deterrence in American Foreign Policy: Theory and Practice* (New York: 1977), 235–241. Ambrose, 51–52. Eisenhower's press conference, April 30, 1953, *Public Papers of the Presidents of the United States: Dwight D. Eisenhower, 1953* (Washington, D.C.: 1960), 209, 239. (Hereafter cited as *EPP, 1953.*) Gaddis, 134. Samuel P. Huntington, *The Common Defense: Strategic Programs in National Politics* (New York: 1961), 66. Douglas Kinnard, *President Eisenhower and Strategy Management: A Study in Defense Politics* (Lexington, Ky.: 1977), 127. Bottome, *The Balance of Terror*, 21. Wittner, 145.

9. Address by Dulles, January 12, 1954, *Department of State Bulletin*, 30 (January 25, 1954), 108. Eisenhower is quoted in Gaddis, 149–150. Guhin, 221–239.

10. Dulles is quoted in James Shepley, "How Dulles Averted War," *Life* 40 (January 16, 1956), 78.

11. Twining is quoted in Alexander, 79. NSC Action No. 1074-a, April 5, 1954, *The Pentagon Papers*, Senator Gravel Edition (Boston: 1971), I, 469–470. Gaddis, 169. Divine, 39–51. Guhin, 239–251. Ambrose, 173–185.

12. Dulles is quoted in Eisenhower, *Mandate for Change*, 476. George and Smoke, 363–389. Gaddis, 170. Jerome H. Kahan, *Security in the Nuclear Age: Developing U.S. Strategic Arms Policy* (Washington, D.C.: 1975), 22–23. Ambrose, 212–214, 231–245.

13. Huntington, 75–76. Russell F. Weigley, *The American Way of War: A History of United States Military Strategy and Policy* (New York: 1973), 401. Divine, 37. Glenn H. Snyder, "The New Look," in Warner R. Schilling, Paul Y. Hammond, and Glenn H. Snyder, *Strategy, Politics, and Defense Budgets* (New York: 1962), 379–524. Eisenhower, *Mandate for Change*, 445–458.

14. Herbert York, *Race to Oblivion: A Participant's View of the Arms Race* (New York: 1970), 86–89, 94–97.

15. Ibid., 98–102.

16. Dulles is quoted in Huntington, 80. Gaddis, 167–168.

17. Henry A. Kissinger, *Nuclear Weapons and Foreign Policy* (New York: 1957), 185. Weigley, 414–417.

18. Weigley, 411–424.

19. George and Smoke, 295–308.

20. Dulles is quoted in Snyder, "The New Look," 465–466. Gaddis, 173.

21. Eisenhower is quoted in Gaddis, 173–174 and in George B. Kistiakowsky, *A Scientist in the White House: The Private Diary of President's Eisenhower's Special Assistant for Science and Technology* (Cambridge, Mass.: 1976), 400.

22. Khrushchev is quoted in Chalmers Roberts, *The Nuclear Years: The Arms Race and Arms Control, 1945–1970* (New York: 1970), 41. Bottome, *The Balance of Terror*, 29. Kahan, 48.

23. Colin S. Gray, " 'Gap' Predictions and America's Defense: Arms Race Behavior in the Eisenhower Years," *Orbis* 16 (Spring 1972), 261. Allen Dulles, *The Craft of Intelligence* (New York: 1963), 149. John Prados, *The Soviet Estimate: U.S. Intelligence Analysis and Russian Military Strength* (New York: 1982), 38–43. Bottome, *The Balance of Terror*, 36. Edgar M. Bottome, *The Missile Gap: A Study of the Formulation of Military and Political Policy* (Rutherford, N.J.: 1971), 58. Arnold L. Horelick and Myron Rush, *Strategic Power and Soviet Foreign Policy* (Chicago: 1966), 27–31. Lincoln P. Bloomfield, Walter C. Clemens, Jr., and Franklyn Griffiths, *Khrushchev and the Arms Race: Soviet Interests in Arms Control and Disarmament, 1954–1964* (Cambridge, Mass.: 1966), 66–67.

24. Gray, 262. Bottome, *The Balance of Terror*, 35–36. Horelick and Rush, 18.

25. Bloomfield et al., 50–58. Kahan, 50. Dulles, 163.

26. Horelick and Rush, 31. Bloomfield et al., 90–99.

27. Bissel is quoted in Charles J. V. Murphy, "Khrushchev's Paper Bear," *Fortune* 70 (December 1970), 227. Dulles, 163–164. Address by Eisenhower, October 6, 1953, *EPP, 1953*, 635. Gray, 262. Kahan, 29–30. Bottome, *The Balance of Terror*, 37. Bottome, *The Missile Gap*, 74–75.

28. Kahan, 29, 32–33. Huntington, 88–105.

29. D. F. Fleming, *The Cold War and Its Origins, 1917–1960* (Garden City,

N.Y.: 1961), II, 885. Horelick and Rush, 42–45. York, 107. Alexander, 210–214, 218. Richard A. Aliano, *American Defense Policy from Eisenhower to Kennedy: The Politics of Changing Military Requirements, 1957–1961* (Athens, Ohio: 1975), 47–49. Prados, 51–57.

30. Murrow is quoted in Fleming, II, 886–887. Eisenhower is quoted in James R. Killian, Jr., *Sputniks, Scientists, and Eisenhower: A Memoir of the First Special Assistant to the President for Science and Technology* (Cambridge, Mass.: 1977), 10.

31. York, 110. Dwight D. Eisenhower, *The White House Years: Waging Peace, 1956–1961* (Garden City, N.Y.: 1965), 209. Aliano, 47. William A. McDougall, "Sputnik, the Space Race, and the Cold War," *Bulletin of the Atomic Scientists* 40 (May 1985), 20–25.

32. Morton H. Halperin, "The Gaither Committee and the Policy Process," *World Politics* 13 (April 1961), 360–384. Huntington, 106–107. Alexander, 227–229. Bottome, *The Missile Gap*, 45–47. Kahan, 40–41. Prados, 71.

33. Eisenhower, *Waging Peace*, 221. Killian, 96–100. Huntington, 111. Kahan, 43. Halperin, 369–371, 379–380. Gaddis, 185. Bottome, *The Balance of Terror*, 43–44. Dulles, 151–154. Bottome, *The Missile Gap*, 178. James C. Dick, "The Strategic Arms Race, 1957–1961: Who Opened a Missile Gap?" *Journal of Politics* 34 (November 1972), 1073–1074.

34. Desmond Ball, *Politics and Force Levels: The Strategic Missile Program of the Kennedy Administration* (Berkeley, Calif.: 1980), 6. Glenn H. Snyder, *Deterrence and Defense: Toward a Theory of National Security* (Princeton, N.J.: 1961), 69. Bottome, *The Missile Gap*, 37, 57–58, 221–234.

35. Ball, 53–58. Prados, 35–36.

36. Dulles, 165. Kahan, 47–53. Horelick and Rush, 36-37, 60–61, 67–70, 105, 107–111. Bottome, *The Missile Gap*, 173–175, 183. Theodore C. Sorensen, *Kennedy* (New York: 1965), 611. Ball, 54.

37. Horelick and Rush, 37, 41, 67, 69. Kahan, 51. Bottome, *The Missile Gap*, 75–77, 173. Bottome, *The Balance of Terror*, 42–43.

38. Khrushchev to Eisenhower, September 7, 1958, U.S. Department of State, *Documents on Disarmament, 1945–1959* (Washington, D.C.: 1960), II. 1121–1125. Khrushchev is quoted in Horelick and Rush, 49.

39. Eisenhower, *Waging Peace*, 336, 342. Kahan, 25.

40. Kahan, 53. Bottome, *The Missile Gap*, 175, 212. Horelick and Rush, 101–102, 127. Bloomfield et al., 90–99.

41. Bottome, *The Missile Gap*, 43, 52–56, 69, 107–110, 157–159. Horelick and Rush, 51. Kahan, 41. Roy E. Licklider, "The Missile Gap Controversy," *Political Science Quarterly* 85 (December 1970), 604–607. Huntington, 384–398. Bottome, *The Balance of Terror*, 59. Aliano, 53–56, 100–140.

42. Licklider, 607. Bottome, *The Missile Gap*, 34, 37–38, 97. Bottome, *The Balance of Terror*, 47.

43. Michael H. Armacost, *The Politics of Weapons Innovation: The Thor-Jupiter Controversy* (New York: 1969), 99. Fred J. Cook, *The Warfare State* (New York: 1962), 22–23.

44. Eisenhower is quoted in Cook, 9 and in Robert L. Branyon and Lawrence H. Lassen, *The Eisenhower Administration, 1953–1961* (New York: 1971), II, 1375.

45. Aliano, 57.

46. Bottome, *The Balance of Terror*, 60. Aliano, 146–174.

47. Johnson is quoted in Bottome, *The Missile Gap*, 52. Kennedy is quoted in Bottome, *The Missile Gap*, 50. Ball, 15.

48. Ball, 15–25, 98, 100–102. Bottome, *The Missile Gap*, 165–168, 203–204. Horelick and Rush, 83–84.

49. Bottome, *The Missile Gap*, 93. Licklider, 605. Ball, 44–45. Horelick and Rush, 46. Ball, 44–46.

50. Alexander, 217. Armacost, 195–196, 199. Harold Karan Jacobsen and Eric Stein, *Diplomats, Scientists, and Politicians: The United States and the Nuclear Test Ban Negotiations* (Ann Arbor: 1966), 36–37. John Simpson, *The Independent Nuclear State: The United States, Britain and the Military Atom* (New York: 1983), 138–141.

Chapter 6. Eisenhower and Nuclear Arms Control, 1953–1961

1. U.S. Department of State, *Documents on Disarmament, 1945–1959* (Washington, D.C.: 1960), I, 393–400. (Hereafter cited as *DD, 1945–1959*.) Lewis L. Strauss, *Men and Decisions* (New York: 1963), 336. Dwight D. Eisenhower, *The White House Years: Waging Peace, 1956–1961* (New York: 1965), 468.

2. *DD, 1945–1959*, I, 393–400. Stephen E. Ambrose, *Eisenhower, Vol. II: The President* (New York: 1984), 149.

3. Robert A. Divine, *Eisenhower and the Cold War* (New York: 1981), 113. Robert J. Donovan, *Eisenhower: The Inside Story* (New York: 1956), 191–197. Dwight D. Eisenhower, *The White House Years: Mandate for Change, 1953–1956* (New York: 1963), 253–254. Gerard H. Clarfield and William M. Wiecek, *Nuclear America: Military and Civilian Nuclear Power in the United States, 1940–1980* (New York: 1984), 185–188.

4. Robert A. Divine, *Blowing on the Wind: The Nuclear Test Ban Debate, 1954–1960* (New York: 1978), 3–4, 25. Stanley A. Blumberg and Gwinn Owens, *Energy and Conflict: The Life and Times of Edward Teller* (New York: 1976), 382. Glenn T. Seaborg, *Kennedy, Khrushchev and the Test Ban* (Berkeley, Calif.: 1981), 3–4.

5. Ralph E. Lapp, *The Voyage of the Lucky Dragon* (New York: 1958), 27–44. Divine, *Blowing on the Wind*, 4–8.

6. Seaborg, 4. *Public Papers of the Presidents of the United States: Dwight D. Eisenhower, 1954* (Washington, D.C.: 1960), 342–346. Strauss's testimony appears in U.S. Congress, Joint Committee on Atomic Energy, *Hearings: Health and Safety Problems and Weather Effects Associated with Atomic Explosions*, 84th Cong., 1st sess., 1955, 3. (Hereafter cited as *Hearings: Health and Safety Problems*.)

7. *U.S. News and World Report* 38 (February 25, 1955), 128–130, 132, 134. Lewis L. Strauss, "The Truth About Radioactive Fallout," ibid., 35–36, 38.

8. Ralph E. Lapp, "Civil Defense Faces New Peril," *Bulletin of the Atomic Scientists* 10 (November 1954), 349–51, and "Fallout and Candor," ibid. 11 (May 1955), 170, 220. Divine, *Blowing on the Wind*, 51–53.

9. *Hearings: Health and Safety Problems*, 1–8. Strauss, "The Truth About Radioactive Fallout," 38.

10. *DD, 1945–1959*, I, 414–423.

11. Ibid., I, 456–467.

12. Ibid.

13. Ibid. Lincoln P. Bloomfield, Walter C. Clemens, Jr., and Franklyn Griffiths, *Khrushchev and the Arms Race: Soviet Interests in Arms Control and Disarmament, 1954–1964* (Cambridge, Mass.: 1966), 24.

14. Bloomfield et al., 17–25.

15. Ibid., 33–49, 51, 54–55, 73–75. Edgar M. Bottome, *The Balance of Terror: A Guide to the Arms Race* (Boston: 1971), 38. David Holloway, *The Soviet Union and the Arms Race* (New Haven, Conn.: 1983), 29–43. Adam B. Ulam, *Expansion and Coexistence: Soviet Foreign Policy, 1917–1973*, 2nd ed. (New York: 1974), 572–576.

16. Bloomfield et al., 45–56, 70–71.

17. Ibid., 19, 25.

18. D. F. Fleming, *The Cold War and Its Origins: 1917–1960* (Garden City, N.Y.: 1961), II, 737, 748.

19. Bloomfield et al., 25–26. John H. Barton and Lawrence D. Weiler, eds., *International Arms Control: Issues and Agreements* (Stanford, Calif.: 1976), 77. *DD, 1945–1959*, I, 481–485.

20. *DD, 1945–1959*, I, 475–480, 489–492, 498, 486–488, 501–503, 523–528. Eisenhower, *Mandate for Change*, 516.

21. Eisenhower, *Mandate for Change*, 516–519. Robert Gilpin, *American Scientists and Nuclear Weapons Policy* (Princeton, N.J., 1962), 162–164. Bernard G. Bechhoefer, *Postwar Negotiations for Arms Control* (Washington, D.C.: 1961), 302–305.

22. Address by Bulganin, August 4, 1955, *DD, 1945–1959*, I, 496. Bulganin to Eisenhower, September 11, 1956, ibid., I, 688–694. Jerome H. Kahan, *Security in the Nuclear Age: Developing U.S. Strategic Arms Policy* (Washington, D.C.: 1975), 56–57. Khrushchev is quoted in Charles E. Bohlen, *Witness to History, 1929–1969* (New York: 1973), 384.

23. *DD, 1945–1959*, I, 593–595, 603–607. Barton and Weiler, 78.

24. Divine, *Blowing on the Wind*, 24.

25. Ibid. Divine, *Eisenhower and the Cold War*, 121. Memorandum on Disarmament Negotiations, *Department of State Bulletin* 35 (November 5, 1956), 712.

26. Divine, *Blowing on the Wind*, 72, 100, 109. John Bartlow Martin, *Adlai Stevenson and the World: The Life of Adlai E. Stevenson* (Garden City, N.Y.: 1977), 308–314, 365–379. Memorandum on Disarmament Negotiations, *Department of State Bulletin* 35 (November 5, 1956), 704–715.

27. Eisenhower is quoted in J. Emmett Hughes, *Ordeal in Power: A Political Memoir of the Eisenhower Years* (New York: 1963), 203. Divine, *Blowing on the Wind*, 118. Ambrose, 350.

28. Khrushchev is quoted in Divine, *Blowing on the Wind*, 120.

29. Ibid., 121, 128. Seaborg, 8. Ralph E. Lapp, *The New Priesthood: The Scientific Elite and the Uses of Power* (New York: 1965), 131–135. George Gallup, *The Gallup Poll: Public Opinion: 1935–1971* (New York: 1972), II, 1487–1488.

30. Divine, *Blowing on the Wind*, 139–140, 143. Sherman Adams, *First Hand Report: The Story of the Eisenhower Administration* (New York: 1961), 326.

31. *DD, 1945–1959*, II, 763–770, 773–774.

32. Radford is quoted in *Time* 69 (June 13, 1957), 38. Eisenhower is quoted in Adams, 326. Divine, *Blowing on the Wind*, 144–145.

33. Harold Macmillan, *Riding the Storm: 1956–1959* (New York: 1971), 300–303. Divine, *Blowing on the Wind*, 144–146.

34. *DD, 1945–1959*, II, 791. *Public Papers of the Presidents of the United States: Dwight D. Eisenhower, 1957* (Washington, D.C.: 1958), 476. (Hereafter cited as *EPP, 1957*.)

35. Divine, *Blowing on the Wind*, 147–149. Divine, *Eisenhower and the Cold War*, 126. Eisenhower's news conference, June 26, 1957, *EPP, 1957*, 498–499.

36. *The New Republic* 137 (July 15, 1957), 3. Divine, *Blowing on the Wind*, 150–152.

37. Divine, *Blowing on the Wind*, 152–153. Dulles' news conference, June 25, 1957, *Department of State Bulletin* 37 (July 15, 1957), 99. Statement by Eisenhower, August 21, 1957, *EPP, 1957*, 627. Eisenhower, *Waging Peace*, 474. *DD, 1945–1959*, II, 845–848.

38. *DD, 1945–1959*, II, 849–868.

39. Divine, *Blowing on the Wind*, 159. Harold Karan Jacobsen and Eric Stein, *Diplomats, Scientists, and Politicians: The United States and the Nuclear Test Ban Negotiations* (Ann Arbor: 1966), 147–151.

40. Divine, *Blowing on the Wind*, 160–161.

41. Nevil Shute, *On the Beach* (New York: 1957), 94. Divine, *Blowing on the Wind*, 162–163.

42. *DD, 1945–1959*, II, 918–926, 932–941. Jacobsen and Stein, 40–41.

43. Divine, *Blowing on the Wind*, 177–178. Samuel P. Huntington, *The Common Defense: Stategic Programs in National Politics* (New York: 1961), 362.

44. U.S. Congress, Subcommittee of the Senate Foreign Relations Committee, *Hearings: Control and Reduction of Armaments*, 84th Cong., 2nd sess., and 85th Cong., 1st and 2nd sess., 1956–1958, 1343.

45. Jacobsen and Stein, 46–48. Divine, *Blowing on the Wind*, 188–192.

46. Divine, *Blowing on the Wind*, 199, 232–235.

47. *DD, 1945–1959*, II, 978–980, 982–985, 1017–1018. Seaborg, *11*.

48. *DD, 1945–1959*, II, 1036–1041. U.S. Disarmament Administration, *Geneva Conference on the Discontinuance of Nuclear Weapons Tests, History and Analysis of Negotiations* (Washington, D.C.: 1961), 19, 271–310. (Hereafter cited as *Geneva Conference*.) Seaborg, 12–13. Gilpin, 186–195, 201–222. Jacobsen and Stein, 54–81.

49. *Geneva Conference*, 310–311, 19–20. Seaborg, 14. *DD, 1945–1959*, II, 1114–1120, 1142–1143, 1153–1171.

50. Divine, *Blowing on the Wind*, 232–234. Clarfield and Wiecek, 201–229.

51. Divine, *Blowing on the Wind*, 238. Jacobsen and Stein, 100. U.S. Department of Defense, Samuel Glasstone, ed., *The Effects of Nuclear Weapons*, rev. ed. (Washington, D.C.: 1962), 671–681.

52. *Geneva Conference*, 22, 29, 31–46. Gilpin, 229.

53. Seaborg, 17. *Geneva Conference*, 331–334. Jacobsen and Stein, 147–151.

54. Wadsworth is quoted in James R. Killian, Jr., *Sputnik, Scientists, and Eisenhower: A Memoir of the First Special Assistant to the President for Science and Technology* (Cambridge, Mass.: 1977), 164. Tsarapkin's statement appears in *Geneva Conference*, 30–31.

55. Teller and Panofsky are quoted in Killian, 167. Divine, *Blowing on the Wind*, 255. Jacobsen and Stein, 145–147.

56. Humphrey is quoted in Divine, *Blowing on the Wind*, 250. Gilpin, 247.

57. Jacobsen and Stein, 126–129, 176.

58. *Geneva Conference,* 354–355. Khrushchev to Eisenhower, April 23, 1959, ibid., 356–358.

59. *Geneva Conference,* 353, 360–365. Gilpin, 232.

60. Seaborg, 18–19. Gilpin, 236. Jacobsen and Stein, 151–154. Killian, 166, 168.

61. *Geneva Conference,* 48–49, 335–354.

62. Ibid., 367–375, 384–413. Jacobsen and Stein, 162–165, 183–197, 210–230. Seaborg, 20–21.

63. *DD, 1945–1959,* II, 1550–1557.

64. *Geneva Conference,* 80–81. Divine, *Eisenhower and the Cold War,* 143. Jacobsen and Stein, 201–204.

65. U.S. Department of State, *Documents on Disarmament, 1960* (Washington, D.C.: 1961), 31–39. (Hereafter cited as *DD, 1960.*) Seaborg, 21. Gilpin, 246. Jacobson and Stein, 235–238.

66. *DD, 1960,* 40–44. *Geneva Conference,* 420–423. Seaborg, 22. Gilpin, 247. Jacobsen and Stein, 238–241.

67. Seaborg, 22–23. George B. Kistiakowsky, *A Scientist in the White House: The Private Diary of President Eisenhower's Special Assistant for Science and Technology* (Cambridge, Mass.: 1976), 281. Eisenhower is quoted in Divine, *Blowing on the Wind,* 256.

68. *Geneva Conference,* 423–424. Divine, *Eisenhower and the Cold War,* 144. Seaborg, 23–25.

69. Seaborg, 23. Eisenhower is quoted in Divine, *Eisenhower and the Cold War,* 147–148.

70. Divine, *Eisenhower and the Cold War,* 149–150. Eisenhower, *Waging Peace,* 543–557. Killian, 81–83. Jacobsen and Stein, 263. *Geneva Conference,* 100–121.

71. Charles C., Alexander, *Holding the Line: The Eisenhower Era, 1952–1961* (Bloomington, Ind.: 1975), 280–288. Seaborg, 23. Divine, *Blowing on the Wind,* 316. Eisenhower, *Waging Peace,* 481.

Chapter 7. Kennedy, Nuclear Weapons, and the Limited Test Ban Treaty, 1961–1963

1. Wiesner's testimony appears in U.S. Congress, Subcommittee on Arms Control, International Law and Organization of the Senate Committee on Foreign Relations, *Hearings: ABM, MIRV, SALT, and the Nuclear Arms Race,* 91st Cong., 2nd sess., 395. (Hereafter cited as *ABM Hearings.*)

2. Arthur M. Schlesinger, Jr., *A Thousand Days: John F. Kennedy in the White House* (Boston: 1965), 301.

3. Theodore C. Sorensen, *Kennedy* (New York: 1965), 608. William W. Kaufmann, *The McNamara Strategy* (New York: 1964), 77. Harold Karan Jacobsen and Eric Stein, *Diplomats, Scientists and Politicians: The United States and the Nuclear Test Ban Negotiations* (Ann Arbor: 1966), 339. Desmond Ball, *Politics and Force Levels: The Strategic Missile Program of the Kennedy Administration* (Berkeley, Calif.: 1980), xix–xx, 50–51. Harland B. Moulton, *From Superiority to Parity: The United States and the Strategic Arms Race, 1961–1971* (Westport, Conn.: 1973), 52. Edgar M. Bottome, *The Balance of Terror: A Guide to the Arms Race* (Boston: 1971), 83.

4. McGeorge Bundy, "The Presidency and the People," *Foreign Affairs* 42 (April 1964), 354. Sorensen, 602.

5. Henry L. Trewhitt, *McNamara, His Ordeal in the Pentagon* (New York: 1971), 6.

6. McNamara is quoted in Edgar M. Bottome, *The Missile Gap: A Study of the Formulation of Military and Political Policy* (Rutherford, N.J.: 1971), 150–151, 157. *Time*, February 17, 1961, 12. Moulton, 61, 63. Ball, 19–22, 89–94.

7. U.S. Arms Control and Disarmament Agency, *Documents on Disarmament, 1961* (Washington, D.C.: 1962), 545. (Hereafter cited as *DD, 1961*). Harold W. Chase and Allen H. Lerman, eds., *Kennedy and the Press* (New York: 1965), 125. McNamara is quoted in *Newsweek*, November 13, 1961, 23, and in Stewart Alsop, "McNamara Thinks the Unthinkable," *Saturday Evening Post*, December 1, 1962, 67.

8. Ball, 93. Moulton, 63.

9. John F. Kennedy, *The Strategy of Peace*, ed. Allan Nevins (New York: 1960), 40. John Lewis Gaddis, *Strategies of Containment: A Critical Appraisal of Postwar American National Security Policy* (New York: 1982), 204. Ball, 109, 112, 257–263.

10. Schlesinger, 310, 311, 317.

11. Maxwell D. Taylor, *The Uncertain Trumpet* (New York: 1959), 146. Kaufmann, 69. Jacobsen and Stein, 340.

12. Address by McNamara, June 16, 1962, *Department of State Bulletin* 47 (July 9, 1962), 64–69. Kaufmann, 92–94. Jerome H. Kahan, *Security in the Nuclear Age: Developing U.S. Strategic Arms Policy* (Washington, D.C.: 1975), 90–91. Ball, 138–139, 186–211.

13. Moulton, 88–93. Michael Brower, "Controlled Thermonuclear War," *The New Republic* 147 (July 30, 1962), 9–15. Bottome, *The Balance of Terror*, 82–83.

14. Schlesinger, 852–853.

15. Kaufmann, 124–126. Ball, 144, 224–231.

16. Ball, 87, 276.

17. Ibid., 140, 173.

18. Ibid., 180.

19. Taylor, 158. Ball, 82–83. *Congressional Record*, August 2, 1963, vol. 109, part 10, 13989.

20. Jerome B. Wiesner, "Arms Control: Current Prospects and Problems," *Bulletin of the Atomic Scientists* 26 (May 1970), 6. Ball, 68–78, 86–87. Bottome, *The Balance of Terror*, 84–85. Schlesinger, 438. *ABM Hearings*, 396. I. F. Stone, "Theater of Delusion," *The New York Review of Books* 14 (April 23, 1970), 23.

21. Kennedy is quoted in Richard J. Walton, *Cold War and Counterrevolution: The Foreign Policy of John F. Kennedy* (New York: 1972), 9. Schlesinger, 298. Sorensen, 515, 509–510.

22. Schlesinger, 506–509, 302–303.

23. Bruce Miroff, *Pragmatic Illusions: The Presidential Politics of John F. Kennedy* (New York: 1976), 64.

24. Ibid., 67, 13. John F. Kennedy, *Profiles in Courage* (New York: 1961), 49.

25. Miroff, 15. *Public Papers of the Presidents of the United States: John F. Kennedy, 1961* (Washington, D.C.: 1962), 19, 22–23. (Hereafter cited as *KPP, 1961*.)

26. Miroff, 44.

27. Louise FitzSimons, *The Kennedy Doctrine* (New York: 1972), 18–71. Herbert S. Parmet, *JFK: The Presidency of John F. Kennedy* (New York: 1983), 157–179. Schlesinger, 233–297. Sorensen, 294–309. Herbert S. Dinerstein, *The Making of a Missile Crisis: October, 1962* (Baltimore: 1976), 129–130. Fitz-Simons, 45, 59–60, 62, 65.

28. *KPP, 1961*, 137–138. Schlesinger, 336–337. Walton, 72. Ralph E. Lapp, *The Weapons Culture* (New York: 1968), 44. FitzSimons, 70.

29. *DD, 1961*, 173–189, 166–173. Kennedy is quoted in Walton, 80. Sorensen, 543–550. Schlesinger, 358–374. FitzSimons, 89. Bottome, *The Balance of Terror*, 87–88. Alexander L. George and Richard Smoke, *Deterrence and American Foreign Policy: Theory and Practice* (New York: 1974), 414–421.

30. FitzSimons, 115, 121. Raymond L. Garthoff, *Soviet Military Policy: A Historical Analysis* (New York: 1966), 115–119.

31. Address by Kennedy, July 25, 1961, *DD, 1961*, 258–267. Walton, 86–88, 92–93. FitzSimons, 123. Sorensen, 613–617. Kennedy is quoted in Schlesinger, 391. George and Smoke, 424–437.

32. Schlesinger, 397–400. George and Smoke, 442–444. Walton, 92.

33. Jacobsen and Stein, 284. Schlesinger, 458–461. Glenn T. Seaborg, *Kennedy, Khrushchev and the Test Ban* (Berkeley, Calif.: 1981), 90.

34. Lincoln P. Bloomfield, Walter C. Clemens, Jr., and Franklyn Griffiths, *Khrushchev and the Arms Race: Soviet Interests in Arms Control and Disarmament, 1954–1964* (Cambridge, Mass.: 1966), 156. Jacobsen and Stein, 28. Dinerstein, 84.

35. Jacobsen and Stein, 277, 281–283. *DD, 1961*, 351. Sorensen, 618. Seaborg, 68–78. Mary Milling Lepper, *Foreign Policy Formulation: A Case Study of the Nuclear Test Ban Treaty of 1963* (Columbus, Ohio: 1971), 51–55.

36. Kennedy is quoted in Schlesinger, 482–483.

37. Jacobsen and Stein, 346. Address by Kennedy, March 2, 1962, *Public Papers of the Presidents of the United States: John F. Kennedy, 1962* (Washington, D.C.: 1963), 190. (Hereafter cited as *KPP, 1962*.) Seaborg, 63–66.

38. Dinerstein, 166. George and Smoke, 447, 450–451. Sorensen, 680. Bottome, *The Balance of Terror*, 90. Nikita S. Khrushchev, *Khrushchev Remembers*, ed. and trans. Strobe Talbott (Boston: 1970), 494. Schlesinger, 794–797. Arnold L. Horelick, "The Cuban Missile Crisis: An Analysis of Soviet Calculations and Behavior," *World Politics* 16 (April 1964), 377. Garthoff, 122. Graham T. Allison, *Essence of Decision: Explaining the Cuban Missile Crisis* (Boston: 1971), 40–56, 102–113, 230–244.

39. Kahan, 110. Bottome, *The Balance of Terror*, 91. Arnold L. Horelick and Myron Rush, *Strategic Power and Soviet Foreign Policy* (Chicago: 1966), 126–127, 141–142. Robin Edmonds, *Soviet Foreign Policy: The Brezhnev Years* (Oxford, Eng.: 1983), 26–34. FitzSimons, 147. George and Smoke, 456–459. Garthoff, 122–124.

40. FitzSimons, 148. Bloomfield et al., 226–233. Kahan, 109–111. Khrushchev, 494. George and Smoke, 461–466. Garthoff, 120-121. Allison, 113–117.

41. Sorensen, 676–678, 683. Schlesinger, 797–802. McNamara is quoted in Roger Hilsman, *To Move A Nation: The Politics of Foreign Policy in the Administration of John F. Kennedy* (Garden City, N.Y.: 1967), 195.

42. Sorensen, 678, 683. Horelick, "The Cuban Missile Crisis," 380. Barton J. Bernstein, "The Week We Almost Went to War," *Bulletin of the Atomic Scientists* 32 (February 1976), 17.

43. Sorensen, 684–688. Schlesinger, 810–811. Elie Abel, *The Missile Crisis,* (New York: 1966), 80. Allison, 141–143.

44. Sorensen, 688–689, 694, 795. *KPP, 1962,* 806–809. Allison, 56–62, 187–210.

45. Allison, 62–66, 210–230. The texts of the Kennedy-Khrushchev correspondence during the Cuban missile crisis appear in *Department of State Bulletin* 69 (November 19, 1973), 635–655.

46. Schlesinger, 841. I. F. Stone, "The Brink," *The New York Review of Books* 6 (April 14, 1966), 12. FitzSimons, 127. Kennedy is quoted in Sorensen, 705.

47. FitzSimons, 172, 161. George and Smoke, 470–472.

48. Henry M. Pachter, *Collision Course: The Cuban Missile Crisis and Coexistence* (New York: 1963), 84. Miroff, 67.

49. Ronald Steel, "Endgame," *The New York Review of Books* 12 (March 13, 1969), 21.

50. Kahan, 112. Edmonds, 35–43.

51. Kennedy interview with Robert Stein, August 1, 1963, *Public Papers of the Presidents of the United States: John F. Kennedy, 1963* (Washington, D.C.: 1964), 609. (Hereafter cited as *KPP, 1963.*) Address by Kennedy, July 26, 1963, ibid., 601–606. Schlesinger, 893, Sorensen, 518.

52. Bernard J. Firestone, *The Quest for Nuclear Stability: John F. Kennedy and the Soviet Union* (Westport, Conn.: 1982), 43, 80, 83. *DD, 1961,* 482–495.

53. Firestone, 67–68. Sorensen, 729.

54. Jacobsen and Stein, 384–386. U.S. Arms Control and Disarmament Agency, *Documents on Disarmament, 1962* (Washington, D.C.: 1963), 633–635. (Hereafter cited as *DD, 1962.*)

55. *DD, 1962,* 714–717, 747–758, 792–819. Jacobsen and Stein, 393–394, 403, 409–413.

56. *DD, 1962,* 776–788, 820–829. Jacobsen and Stein, 389–390, 413–414.

57. *DD, 1962,* 1239–1242, 863–865, 1144–1153. Kennedy to Khrushchev, December 28, 1962, ibid., 1277–1279. Jacobsen and Stein, 425–426, 430–432. Seaborg, 188.

58. Seaborg, 187. Jacobsen and Stein, 436–437. Firestone, 121.

59. Khrushchev is quoted in Norman Cousins, "Notes on a 1963 Visit with Khrushchev," *The Saturday Review* 47 (November 11, 1964), 21.

60. Seaborg 180, 192. Jacobsen and Stein, 432–435.

61. Sorensen, 733. U.S. Arms Control and Disarmament Agency, *Documents on Disarmament, 1963* (Washington, D.C.: 1964), 215–222. (Hereafter cited as *DD, 1963.*) Sorensen, 730–733.

62. *DD, 1963,* 222–228, 236–238. Seaborg, 216–218.

63. *DD, 1963,* 244–246. Harriman is quoted in Seaborg, 242.

64. Seaborg, 220–223, 242.

65. *DD, 1963,* 291–293. Seaborg, 254–255. Jacobsen and Stein, 454–458.

66. *DD, 1963,* 299–302. Seaborg, 263–264. Jacobsen and Stein, 458.

67. Teller's testimony is in U.S. Congress, Senate Committee on Foreign Relations, *Hearings: Nuclear Test Ban Treaty,* 88th Cong., 1st sess., 1963, 417–506, and U.S. Congress, Senate Committee on Armed Services, *Hearings: Military Aspects and Implications of Nuclear Test Ban Proposals and Related Matters,* 88th Cong., 1st sess., 1964, vol. I, 542–584.

68. *Hearings: Nuclear Test Ban,* 528–578, 758–791, 580. Seaborg, 272–273.

69. *Hearings: Military Aspects*, 796. *Hearings: Nuclear Test Ban*, 275.

70. Seaborg, 269–273.

71. Ibid., 265. Lepper, 55–121.

72. Lepper, 51, 54. Firestone, 135. Sorensen, 740.

73. Seaborg, 285–288.

74. Walton, 158–159.

75. Firestone, 149–150.

Chapter 8. Johnson, Nuclear Weapons, and the Pursuit of SALT, 1963–1969

1. Johnson is quoted in Eric Goldman, *The Tragedy of Lyndon Johnson* (New York, 1968), 378.

2. Lyndon Baines Johnson, *The Vantage Point: Perspectives of the Presidency, 1963–1969* (New York: 1971), 467, 471, 473, 476.

3. Philip L. Geyelin, *Lyndon B. Johnson and the World* (New York: 1966), 31–32.

4. U.S. Department of Defense, *Statement of Secretary of Defense Robert S. McNamara before the Senate Armed Services Committee on the Fiscal 1969–1973 Defense Program and 1969 Defense Budget* (Washington, D.C.: 1968), 55. (Hereafter cited as *Fiscal 1969–1973 Defense Program.*) Thomas W. Wolfe, *Soviet Power and Europe: 1945–1970* (Baltimore: 1970), 432–433. John Prados, *The Soviet Estimate: U.S. Intelligence Analysis and Russian Military Strength* (New York: 1982), 183–199.

5. *Fiscal 1969–1973 Defense Program*, 54. Address by McNamara, "The Dynamics of Nuclear Strategy," September 18, 1967, *Department of State Bulletin* 57 (October 9, 1967), 445.

6. McNamara, "The Dynamics of Nuclear Strategy," 446.

7. Statement of Robert S. McNamara, U.S. Congress, House Committee on Armed Services, *Hearings on Military Posture and H.R. 2440 to Authorize Appropriations During Fiscal Year 1964 for Procurement, Research, Development, Testing, and Evaluation of Aircraft, Missiles, and Naval Vessels for the Armed Forces, and for Other Purposes*, 88th Cong., 1st sess., 1963, 310.

8. Alain C. Enthoven and K. Wayne Smith, *How Much Is Enough? Shaping the Defense Program, 1961–1969* (New York: 1971), 207–208. *Fiscal 1969–1973 Defense Program*, 57–58.

9. Desmond Ball, *Politics and Force Levels: The Strategic Missile Program of the Kennedy Administration* (Berkeley, Calif.: 1980), 157, 161. Statement of Robert S. McNamara, U.S. Congress, House Committee on Armed Services, *Hearings on Military Posture and H.R. 4016 to Authorize Appropriations During Fiscal Year 1966 for Procurement of Aircraft, Missiles, and Naval Vessels, and Research, Development, Testing, and Evaluation, for the Armed Forces, and for Other Purposes*, 89th Cong., 1st sess., 1965, 211. (Hereafter cited as *Hearings on Military Posture Fiscal 1966.*)

10. McNamara, "The Dynamics of Nuclear Strategy," 445. Herbert York, *Race to Oblivion: A Participant's View of the Arms Race* (New York: 1970), 173–187. Ronald L. Tammen, *MIRV and the Arms Race: An Interpretation of Defense Strategy* (New York: 1973), 71–108.

11. U.S. Department of Defense, *Annual Report for Fiscal Year 1965* (Wash-

ington, D.C.: 1967), 13–14. U.S. Department of Defense, *Annual Report of the Secretary of the Air Force for Fiscal Year 1966* (Washington, D.C.: 1968), 349.

12. Ted Greenwood, *Making the MIRV: A Study of Defense Decision Making* (Cambridge, Mass.: 1975), 96–101, 171–177. Jerome H. Kahan, *Security in the Nuclear Age: Developing U.S. Strategic Arms Policy* (Washington, D.C.: 1975), 98. York, 178. Prados, 151–171.

13. James R. Kurth, "Why We Buy the Weapons We Do," *Foreign Policy*, No. 11 (Summer 1973), 48. Greenwood, 100, 103–104, 58–59.

14. Greenwood, 70–72.

15. Ibid., 72, 60, 62. Tammen, 119–122. Kosta Tsipis, *Arsenal: Understanding Weapons in the Nuclear Age* (New York: 1983), 130–146.

16. Harland B. Moulton, *From Superiority to Parity: The United States and the Strategic Arms Race, 1961–1971* (Westport, Conn.: 1973), 285. *Hearings on Military Posture Fiscal 1966*, 209. Greenwood, 75–76. Enthoven and Smith, 243–251. Ball, 214–224.

17. Moulton, 123, 154–176, 218–228. William W. Kaufmann, *The McNamara Strategy* (New York: 1964), 220–229.

18. Greenwood, 74–76.

19. Ibid., 75, 100, 107, 73–77. Kahan, 100, 107.

20. Kahan, 133–134. Enthoven and Smith, 184–196. Morton H. Halperin, "The Decision to Deploy the ABM: Bureaucratic and Domestic Politics in the Johnson Administration," *World Politics* 25 (October 1972), 62–95.

21. U.S. Department of Defense, *Statement by Secretary of Defense Robert S. McNamara before the House Armed Services Committee on the FY 1965–1969 Defense Program and 1965 Defense Budget*, January 27, 1964 (Washington, D.C.: 1964), 52–54. (Hereafter cited as *FY 1965–1969 Defense Program*.) Kaufmann, 229–232.

22. *FY 1965–1969 Defense Program*, 137. Moulton, 235–241. Kaufmann, 55–58.

23. McNamara, "The Dynamics of Nuclear Strategy," 450. Robert S. McNamara, *The Essence of Security: Reflections in Office* (New York: 1968), 60–67.

24. Halperin, 70–71.

25. *Congress and the Nation, Vol. II: 1965–1968* (Washington, D.C.: 1969), 873. Adam Yarmolinsky, "The Problem of Momentum," in Abram Chayes and Jerome B. Wiesner, *ABM: An Evaluation of the Decision to Deploy an Antiballistic Missile System* (New York: 1969), 146.

26. Moulton, 287. Greenwood, 73–75. Edgar M. Bottome, *The Balance of Terror: A Guide to the Arms Race* (Boston: 1971), 127. Halperin, 72–78, 82–83.

27. McNamara, "The Dynamics of Nuclear Strategy," 447–450. Greenwood, 76–81. Moulton, 288–289. Halperin, 84–86.

28. John Newhouse, *Cold Dawn: The Story of SALT* (New York: 1973), 98. Halperin, 86–87.

29. McNamara, "The Dynamics of Nuclear Strategy," 446. Kahan, 132.

30. *Fiscal 1969–1973 Defense Program*, 55. Greenwood, 174–177.

31. *Fiscal 1969–1973 Defense Program*, 56, Allen S. Whiting, "The Chinese Nuclear Threat," in Chayes and Wiesner, 160–170. Wiesner is quoted in Louis Fisher, *President and Congress: Power and Policy* (New York: 1972), 213.

32. U.S. Arms Control and Disarmament Agency, *Documents on Disarma-*

ment, 1964 (Washington, D.C.: 1965), 294–295. (Hereafter cited as *DD, 1964.*) U.S. Arms Control and Disarmament Agency, *Documents on Disarmament, 1967* (Washington, D.C.: 1968), 69–83. (Hereafter cited as *DD, 1967.*) Statement by Johnson, February 14, 1968, U.S. Arms Control and Disarmament Agency, *Documents on Disarmament, 1968,* 71. (Hereafter cited as *DD, 1968.*)

33. *DD, 1967,* 38–43.

34. *DD, 1964,* 276–278, 381–382.

35. *Seventh Annual Report of the U.S. Arms Control and Disarmament Agency,* January 30, 1968, *DD, 1967,* 741–742. Thomas B. Larson, *Disarmament and Soviet Policy, 1964–1968* (Englewood Cliffs, N.J.: 1969), 150–151.

36. *DD, 1968,* 110–118. John H. Barton and Lawrence D. Weiler, eds., *International Arms Control: Issues and Agreements* (Stanford, Calif.: 1976), 295–298.

37. *DD, 1968,* 444, 461–465.

38. Ibid., 655. U.S. Arms Control and Disarmament Agency, *Documents on Disarmament, 1965* (Washington, D.C.: 1966), 462–464. (Hereafter cited as *DD, 1965.*) U.S. Arms Control and Disarmament Agency, *Documents on Disarmament, 1966,* 355–359. (Hereafter cited as *DD, 1966.*)

39. *DD, 1965,* 338. Larson, 152.

40. *DD, 1964,* 4, 44–48, 235–238, 166–168.

41. Ibid., 101–105, 289–293, 137–139, 12–17.

42. Ibid., 7–9, 539. Barton and Weiler, 172. *DD, 1964,* 72. Greenwood, 109. Kahan, 119–121.

43. *DD, 1965,* 538–554.

44. Newhouse, 87, 91.

45. Barton and Weiler, 174–176. Newhouse, 103.

46. Johnson, 489.

47. Barton and Weiler, 177. Robin Edmonds, *Soviet Foreign Policy: The Brezhnev Years* (Oxford, Eng.: 1983), 65–73.

48. Newhouse, 134–137. *DD, 1968,* 709. Fulbright is quoted in Newhouse, 136. Henry A. Kissinger, *White House Years* (Boston: 1979), 49–50.

Chapter 9. Nixon and SALT I, 1969–1972

1. Gerard C. Smith, *Doubletalk: The Story of the First Strategic Arms Limitations Talks* (New York: 1980), 1.

2. Ibid., 22. "U.S. Foreign Policy for the 1970s: A New Strategy for Peace," A Report to the Congress by Richard M. Nixon, President of the United States, February 18, 1970, in *Department of State Bulletin* 62 (March 9, 1970), 323–327. (Hereafter cited as Nixon, "A New Strategy for Peace.")

3. Smith, 22. Jerome H. Kahan, *Security in the Nuclear Age: Developing U.S. Strategic Arms Policy* (Washington, D.C.: 1975), 144–145. Nixon, "A New Strategy for Peace," 327–329.

4. Henry Kissinger, *White House Years* (Boston: 1979), 55–70, 128, 202–203. Alexander L. George, "Détente: The Search for a 'Constructive Relationship,'" in Alexander L. George, ed., *Managing U.S.-Soviet Rivalry: Problems of Crisis Prevention* (Boulder, Colo: 1983), 17–29.

5. Joseph G. Berliner and Franklyn D. Holzman, "The Soviet Economy: Domestic and International Issues," in William E. Griffith, ed., *The Soviet Em-*

pire: Expansion and Détente, Critical Choices for Americans (Lexington, Mass.: 1976), 85–144. Robin Edmonds, *Soviet Foreign Policy: The Brezhnev Years* (Oxford, Eng: 1983), 80–86. Marshall D. Schulman, "SALT and the Soviet Union," in Mason Willrich and John B. Rhinelander, eds., *SALT: The Moscow Agreements and Beyond* (New York: 1974), 101–110.

6. Adam B. Ulam, *Dangerous Relations: The Soviet Union in World Politics, 1970–1982* (New York: 1983), 39–41, 43–46. Edmonds, 17–19. Thomas W. Robinson, "The Sino-Soviet Border Conflict," in Stephen S. Kaplan, ed., *Diplomacy of Power: Soviet Armed Forces as a Political Instrument* (Washington, D.C.: 1981), 265–313. Kissinger, *White House Years*, 171–194.

7. Smith, 31–35. Kahan, 167–169. Raymond L. Garthoff, "SALT and the Soviet Military," *Problems of Communism* 24 (January–February 1975), 26. Address by Gromyko, July 10, 1969, U.S. Arms Control and Disarmament Agency, *Documents on Disarmament, 1969* (Washington, D.C.: 1970), 314. (Hereafter cited as *DD, 1969.*)

8. Kissinger, *White House Years*, 127–130.

9. "U.S. Foreign Policy for the 1970's: Building for Peace," A Report to the Congress by Richard Nixon, President of the United States, February 25, 1971, in *Department of State Bulletin* 64 (March 22, 1971), 345–346, 415. (Hereafter cited as Nixon, "Building for Peace.") Congressional briefing by Kissinger, June 15, 1972, in U.S. Congress, Senate Committee on Foreign Relations, *Hearings: Strategic Arms Limitations Agreements*, 92nd Cong., 2nd sess., 1972, 396. Thomas W. Wolfe, *The SALT Experience* (Cambridge, Mass.: 1979), 29–36. Smith, 109–111.

10. Kissinger, *White House Years*, 127–130. Henry Kissinger, *The Necessity for Choice: Prospects of American Foreign Policy* (New York: 1961), 213. Smith, 25–26.

11. Kissinger, *White House Years*, 202. Kahan, 174–175.

12. Kahan, 165.

13. Ibid., 165, 168. Garthoff, "SALT and the Soviet Military," 26, 29–33.

14. Nixon, "A New Strategy for Peace," 318–322. Nixon, "Building for Peace," 345–346. Kissinger, *White House Years*, 220–225.

15. Nixon, "Building for Peace," 408–410.

16. Kahan, 155, 160. Kissinger, *White House Years*, 212.

17. *DD, 1969*, 102–105. Kissinger, *White House Years*, 208.

18. Kissinger, *White House Years*, 206–207. Ralph E. Lapp, *Arms Beyond Doubt: The Tyranny of Weapons Technology* (New York: 1970), 54–55, 65–66. Kahan, 152.

19. Kissinger, *White House Years*, 208.

20. Ibid., 209–210. *The New York Times*, November 25, 1975.

21. Smith, 75–107.

22. Ibid., 86–96. Kissinger, *White House Years*, 208.

23. John H. Barton and Lawrence Weiler, eds., *International Arms Control: Issues and Agreements* (Stanford, Calif.: 1976), 180–181. Smith, 90–93, 101. *The New York Times*, June 18, 1972.

24. Alan Platt, *The United States Senate and Strategic Arms Policy, 1969–1977* (Boulder, Colo.: 1978), 14. U.S. Arms Control and Disarmament Agency, *Documents on Disarmament, 1970* (Washington, D.C.: 1971), 132. (Hereafter cited as *DD, 1970.*)

25. Smith, 159–160, 479–482.

26. Kahan, 153. Smith, 161.

27. Kissinger, *White House Years*, 212–215. Ted Greenwood, *Making the MIRV: A Study in Decision Making* (Cambridge, Mass.: 1975), 129–134.

28. Kissinger, *White House Years*, 544. Smith, 477–478.

29. Kissinger, *White House Years*, 541.

30. Smith, 174–175.

31. Barton and Weiler, 183.

32. Smith, 123–124.

33. Kissinger, *White House Years*, 545. Smith, 157–158.

34. Smith, 156, 177–178. Kissinger is quoted in Smith, 177.

35. Platt, 13.

36. Smith, 146–150. Kissinger, *White House Years*, 548, 551. Platt, 16.

37. John Newhouse, *Cold Dawn: The Story of SALT* (New York: 1973), 195. Smith, 148, 180. Semenov is quoted in Smith, 186. Kahan, 182–183.

38. Kahan, 183. Smith, 216.

39. Kissinger, *White House Years*, 1216–1217, 802–803.

40. U.S. Arms Control and Disarmament Agency, *Documents on Disarmament, 1971,* (Washington, D.C.: 1973), 7–11. (Hereafter cited as *DD, 1971*).

41. *DD, 1971,* 298. Kahan, 184. Barton and Weiler, 189.

42. Smith, 223–225, 233, 235, 243, 228.

43. Ibid., 250–252, 255. Kahan, 184.

44. Kahan, 184. Barton and Weiler, 191.

45. Smith, 256, 258, 261.

46. Ibid., 264–265.

47. Kissinger, *White House Years*, 837. *DD, 1971,* 634–641.

48. Barton and Weiler, 190. Smith, 272–274, 323.

49. Smith, 325–326, 329, 336.

50. Kissinger, *White House Years*, 1129. Seymour M. Hersh, *The Price of Power: Kissinger in the White House* (New York: 1983), 536. Smith, 340.

51. Smith, 341–343.

52. Ibid., 370–371. Kissinger, *White House Years*, 1149–1150. News conference remarks by Kissinger, May 27, 1972, *DD, 1972,* 217–230. Kahan, 186. Hersh, 334–335, 343–348. Newhouse, 245–248.

53. Kissinger, *White House Years*, 1152.

54. Smith, 371–377.

55. Ibid., 382–383.

56. Kissinger, *White House Years*, 1202–1257. Smith, 408, 410.

57. U.S. Arms Control and Disarmament Agency, *Documents on Disarmament, 1972* (Washington, D.C.: 1973), 197–201. (Hereafter cited as *DD, 1972*.)

58. Ibid.

59. Ibid.

60. Ibid., 202–207. Barton and Weiler, 200–201. News conference remarks by Kissinger and Smith, May 26, 1972, in *DD, 1972,* 207–217.

61. *DD, 1972,* 202–217.

62. Hersh, 544–548, Smith, 413.

63. *DD, 1972,* 237–240.

64. Congressional briefing by Kissinger, June 15, 1972, *Hearings: Strategic Arms Limitations Agreements*, 394.

65. Testimony of Senators Edward Kennedy and James Buckley, ibid., 245–271.

66. Congressional briefing by Kissinger, June 15, 1972, ibid., 403.

67. Ibid., 401.

68. *DD, 1972*, 546–547, 652–653. Platt, 26–28. Roger P. Labrie, ed., *SALT Hand Book: Key Documents and Issues, 1972–1979* (Washington, D.C.: 1979), 144–155. Smith, 442.

69. Kissinger, *White House Years*, 1240. Milton Leitenberg, "The Race to Oblivion," *Bulletin of the Atomic Scientists* 30 (September 1974), 9–10.

70. Platt, 28–29. Labrie, 141–143.

Chapter 10. Nixon, Ford, and the Decline of Détente, 1972–1977

1. Peter J. Ognibene, *Scoop: The Life and Politics of Henry M. Jackson* (New York: 1975), 202–203, 210–213. Duncan L. Clarke, *Politics of Arms Control: The Role and Effectiveness of the U.S. Arms Control and Disarmament Agency* (New York: 1979), 51. Gerard C. Smith, *Doubletalk: The Story of the First Strategic Arms Limitations Talks* (New York: 1980), 444.

2. Clarke, 55, 47. Alan Platt, *The U.S. Senate and Strategic Arms Policy, 1969–1977* (Boulder, Colo.: 1978), 39–40.

3. Henry Kissinger, *Years of Upheaval* (Boston: 1982), 267.

4. Ibid., 1012–1013.

5. Ibid., 270, 1013.

6. Ibid., 271, 268.

7. U.S. Arms Control and Disarmament Agency, *Documents on Disarmament, 1973* (Washington, D.C.: 1974), 271–283. (Hereafter cited as *DD, 1973*.)

8. Ibid., 283–285.

9. Platt, 49–50. Kissinger, *Years of Upheaval*, 1016.

10. Robert S. Litwak, *Détente and the Nixon Doctrine: American Foreign Policy and the Pursuit of Stability, 1969–1976* (Cambridge, Eng.: 1984), 156–167. Kissinger, *Years of Upheaval*, 450–613.

11. Kissinger, *Years of Upheaval*, 1001–1004. Barry Carter, "Nuclear Strategy and Nuclear Weapons," *Scientific American* 230 (May 1974), 25. Desmond Ball, "Targeting for Strategic Deterrence," *Adelphi Paper* No. 185 (London: 1983), 18–19. Raymond L. Garthoff, *Détente and Confrontation: American-Soviet Relations from Nixon to Reagan* (Washington, D.C.: 1985), 417.

12. James R. Schlesinger, *Annual Defense Department Report, FY 1975* (Washington, D.C.: 1974), 45–46. Thomas W. Wolfe, *The SALT Experience* (Cambridge, Mass.: 1979), 162–171.

13. Kissinger, *Years of Upheaval*, 1018–1019. News conference remarks by Kissinger, March 28 and April 26, 1974, U.S. Arms Control and Disarmament Agency, *Documents on Disarmament, 1974* (Washington, D.C.: 1975), 73–74, 94–99. (Hereafter cited as *DD, 1974*.)

14. Platt, 57.

15. *DD, 1974*, 225–231. U.S. Arms Control and Disarmament Agency, *Documents on Disarmament, 1976* (Washington, D.C.: 1977), 328–348, 493–495, 529–534. (Hereafter cited as *DD, 1976*.)

16. Platt, 56, 59. Garthoff, 435–437.

17. Schlesinger is quoted in Platt, 60. Kissinger, *Years of Upheaval*, 1156–1159.

18. *DD, 1974*, 746–750.

19. Kissinger's press conferences, November 24 and December 7, 1974, *DD, 1974,* 750–761, 779–784. *The Washington Post,* December 7, 1974.

20. Platt, 61–62. Goldwater is quoted in Platt, 62.

21. Platt, 62, 65.

22. Brezhnev is quoted in Philip S. Gillette, "American-Soviet Trade in Perspective," *Current History* 65 (October 1973), 158. Henry Kissinger, *White House Years* (Boston: 1979), 152–153. Kissinger, *Years of Upheaval,* 986.

23. Gillette, 161–162.

24. Jackson is quoted in *Time,* January 27, 1975, 35.

25. Ibid. Kissinger, *Years of Upheaval,* 248–249.

26. For testimony for and against détente, see U.S. Congress, Subcommittee on Europe, House Committee on Foreign Affairs, *Hearings: Détente,* 93rd Cong., 2nd sess., 1974.

27. John Lewis Gaddis, *Strategies for Containment: A Critical Appraisal of Postwar American National Security Policy* (New York: 1982), 321. Reagan is quoted in Gerald Ford, *A Time to Heal: The Autobiography of Gerald R. Ford* (New York: 1979), 373.

28. Kissinger is quoted in *The New York Times,* July 16, 1975.

29. Paul H. Nitze, "Assuring Strategic Stability in an Era of Détente," *Foreign Affairs* 54 (January 1976), 209.

30. Alexander L. George, "The Arab-Israeli War of October 1973: Origins and Impact," in Alexander L. George, ed., *Managing U.S.-Soviet Rivalry: Problems of Crisis Prevention* (Boulder, Colo.: 1983), 139–154. George W. Breslauer, "Why Détente Failed: An Interpretation," in George, *Managing U.S.-Soviet Rivalry,* 319–340.

31. *Izvestia,* December 2, 1975, quoted in Theodore Draper, "Appeasement and Détente," *Commentary* 61 (February 1976), 31. Wolfe, 203–205. Coit D. Blacker, "The Kremlin and Détente: Soviet Conceptions, Hopes, and Expectations," in George, *Managing U.S.-Soviet Rivalry,* 119–137.

32. Address by Kissinger, "The Permanent Challenge of Peace: U.S. Policy Toward the Soviet Union," February 3, 1976, *Department of State Bulletin* 74 (February 23, 1976), 201–212. Statement by Kissinger to the Senate Committee on Foreign Relations, "Détente with the Soviet Union: The Reality of Competition and the Imperative of Cooperation," September 19, 1974, *Department of State Bulletin* 71 (October 14, 1974), 518.

33. Kissinger, *Years of Upheaval,* 237, 998. Gaddis, 323–325.

34. Kissinger's press conference, July 3, 1974, *Department of State Bulletin* 71 (July 29, 1974), 215. Kissinger, "The Permanent Challenge of Peace," 206–208.

35. Kissinger, "The Permanent Challenge of Peace," 206–208.

36. Kissinger, "Détente with the Soviet Union," 518. Kissinger, *Years of Upheaval,* 243–245.

37. Kissinger, *Years of Upheaval,* 985.

38. Ibid., 243–245.

39. Roger P. Labrie, ed., *SALT Hand Book: Key Documents and Issues, 1972–1979* (Washington, D.C.: 1979), 167. John H. Barton and Lawrence D. Weiler, eds., *International Arms Control: Issues and Agreements* (Stanford, Calif.: 1976), 221.

40. F. A. Long, "Should We Buy the Vladivostok Accord?" *Bulletin of the Atomic Scientists* 31 (February 1975), 5. Wolfe, 199–202.

41. *The New York Times,* December 3, 1975. Kissinger's press conference,

November 10, 1975, *Department of State Bulletin* 73 (December 1, 1975), 776–784. Kissinger is quoted in *The New York Times*, February 17, 1976.

42. Jackson is quoted in *Congressional Quarterly: Weekly Report* 33 (November 29, 1975), 2587.

43. Ibid., 33 (December 6, 1975), 2645. Kissinger's press conference, December 9, 1975, *Department of State Bulletin* 74 (January 5, 1976), 1–12. W. H. Kincade and J. D. Porro, *Negotiating Security: An American Arms Control Reader* (Washington, D.C.: 1979), 37–41. Charles William Maynes, Daniel Yankelovich, and Richard Lawrence Cohen, *U.S. Foreign Policy: Principles for Defining The National Interest* (New York: 1976), 73. Litwak, 214.

44. *The New York Times*, February 17, 1976. Labrie, 168. Ford, 357–358. Wolfe, 211–217.

45. *DD, 1974,* 146. William Epstein, *The Last Chance: Nuclear Proliferation and Arms Control* (New York: 1976), 227. Ashok Kapur, *International Nuclear Proliferation: Multilateral Diplomacy and Regional Aspects* (New York: 1979), 38–54.

46. Barton and Weiler, 303–305. Epstein, 221–224. Michael J. Brenner, *Nuclear Power and Non-Proliferation: The Remaking of U.S. Policy* (Cambridge, Eng: 1981), 6, 70.

47. Address by Kissinger, September 23, 1974, *Department of State Bulletin* 71 (October 14, 1974), 501. Epstein, 227. Brenner, 68–69.

48. Statement by Kissinger to the Senate Committee on Government Operations, March 9, 1976, *Department of State Bulletin* 74 (March 29, 1976), 405–411. Brenner, 14, 80, 93, 113–115, 269–280.

49. Statement by President Ford, October 28, 1976, *DD, 1976,* 703–716. Kapur, 67–86.

50. *The New York Times*, October 3, 1977. Epstein, 255. Brenner, 97.

51. Epstein, 248–249. U.S. Arms Control and Disarmament Agency, *Documents on Disarmament, 1975* (Washington, D.C.: 1976), 146–156.

52. Epstein, 249.

53. Ibid., 250.

Chapter 11. Carter and SALT II, 1977–1981

1. Inaugural address by Carter, January 20, 1977, *Public Papers of the Presidents of the United States: Jimmy Carter, 1977* (Washington, D.C.: 1977), I, 3. Zbigniew Brzezinski, *Power and Principle: Memoirs of the National Security Adviser, 1977–1981* (New York: 1983), 159. *The New York Times*, January 25 and October 5, 1977. U.S. Arms Control and Disarmament Agency, *Documents on Disarmament, 1977* (Washington, D.C.: 1979), 604–609. (Hereafter cited as *DD, 1977.*)

2. Brzezinski, 18. Strobe Talbott, *Endgame: The Inside Story of Salt II* (New York: 1979), 2. Raymond L. Garthoff, *Détente and Confrontation: American Soviet Relations from Nixon to Reagan* (Washington, D.C.: 1985), 563.

3. Cyrus Vance, *Hard Choices: Critical Years in America's Foreign Policy* (New York: 1983), 27–28.

4. Talbott, 48–49. Brzezinski, 519–520. Simon Serfaty, "Brzezinski: Play It Again, Zbig," *Foreign Policy*, No. 32 (Fall 1978), 3–21.

5. Brzezinski, 520.

6. Vance, 35–36.

7. Ibid., 51. Talbott, 52–53.

8. Vance, 51. Jackson is quoted in Talbott, 53 and 56. Jerry W. Sanders, *Peddlers of Crisis: the Committee on the Present Danger and the Politics of Containment* (Boston: 1983), 204–210. Duncan L. Clarke, *Politics of Arms Control: The Role and Effectiveness of the U.S. Arms Control and Disarmament Agency* (New York: 1979), 221–232.

9. Talbott, 58–59.

10. Vance, 50, 49.

11. Ibid., 52. News conference remarks by Brzezinski, April 1, 1977, *DD, 1977*, 199–209. Herbert Scoville, Jr., "The SALT Negotiations," *Scientific American* 237 (August 1977), 24–31.

12. Alan F. Geyer, *The Idea of Disarmament: Rethinking the Unthinkable* (Elgin, Ill.: 1982), 115. *The New York Times*, April 12, 1977. Talbott, 70. Garthoff, 806–807. Gromyko's news conference, March 31, 1977, *DD, 1977*, 180–191.

13. Kornienko is quoted in Talbott, 73.

14. Vance, 54.

15. Ibid., 55. Roger P. Labrie, ed., *SALT Hand Book: Key Documents and Issues, 1972–1979* (Washington, D.C.: 1979), 384–386.

16. Vance, 57. Labrie, 386–387. News conference remarks by Vance, May 21, 1977, *Department of State Bulletin* 56 (June 13, 1977), 628–633.

17. Talbott, 90, 103. *The New York Times*, June 13, 1977.

18. Carter's press conference, June 30, 1977, in Labrie, 468–469. Talbott, 104.

19. Talbott, 104–106. Vance, 58.

20. U.S. Congress, Senate Committee on Foreign Relations, *Hearings: The SALT II Treaty*, 96th Cong., 1st sess., 1979, part 5, 280.

21. Vance, 61. Labrie, 388–389. Vance to Senator John J. Sparkman, September 21, 1977, *DD, 1977*, 576–577.

22. Labrie, 389–390. Talbott, 133–135.

23. Perle is quoted in Talbott, 136. Vance, 61–62.

24. Talbott, 141. Brzezinski, 307.

25. Statement by Carter, April 7, 1978, U.S. Arms Control and Disarmament Agency, *Documents on Disarmament, 1978* (Washington, D.C.: 1980), 230. (Hereafter cited as *DD, 1978*.) Jimmy Carter, *Keeping Faith: Memoirs of a President* (New York: 1982), 225–228. Vance, 68–69, 93–96. Brzezinski, 304–306. Fred M. Kaplan, "Enhanced Radiation Weapons," *Scientific American* 238 (May 1978), 44–51. Garthoff, 851–853.

26. Talbott, 141–142. Brzezinski, 307–309. Vance, 98. Garthoff, 853–886.

27. Larry C. Napper, "The Ogaden War: Some Implications for Crisis Prevention," in Alexander L. George, ed., *Managing U.S.-Soviet Relations: Problems of Crisis Prevention* (Boulder, Colo.: 1983), 225–253.

28. Brzezinski, 181–182, 186.

29. Vance, 84–85, 88.

30. Brzezinski, 186, 189.

31. Vance is quoted in ibid., 185. Carter's news conference, March 2, 1978, *Department of State Bulletin*, 78 (April 1978), 20–21. Carter is quoted in *The New York Times*, March 18, 1978.

32. *Tass* is quoted in *The New York Times*, March 18, 1978.

33. Vance, 91, 88.

34. Ibid., 100, 102–103. Brzezinski, 326.

35. Labrie, 393–394.

36. Ibid., 393.

37. Ibid., 394–395. Garthoff, 609–614.

38. Vance, 104.

39. Labrie, 399.

40. Vance, 109; Vance's italics. Talbott, 194–195.

41. Talbott, 203–204.

42. Ibid., 215–216.

43. Vance, 111. Talbott, 250–251. Garthoff, 619–621.

44. Talbott, 237, 241–244.

45. Ibid., 252–254.

46. Labrie, 412. *Hearings: The SALT II Treaty*, 96th Cong., 1st sess., 1979, part 1, 81. Talbott, 256–259.

47. Labrie, 411, 413. Vance, 134–135.

48. Statement by Carter, September 9, 1979, U.S. Arms Control and Disarmament Agency, *Documents on Disarmament, 1979* (Washington, D.C.: 1982), 565–567. (Hereafter cited as *DD, 1979*.)

49. Carter, 241. Carter is quoted in Brzezinski, 334, 336.

50. Brzezinski, 332. U.S. Department of Defense, *Annual Report for Fiscal 1979* (Washington, D.C.: 1978), 106–107. Vance, 365. Brzezinski, 336–337.

51. *The New York Times*, March 3, 1980. Bernard T. Feld and Kosta Tsipis, "Land-Based Intercontinental Ballistic Missiles," *Scientific American* 241 (November 1979), 51–61.

52. Talbott, 169–170, 173–174.

53. Brzezinski, 337.

54. *Hearings: The SALT II Treaty*, part 1, 4–81.

55. Ibid.

56. Ibid., 54–74.

57. Ibid., 15–16.

58. Carter, 246.

59. Ibid., 251, 253, 251; Carter's italics.

60. Ibid., 252, 255.

61. Vance, 135.

62. *The New York Times*, April 13, 1979.

63. Eugene V. Rostow, "The Case Against SALT II," *Commentary* 67 (February 1979), 27. *Hearings: SALT II Treaty*, part 2, 372–412, and part 1, 433–598.

64. *Hearings: SALT II Treaty*, part 1, 547, 446.

65. Ibid., part 1, 447–448.

66. Rostow, 24.

67. *The New York Times*, August 4, 1979. Vance, 137.

68. *DD, 1979*, 231–238.

69. Brzezinski, 348–350. Vance, 362. Gloria Duffy, "Crisis Prevention in Cuba," in George, 296–318. Garthoff, 828–848.

70. Carter is quoted in *The New York Times*, October 2, 1979.

71. Ibid., October 9, 1979. Garthoff, 847–848.

72. U.S. Congress, Senate Committee on Foreign Relations, *Report: The SALT II Treaty* (Senate Executive Report 96-14), 96th Cong., 1st sess., 1979. Vance, 366–367.

73. Vance, 314–333, 368–383, 398–413.

74. Ibid., 386–388. Garthoff, 887–937.

75. Brzezinski, 429, 432. Milton Rosenberg, "The Decline and Rise of the Cold War Consensus," *Bulletin of the Atomic Scientists* 37 (March 1981), 8. Vance, 389.

76. William H. Kincade, "Banning Nuclear Tests: Cold Feet in the Carter Administration," *Bulletin of the Atomic Scientists* 34 (November 1978), 8–11. Labrie, 398. *The New York Times*, November 17, 1980. U.S. Arms Control and Disarmament Agency, *Documents on Disarmament, 1980* (Washington, D.C.: 1983), 157–164, 317–321.

77. Jo Pomerance, "The Comprehensive Test Ban at Last," *Bulletin of the Atomic Scientists* 35 (September 1979), 10.

78. Brzezinski, 129–134. Pierre Lellouche, "International Nuclear Politics," *Foreign Affairs* 58 (Winter 1979–1980), 336–350. Ashok Kapur, *International Nuclear Proliferation: Multilateral Diplomacy and Regional Aspects* (New York: 1979), 86–119. Michael J. Brenner, *Nuclear Power and Non-Proliferation* (Cambridge, Eng.: 1981), 116–144. Address by Joseph S. Nye, Jr., June 30, 1977, *Department of State Bulletin* 57 (August 8, 1977), 183–188.

79. *DD, 1978*, 118–164.

80. Brenner, 145–147.

81. Ibid., 199–204.

Chapter 12. Reagan and the "Rearmament" of America, 1981–1983

1. Robert Dallek, *Ronald Reagan: The Politics of Symbolism* (Cambridge, Mass.: 1983), 3–29. Lou Cannon, *Reagan* (New York: 1982), 22–26. Ronnie Dugger, *On Reagan: The Man and His Presidency* (New York: 1983), 2–24.

2. Cannon, 386–387, 393–396. Dallek, 134, 136–137. Nicholas Lemann, "The Peacetime War," *The Atlantic* 254 (October 1984), 71–94. *The New York Times*, July 21, 1981. Seyom Brown, *The Faces of Power: Constancy and Change in United States Foreign Policy from Truman to Reagan* (New York: 1983), 624.

3. *The New York Times*, November 11, 1981.

4. Address by Reagan, March 8, 1983, *Public Papers of the Presidents of the United States: Ronald W. Reagan, 1983* (Washington, D.C.: 1984), 363. (Hereafter cited as *RPP, 1983*.)

5. Dallek, 133. Jerry W. Sanders, *Peddlers of Crisis: The Committee on the Present Danger and the Politics of Containment* (Boston: 1983), 222.

6. Reagan's news conference, January 29, 1981, *Public Papers of the Presidents of the United States: Ronald W. Reagan, 1981* (Washington, D.C.: 1982), 57. (Hereafter cited as *RPP, 1981*.) Reagan is quoted in Dugger, 352.

7. Robert Sheer, *With Enough Shovels: Reagan, Bush, and Nuclear War* (New York: 1982), 7. Clark is quoted in Brown, 607, and in *The New York Times*, May 22, 1982. *The New York Times*, May 30, June 17, 19, and 22, 1982.

8. *Public Papers of the Presidents of the United States: Ronald W. Reagan, 1982* (Washington, D.C.: 1983), 487. (Hereafter cited as *RPP, 1982*.)

9. Paul H. Nitze, "Deterring Our Deterrent," *Foreign Policy*, No. 25 (Winter 1976–1977), 195–210. Fred Charles Iklé, "Can Nuclear Deterrence Last Out

the Century?" *Foreign Affairs* 51 (January 1973), 267–285. Edward N. Luttwak, "Nuclear Strategy: The New Debate," *Commentary* 57 (April 1974), 53–59.

10. Richard Pipes, "Why the Soviet Union Thinks It Could Fight and Win a Nuclear War," *Commentary* 64 (July 1977), 30–31.

11. Colin S. Gray and Keith Payne, "Victory Is Possible," *Foreign Policy*, No. 39 (Summer 1980), 20–21.

12. *The New York Times*, May 30 and June 10, 1982. Dugger, 402.

13. Caspar W. Weinberger, *Report of Secretary of Defense to the Congress on the FY 1983 Budget, FY Authorization Request and FY 1983–1987 Defense Programs* (Washington, D.C.: 1982), III-57 to III-63 and III-77 to III-89. U.S. Congress, Senate Committee on Foreign Relations, *Hearings: Strategic Weapons Proposals*, 97th Cong., 1st sess., 1981, 8–14. U.S. Congress, Congressional Budget Office, *Modernizing U.S. Strategic Offensive Forces: The Administration's Program and Alternatives* (Washington, D.C.: 1983), xiii–xiv. Christopher Paine, "Nuclear Combat: The Five-Year Defense Plan," *Bulletin of the Atomic Scientists* 38 (November 1982), 5–12. Michael D. Wormser, ed., *U.S. Defense Policy*, 3rd ed. (Washington, D.C.: 1983), 26–30, 33–35, 63–68.

14. Congressional Budget Office, *Modernizing U.S. Strategic Offensive Forces*, xiii–xiv. Christopher Paine, "Reagatomics, or How to 'Prevail,'" *The Nation* 236 (April 9, 1983), 423–433.

15. Gray and Payne, 25. White is quoted in Sheer, 18.

16. J. Edward Anderson, "First Strike: Myth or Reality," *Bulletin of the Atomic Scientists* 37 (November 1981), 6–11. Michael E. Howard, "On Fighting a Nuclear War," in Bernard Brodie, Michael D. Intriligator and Roman Kolkowicz, *National Security and International Stability* (Cambridge, Mass.: 1983), 23–35. Matthew Bunn and Kosta Tsipis, "The Uncertainties of a Preemptive Nuclear Attack," *Scientific American* 249 (November 1983), 38–47.

17. U.S. Congress, Office of Technology Assessment, *The Effects of Nuclear War* (Washington, D.C.: 1979), 84. Jones is quoted in *The New York Times*, June 19, 1982.

18. William M. Arkin, Thomas B. Cochran, and Milton M. Hoenig, *Nuclear Weapons Databook, Vol. I: U.S. Nuclear Forces and Capabilities* (New York: 1984), 102. Henry W. Kendall, "Second Strike," *Bulletin of the Atomic Scientists* 35 (September 1979), 33. Jones is quoted in John Newhouse, "Arms and Orthodoxy," *The New Yorker*, January 7, 1982, 93.

19. Lloyd J. Dumas, "Human Fallibility and Weapons," *Bulletin of the Atomic Scientists* 36 (November 1980), 16. *Newsweek*, October 5, 1981, 37. *The New York Times*, June 22, 1980 and October 10, 1980. U.S. Congress, Senate Committee on Armed Services, Report of Senator Gary Hart and Senator Barry Goldwater, *Recent False Alerts from the Nation's Missile Attack Warning System*, Committee Print, 96th Cong., 2nd sess., 1980. Stephen Talbot, "The H-Bombs Next Door," *The Nation* 232 (February 7, 1981). 145–146.

20. Les Aspin, "Soviet Civil Defense: Myth and Reality," in William H. Kincade and J. D. Porro, *Negotiating Security: An Arms Control Reader* (Washington, D.C.: 1979), 104–111. Fred M. Kaplan, "The Soviet Civil Defense Myth," *Bulletin of the Atomic Scientists* 34 (March 1978), 14–20.

21. Louis René Beres, *Mimicking Sisyphus: America's Countervailing Nuclear Strategy* (Lexington, Mass.: 1983), 40. *The New York Times*, March 4, 1985.

22. Richard Stubbing, "The Defense Program: Buildup or Binge?" *Foreign*

Affairs 63 (Spring 1985), 848–872. Anthony Lewis, "State of the Union," *The New York Times*, February 6, 1986.

23. Dugger, 419–420. *The New York Times*, August 9–11, 1981.

24. U.S. Congress, Senate Committee on Foreign Relations and House Committee on Foreign Affairs, *U.S. Arms Control and Disarmament Agency 1983 Annual Report*, Joint Committee Print, 98th Cong., 2nd sess., 1984, 20.

25. *RPP, 1981*, 956–957. *The New York Times*, October 22, 1981. Reagan's news conference, November 10, 1981, *RPP, 1981*, 1033.

26. Thompson is quoted in Coalition for a New Foreign and Military Policy, *Newsletter*, November 1983, 3.

27. Sanders, 325–326. Michael Lucas, "E.N.D. of the Beginning," *The Nation* 233 (October 10, 1981), 336–338.

28. *The New York Times*, February 27, 1983. C. F. Weizsacker, "European Armaments in the 1980s," *Bulletin of the Atomic Scientists* 36 (December 1980), 8–11. William M. Arkin, "Pershing II and U.S. Nuclear Strategy," *Bulletin of the Atomic Scientists* 39 (June–July 1983), 12–13.

29. Eugenia V. Osgood, "Euromissiles: Historical and Political Realities," *Bulletin of the Atomic Scientists* 39 (December 1983), 15.

30. Laurence I. Barrett, *Gambling with History: Reagan in the White House* (New York: 1983), 308. Kennan and Rickover are quoted in L. Bruce Van Voorst, "The Critical Masses," *Foreign Policy*, No. 48 (Fall 1982), 86.

31. *Time*, December 5, 1983, 38–40.

32. The *Jerusalem Post* is quoted in *The New York Times*, October 24, 1984. Ibid.

33. Peter Pringle, "Disarming Proposals," *The New Republic*, April 21, 1982, 14. Fox Butterfield, "Anatomy of the Nuclear Protest," *The New York Times Magazine*, July 11, 1982, 14–17, 32ff. Randall Forsberg, "A Bilateral Nuclear Weapon Freeze," *Scientific American* 247 (November 1982), 52–61.

34. Sanders, 330–331. Wormser, 107–108.

35. National Conference of Catholic Bishops, *The Challenge of Peace: God's Promise and Our Response* (Washington, D.C.: 1983). Testimony of Cardinal Joseph L. Bernardin and Archbishop John J. O'Connor in U.S. Congress, House Committee on Foreign Affairs, *Hearings: The Role of Arms Control in U.S. Defense Policy*, 98th Cong., 2nd sess., 1984, 131–178. Van Voorst, 82, 86, 87. *The New York Times*, April 16, 1982.

36. *The New York Times*, May 30, 1982.

37. Ibid., March 21, 1982.

38. U.S. Congress, House Committee on Foreign Affairs, *Congress and Foreign Policy, 1983*, Committee Print, 98th Cong., 2nd sess., 1984, 87.

39. *RPP, 1983*, 363.

40. Reagan is quoted in *The New York Times*, October 5, 1982. *The New York Times*, March 26, 1983. Halperin is quoted in *The New York Times*, November 11, 1982. Frank Donner, "But Will They Come? The Campaign to Smear the Nuclear Freeze Movement," *The Nation* 235 (November 6, 1982), 456–465.

41. *The New York Times*, March 31, 1982. Statements by Jackson and Warner, U.S. Congress, Senate Committee on Foreign Relations, *Hearings: Nuclear Arms Reduction Proposals*, 97th Cong., 2nd sess., 1982, 190–202.

42. *The New York Times*, March 9, 1983. Wormser, 110–112. *Congressional Quarterly*, May 7, 1983, 867–868.

43. Dallek, 152–153.

Chapter 13. Reagan and Nuclear Arms Talks, 1981 to the Present

1. Strobe Talbott, *Deadly Gambits: The Reagan Administration and the Stalemate in Nuclear Arms Control* (New York: 1983), 56–57, 59–61. Seyom Brown, *The Faces of Power: Constancy and Change in United States Foreign Policy from Truman to Reagan* (New York: 1983), 588–589.

2. Alexander, M. Haig, Jr., *Caveat: Realism, Reagan, and Foreign Policy* (New York: 1984). Talbott, *Deadly Gambits*, 50, 56, 59–64, 70–79, 83. Reagan is quoted in Laurence I. Barrett, *Gambling with History: Reagan in the White House* (New York: 1983), 313.

3. Talbott, *Deadly Gambits*, 198. U.S. Congress, Senate Committee on Foreign Relations and House Committee on Foreign Affairs, *U.S. Arms Control and Disarmament Agency 1983 Annual Report*, Joint Committee Print, 98th Cong., 2nd sess., 1984, 26–27, 157, 162. (Hereafter cited as *ACDA, 1983 Report.*) *The New York Times*, November 30, 1981.

4. George M. Seignious II and Jonathan Paul Yates, "Europe's Nuclear Superpowers," *Foreign Policy*, No. 55 (Summer 1984), 40–53. Talbott, *Deadly Gambits*, 88–89. *The New York Times*, February 27, 1983.

5. David S. Yost, "European-American Relations and NATO's Initial Missile Deployments," *Current History* 83 (April 1984), 147. *The New York Times*, February 2, 11, and March 17, 1982.

6. *ACDA, 1983 Report*, 25–26. Talbott, *Deadly Gambits*, 4, 125, 128. Paul H. Nitze, "The U.S. Negotiator's View of the Geneva Talks," *The New York Times*, January 19, 1984.

7. Talbott, *Deadly Gambits*, 132, 135–137, 140, 142, 144, 146–147, 151. Strobe Talbott, "Buildup and Breakdown," *Foreign Affairs: America and the World, 1983* 62 (1983), 597–598.

8. *The New York Times*, December 22, 1982. *ACDA, 1983 Report*, 27–28. Talbott, *Deadly Gambits*, 161–162. Talbott, "Buildup and Breakdown," 598. *The New York Times*, January 16, 20, 24 and February 14, 1983.

9. *The New York Times*, March 31, 1983. *ACDA, 1983 Report*, 21–22. Gromyko is quoted in *The New York Times*, April 3, 1983.

10. *The New York Times*, August 27, 1983. *ACDA, 1983 Report*, 28. Talbott, *Deadly Gambits*, 192.

11. *The New York Times*, September 27 and 29, 1983. *ACDA, 1983 Report*, 22–23, 28–30. Talbott, "Buildup and Breakdown," 601.

12. *The New York Times*, November 15, 1983 and January 19, 1984. Talbott, "Buildup and Breakdown," 601. Talbott, *Deadly Gambits*, 199.

13. *ACDA, 1983 Report*, 30–31. Yuli Kvitsinsky, "Soviet View of Geneva," *The New York Times*, January 12, 1984. *The New York Times*, November 22 and 23, 1983. Talbott, *Deadly Gambits*, 200–205.

14. *ACDA, 1983 Report*, 37. *The New York Times*, November 23, 1983. Kvitsinsky is quoted in *The New York Times*, January 12, 1984.

15. Robert Sheer, *With Enough Shovels: Reagan, Bush, and Nuclear War* (New York: 1982), 84. Rostow is quoted in *The New York Times*, June 28, 1981. Haig, 222.

16. Haig, 222–223. Barrett, 315–320.

17. *Time*, April 18, 1983, 29.

18. Haig, 223.

19. *The New York Times,* May 10, 1982. *ACDA, 1983 Report,* 6, 154–155. Haig, 222–223. Talbott, *Deadly Gambits,* 233–276.

20. *The New York Times,* May 10, 1982. U.S. Congress, Congressional Budget Office, *Modernizing Strategic Forces: The Administration's Programs and Alternatives* (Washington, D.C.: 1983), 29–40.

21. *The New York Times,* May 11, 1982. Congressional Budget Office, *Modernizing Strategic Forces,* 29–40. Talbott, *Deadly Gambits,* 263–265.

22. Sheer, 167–168. *The New York Times,* May 19, 1982. Christopher Paine, "A False START," *Bulletin of the Atomic Scientists* 38 (August–September 1982), 11–13.

23. *The New York Times,* May 19, 1982. Talbott, "Buildup and Breakdown," 605. *ACDA, 1983 Report,* 12, 167.

24. *ACDA, 1983 Report,* 12. Talbott, *Deadly Gambits,* 280. Talbott, "Buildup and Breakdown," 606.

25. *The New York Times,* June 1, 1982. Talbott, *Deadly Gambits,* 224–226.

26. *The New York Times,* July 21, 1982. Lynn R. Sykes and Jack F. Evernden, "The Verification of a Comprehensive Nuclear Test Ban," *Scientific American* 247 (October 1982), 47–55.

27. *The New York Times,* June 6, 1984. Paul Leventhal, "Getting Serious About Proliferation," *Bulletin of the Atomic Scientists* 40 (March 1984), 7–8. Reagan is quoted in *The New York Times,* July 7, 1981.

28. Michael D. Wormser, ed., *U.S. Defense Policy,* 3rd ed. (Washington, D.C.: 1983), 82–92. Kosta Tsipis, "Not Such a Bargain After All," *Bulletin of the Atomic Scientists* 39 (November 1983), 54–55.

29. Wormser, 89. *Congressional Quarterly,* July 17, 1982, 1702–1704.

30. U.S. President's Commission on Strategic Forces, *Report of the President's Commission on Strategic Forces* (Washington, D.C.: 1983). *The New York Times,* April 20, 1983.

31. The testimony of Stone and Scoville appears in U.S. Congress, Senate Committee on Armed Services, *Hearings: MX Missile Basing System and Related Issues,* 98th Cong., 1st sess., 1983, 288, 293.

32. *The New York Times,* May 25, 1983. John D. Isaacs, "What Happened to the 98th Congress," *Bulletin of the Atomic Scientists* 39 (August–September 1983), 11–12. U.S. Congress, House Committee on Foreign Affairs, *Congress and Foreign Policy, 1983,* Committee Print, 1984, 95–98. U.S. Congress, Senate Committee on Foreign Relations, *Hearings: United States-Soviet Relations,* 98th Cong., 1st sess., 1983, 44–45.

33. *The New York Times,* June 9 and August 11, 1983. *ACDA, 1983 Report,* 11–12, 166. *Congress and Foreign Policy, 1983,* 98.

34. *The New York Times,* June 10, 1983.

35. Ibid., July 14, 1983. Talbott, *Deadly Gambits,* 326.

36. *Congress and Foreign Policy, 1983,* 93–95. U.S. Congress, Congressional Budget Office, *An Analysis of Administration Strategic Arms Reduction and Modernization Proposals* (Washington, D.C.: 1984), 1–7. *Hearings: United States-Soviet Relations,* 29–41. Alton Frye, "Strategic Build-Down: A Context for Restraint," *Foreign Affairs* 52 (Winter 1983–1984), 293–317.

37. Frye, 303–307, 315–316. *Congress and Foreign Policy, 1983,* 105.

38. Talbott, *Deadly Gambits,* 334–339. Reagan to Cohen, May 12, 1983, *Hearings: United States-Soviet Relations,* 43–44. *ACDA, 1983 Report,* 13–15. *Congress and Foreign Policy, 1983,* 100.

39. *ACDA, 1983 Report*, 166–167. Talbott, *Deadly Gambits*, 338–339.

40. Senator Mark O. Hatfield, "This Is Arms Control?" *The New York Times*, September 22, 1983. *Congress and Foreign Policy, 1983*, 94–95, 99, 105–106.

41. *Pravda* is quoted in *The New York Times*, October 24, 1983. Talbott, *Deadly Gambits*, 339–342.

42. Richard Burt, "The Yearlong Shadow of K.A.L. Flight 7," *The New York Times*, August 31, 1984, and David Pearson, "K.A.L. 007: What the U.S. Knew and When We Knew It," *The Nation* 239 (August 18–25, 1984), 105–124.

43. *ACDA, 1983 Report*, 169. *The Washington Post*, June 17, 1984.

44. *The New York Times*, December 1, 1983. *Public Papers of the Presidents of the United States: Ronald W. Reagan, 1983* (Washington, D.C.: 1984), 437–443.

45. Weinberger is quoted in Richard L. Garwin and John Pike, "Space Weapons: History and Current Debate," *Bulletin of the Atomic Scientists* 40 (May 1984), 5s.

46. *The New York Times*, March 24, 1983. Hans A. Bethe, Richard L. Garwin, Kurt Gottfried, and Henry W. Kendall, "Space-based Ballistic-Missile Defense," *Scientific American* 251 (October 1984), 41. Keith B. Payne and Colin S. Gray, "Nuclear Policy and the Defense Transition," *Foreign Affairs* 60 (Spring 1984), 820–842.

47. Graham is quoted in *World Press Review* 30 (June 1983), 22. Teller is quoted in *The New York Times*, November 5, 1983.

48. U.S. Department of Defense, "Defense Against Ballistic Missiles," in U.S. Congress, Senate Committee on Foreign Relations, *Hearings: Strategic Defense and Anti-satellite Weapons*, 98th Cong., 2nd sess., 1984, 94. *The New York Times*, January 17, 1985.

49. *The New York Times*, November 5, 1983 and February 5, 1985.

50. Excerpts from a report by the Union of Concerned Scientists, in *The New York Review of Books* 31 (April 26, 1984), 47–52.

51. Michio Kaku, "Wasting Space," *The Progressive* 47 (June 1983), 22.

52. McGeorge Bundy, George F. Kennan, Robert S. McNamara, and Gerard Smith, "The President's Choice: Star Wars or Arms Control," *Foreign Affairs* 63 (Winter 1984–1985), 277.

53. *Congressional Quarterly*, January 28, 1984, 150–152. "Report on Soviet Noncompliance with Arms Control Agreements," *Department of State Bulletin* 85 (April 1985), 29–34.

54. *Congressional Quarterly*, April 7, 1984, 797–800. *The New York Times*, August 21 and September 14, 1985.

55. Garwin and Pike, "Space Weapons," 3s. Richard L. Garwin, Kurt Gottfried, and Donald L. Hafner, "Antisatellite Weapons," *Scientific American* 250 (June 1984), 45–55.

56. *Congressional Digest* 63 (March 1984), 69.

57. *Congressional Quarterly*, April 7, 1984, 797–800.

58. *The New York Times*, September 6, 1984.

59. *Congressional Quarterly*, June 23, 1984, 1482, and September 22, 1984, 2291.

60. *The New York Times*, July 28 and September 25, 1984.

61. Ibid., November 10, 1984.

62. Ibid., January 9, 1985. *Newsweek*, January 21, 1985, 22–25. *The New York Times*, January 19 and 23, 1985.

63. R. Jeffrey Smith, "Negotiators Report No Progress at Arms Talks," *Science* 228 (May 24, 1985), 971–972. *Department of State Bulletin* 85 (July 1985), 44–47. *The New York Times*, May 27, 1985.

64. *The New York Times*, November 1, 1985. *Congressional Quarterly*, January 4, 1986, 31.

65. *The New York Times*, April 10 and September 13, 1985.

66. Ibid., October 1, 5, 18 and November 1, 1985. Address by Paul H. Nitze, "The Nuclear and Space Arms Talks: Where We Are After the Summit," December 5, 1985, *Department of State Bulletin* 86 (February 1986), 58.

67. Nitze, "The Nuclear and Space Arms Talks," 58–59. *The New York Times*, November 1, 2, 4, 7, and 8, 1985.

68. Nitze, "The Nuclear and Space Arms Talks," 59–60. *The New York Times*, November 22, 1985 and April 5, 1986.

69. The text of Gorbachev's proposal appears in *The New York Times*, February 5, 1986.

70. Ibid., January 17, February 25, 27, 1986. Reagan's statement, January 15, 1986, *Department of State Bulletin* 86 (March 1986), 27.

71. Statement by President Reagan, May 27, 1986, *Department of State Bulletin* 86 (August 1986), 36–43. *The New York Times*, May 28, June 3 and 17, July 31, 1986.

72. *The New York Times*, June 5 and 20, 1986.

73. Ibid., June 17, 1986.

74. Ibid., June 25 and July 13, 1986.

75. Ibid., July 25, 26, August 2, 9, and September 25, 1986.

76. Ibid., August 9 and 16, 1986.

77. Ibid., October 1, 1986.

78. Ibid., October 11, 1986.

79. Ibid., October 13, 1986. Michael Mandelbaum and Strobe Talbott, "Reykjavik and Beyond," *Foreign Affairs* 65 (Winter 86/87), 228.

80. *The New York Times*, October 14, 1986.

81. Ibid., October 14, 16, and 18, 1986.

82. Ibid., October 14 and 16, 1986.

83. Ibid., October 14 and September 23, 1986. "U.S. Policy Regarding Limitations on Nuclear Testing," *Department of State Bulletin* 86 (October 1986), 14–15.

84. *The New York Times*, October 13, 16, and 17, 1986.

85. Gorbachev is quoted in ibid., October 15, 1986.

86. Buchanan is quoted in ibid., October 16, 1986. Ibid.

87. Nunn is quoted in *The (Cleveland) Plain Dealer*, October 15, 1986. Raimon is quoted in *The New York Times*, October 22, 1986.

88. *The New York Times*, October 21 and 19, 1986.

89. Ibid., October 21, 1986.

90. Ibid., November 29, 1986 and January 16, 1987.

91. Ibid., January 18, 1987.

SUGGESTED READINGS

General Sources

Among the few general sources dealing with the history of the nuclear arms race are Gerard H. Clarfield and William M. Wiecek, *Nuclear America: Military and Civilian Power in the United States, 1940–1980* (New York: 1984) and Michael Mandelbaum, *The Nuclear Question: The United States and Nuclear Weapons, 1946–1976* (New York: 1979). A very useful documentary history is Robert C. Williams and Philip L. Cantelon, eds., *The American Atom: A Documentary History of Nuclear Policies from the Discovery of Fission to the Present, 1939– 1984* (Philadelphia: 1985).

Two periodicals were relied upon heavily in this book. The *Department of State Bulletin* provides the texts of important diplomatic statements and excerpts from the news conferences of the secretary of state. The *Bulletin of the Atomic Scientists* contain cogent articles dealing with the diplomatic and military aspects of the nuclear arms race.

Among the works dealing with nuclear weapons technology are William M. Arkin, Thomas B. Cochran, and Milton M. Hoenig, *Nuclear Weapons Databook, Vol. I: U.S. Nuclear Forces and Capabilities* (New York: 1984) and Kosta Tsipis, *Arsenal: Understanding Weapons in the Nuclear Age* (New York: 1983). A recent work on the history of American strategic doctrine is Gregg Herken's *Counsels of War* (New York: 1985).

Roosevelt, Truman, and the Manhattan Project, 1939–1945 (Chapters 1–2)

The official account of the Manhattan Project is Henry De Wolf Smyth's *Atomic Energy for Military Purposes: The Official Report on the Development of the Atomic Bomb under the Auspices of the United States Government, 1940–1945* (Princeton, N.J.: 1945). The Manhattan Project is also discussed in the official history of the Atomic Energy Agency by Richard G. Hewlett and Oscar E. Anderson, Jr., *A History of the United States Atomic Energy Commission, Vol. I: The New World, 1939–1946* (University Park, Pa.: 1962).

Among the many personal accounts of the Manhattan Project are Arthur H. Compton, *Atomic Quest: A Personal Narrative* (New York: 1956); Spencer R. Weart and Gertrud Weiss Szilard, eds., *Leo Szilard: His Version of the Facts, Selected Recollections and Correspondence* (Cambridge, Mass.: 1978); Leslie R. Groves, *Now It Can Be Told: The Story of the Manhattan Project* (New York: 1962); James F. Brynes, *All in One Lifetime* (New York: 1958); and Henry L. Stimson and McGeorge Bundy, *On Active Service in Peace and War* (New York: 1947). An unpublished manuscript that proved invaluable was the Henry L. Stimson Diary, 1944–1945 (Sterling Memorial Library, Yale University, New Haven, Conn.).

There is an abundance of secondary accounts about the Manhattan Project. Among the most useful were Martin J. Sherwin, *A World Destroyed: The Atomic Bomb and the Grand Alliance* (New York: 1975); Walter Smith Schoenberger, *Decision of Destiny* (Athens, Ohio: 1969); Alice Kimball Smith, *A Peril and a Hope* (Chicago: 1965); Robert C. Batchelder, *The Irreversible Decision, 1939–1950* (Boston: 1962); Robert Jungk, *Brighter Than a Thousand Suns: A Personal History of the Atomic Scientists* (New York: 1958); Herbert Feis, *The Atomic Bomb and the End of World War II*, rev. ed. (Princeton, N.J.: 1960); Len Giovannitti and Fred Freed, *The Decision to Drop the Bomb* (New York: 1965); and Anthony Cave Brown and Charles B. MacDonald eds., *The Secret History of the Atomic Bomb* (New York: 1977).

For the actions of Roosevelt and Truman with respect to the Manhattan Project and American military and diplomatic policy, see James MacGregor Burns, *Roosevelt: The Soldier of Freedom, 1940–1945* (New York: 1970); Robert Dallek, *Franklin D. Roosevelt and American Foreign Policy, 1932–1945* (New York: 1979); Harry S. Truman, *Memoirs*, 2 vols. (Garden City, N.Y.: 1955–1956); and Robert J. Donovan, *Conflict and Crisis: The Presidency of Harry S. Truman* (New York: 1977).

For an introduction to the historiography of the Manhattan Project, see Barton J. Bernstein, ed., *The Atomic Bomb: The Critical Issues* (Boston: 1976) and Paul R. Baker, ed., *The Atomic Bomb: The Great Decision* (Hinsdale, Ill.: 1976). A leading revisionist interpretation is provided by Gar Alperovitz, *Atomic Diplomacy: Hiroshima and Potsdam* (New York: 1965).

Three excellent accounts of the British atomic energy program are Margaret Gowing, *Britain and Atomic Energy* (New York: 1964); Ronald W. Clark, *The Birth of the Bomb* (New York: 1961); and John Simpson, *The Independent Nuclear State: The United States and Britain and the Military Atom* (New York: 1983).

For Japanese-American relations, see Joseph C. Grew, *Turbulent Era: A Diplomatic Record of Forty Years, 1904–1945* (Boston: 1952); Togo Shigenori, *The Cause of Japan* (New York: 1956); Robert J. C. Butow, *Japan's Decision to Surrender* (Stanford, Calif.: 1954); and Kase Toshikazu, *Journey to the Missouri* (New Haven, Conn.: 1950).

For Soviet-American relations, see William Hardy McNeill, *America, Britain, and Russia: Their Cooperation and Conflict, 1941–1946* (London: 1953); John R. Deane, *The Strange Alliance* (New York: 1947); Herbert Feis, *Churchill, Roo-*

sevelt, Stalin: The War They Waged and the Peace They Sought (Princeton, N.J.: 1957); Adam B. Ulam, *Expansion and Coexistence: Soviet Foreign Policy, 1917–1973*), 2nd ed. (New York: 1974); and John Lewis Gaddis, *The United States and the Origins of the Cold War, 1941–1947* (New York: 1972).

Truman, the Cold War, Nuclear Arms Control, and the Hydrogen Bomb, 1945–1952 (Chapters 3–4)

An important primary source for this period is U.S. Department of State, *Foreign Relations of the United States, Diplomatic Papers: The Conference of Berlin (The Potsdam Conference), 1945* (Washington, D.C.: 1960). See also the *Public Papers of the Presidents of the United States: Harry S. Truman, 1945–1953* (Washington, D.C.: 1961–1966).

Among the most useful personal accounts during this period were Henry L. Stimson Dairy, *1944–1945* (unpublished MS, Sterling Memorial Library, Yale University, New Haven, Conn.); Harry S. Truman, *Memoirs, Vol. I: Year of Decisions* (Garden City, N.Y.: 1955); David E. Lilienthal, *Journals of David E. Lilienthal, Vol. II: The Atomic Energy Years, 1945–1950* (New York: 1964); and Dean Acheson, *Present at the Creation: My Years in the State Department* (New York: 1969).

Secondary works dealing with the atomic energy policies of the Truman administration are Gregg Herken, *The Winning Weapon: The Atomic Bomb and the Cold War, 1945–1950* (New York: 1980); Joseph I. Lieberman, *The Scorpion and the Tarantula: The Struggle to Control Atomic Weapons, 1945–1949* (Boston: 1970); Richard G. Hewlett and Oscar E. Anderson, Jr., *A History of the United States Atomic Energy Commission, Vol. I: The New World, 1939–1946* (University Park, Pa.: 1962); and Richard G. Hewlett and Francis Duncan, *A History of the United States Atomic Energy Commission, Vol. II: Atomic Shield, 1947–1952* (University Park, Pa.: 1969). Herbert F. York, *The Advisors: Oppenheimer, Teller, and the Super-bomb* (San Francisco: 1976); Stanley A. Blumberg and Gwinn Owens, *Energy and Conflict: The Life and Times of Edward Teller* (New York: 1976).

Among the works dealing with the early Cold War are Daniel Yergin, *Shattered Peace: The Origins of the Cold War and the National Security State* (Boston: 1977); John Lewis Gaddis, *The United States and the Origins of the Cold War, 1941–1947* (New York: 1972); Bruce Robellet Kuniholm, *The Origins of the Cold War in the Near East: Great Power Conflict and Diplomacy in Iran, Greece, and Turkey* (Princeton, N.J.: 1980); Robert Messer, *The End of an Alliance: James F. Byrnes, Roosevelt, Truman, and the Origins of the Cold War* (Chapel Hill, N.C.: 1982); Walter LaFeber, *America, Russia and the Cold War*, 3rd ed. (New York: 1976); D. F. Fleming, *The Cold War and Its Origins, 1917–1960* 2 vols. (Garden City, N.Y.: 1961); and Barton J. Bernstein, ed., *Politics and Policies of the Truman Administration* (Chicago: 1972). For an introduction to the historiography of the Cold War, see J. Samuel Walker, "Historians and Cold War Origins: The New Consensus," in Gerald K. Haines and J. Samuel Walker, eds., *American Foreign Relations: A Historiographical Review* (Westport, Conn.: 1981), 207–236.

The Eisenhower Years, 1953–1961 (Chapters 5–6)

For an overview of the diplomatic and military policies of the Eisenhower administration, see the President's memoirs, *The White House Years: Mandate for Change, 1953–1956* (Garden City, N.Y.: 1963) and *The White House Years: Waging Peace, 1956–1961* (Garden City, N.Y.: 1965). The *Public Papers of the Presidents of the United States: Dwight D. Eisenhower, 1953–1961*, 8 vols. (Washington, D.C.: 1960–1961) contain the texts of presidential statements and news conferences dealing with nuclear arms policy. Important documentation also appears in Robert L. Branyon and Lawrence H. Lassen, *The Eisenhower Administration, 1953–1961*, 2 vols. (New York: 1971).

Secondary sources covering the diplomatic and military policies of the administration are Robert A. Divine, *Eisenhower and the Cold War* (New York: 1981); Charles C. Alexander, *Holding the Line: The Eisenhower Era, 1952–1961* (Bloomington, Ind.: 1975); D. F. Fleming, *The Cold War and Its Origins: 1917–1960, Vol. II: 1950–1960* (Garden City, N.Y.: 1961); and Stephen E. Ambrose, *Eisenhower, Vol. II: The President* (New York: 1984).

The defense policies of the Eisenhower administration are discussed in Desmond Ball, *Politics and Force Levels: The Strategic Missile Programs of the Kennedy Administration* (Berkeley, Calif.: 1960); John Lewis Gaddis, *Strategies of Containment: A Critical Appraisal of Postwar American National Security Policy* (New York: 1982); Samuel P. Huntington, *The Common Defense: Strategic Programs in National Politics* (New York: 1961); Glenn H. Snyder, "The New Look," in Warner R. Schilling, Paul Y. Hammond, and Glenn H. Snyder, *Strategy, Politics, and Defense Budgets* (New York: 1962); Richard A. Aliano, *American Defense Policy from Eisenhower to Kennedy: The Politics of Changing Military Requirements, 1957–1961* (Athens, Ohio: 1975); and Michael H. Armacost, *The Politics of Weapons Innovation: The Thor-Jupiter Controversy* (New York: 1969).

Several works cover the role played by scientists in the development of the administration's defense and arms control policies. Among them are Herbert York, *Race to Oblivion: A Participant's View of the Arms Race* (New York: 1970); Ralph E. Lapp, *The New Priesthood: The Scientific Elite and the Uses of Power* (New York: 1965); Robert Gilpin *American Scientists and Nuclear Weapons Policy* (Princeton, N.J.: 1962); James P. Killian, Jr., *Sputnik, Scientists, and Eisenhower: A Memoir of the First Special Assistant to the President for Science and Technology* (Cambridge, Mass.: 1977); and George B. Kistiakowsky, *A Scientist in the White House: The Private Diary of President Eisenhower's Special Assistant for Science and Technology* (Cambridge, Mass.: 1976).

The bomber and missile gaps are discussed in Edgar M. Bottome's *The Missile Gap: A Study of the Formulation of Military and Political Policy* (Rutherford, N.J.: 1971) and Bottome's *The Balance of Terror: A Guide to the Arms Race* (Boston: 1971); Edmund Beard, *Developing the ICBM: A Study in Bureaucratic Politics* (New York: 1976); and John Prados, *The Soviet Estimate: U.S. Intelligence Analysis and Russian Military Strength* (New York: 1982).

Documents relating to the test ban negotiations can be found in U.S. Department of State, *Documents on Disarmament, 1945–1959*, 2 vols. (Washington, D.C.: 1960); U.S. Department of State, *Documents on Disarmament, 1960* (Washington, D.C.: 1961); and U.S. Disarmament Administration, *Geneva Con-*

ference on the Discontinuance of Nuclear Weapons Tests, History and Analysis of Negotiations (Washington, D.C.: 1961).

Secondary sources dealing with the test ban negotiations are Robert A. Divine, *Blowing on the Wind: The Nuclear Test Ban Debate, 1954–1960* (New York: 1978); Harold Karan Jacobsen and Eric Stein, *Diplomats, Scientists and Politicians: The United States and the Nuclear Test Ban Negotiations* (Ann Arbor: 1966); and Bernard G. Bechhoefer, *Postwar Negotiations for Arms Control* (Washington, D.C.: 1961).

Interpretations of Soviet diplomatic and military policies during the Eisenhower years can be found in Arnold L. Horelick and Myron Rush, *Strategic Power and Soviet Foreign Policy* (Chicago: 1966); Lincoln P. Bloomfield, Walter C. Clemens, Jr., and Franklyn Griffiths, *Khrushchev and the Arms Race: Soviet Interests in Arms Control and Disarmament, 1954–1964* (Cambridge, Mass.: 1966); and Adam C. Ulam, *Expansion and Coexistence: Soviet Foreign Policy, 1917–1973*, 2nd ed. (New York: 1974).

The Kennedy-Johnson Years, 1961–1969 (Chapters 7–8)

A significant primary source for the Kennedy years is the *Public Papers of the Presidents of the United States: John F. Kennedy, 1961–1963* (Washington, D.C.: 1962–1964).

For general histories of the Kennedy presidency by two of the President's aides, see Arthur M. Schlesinger, Jr., *A Thousand Days: John F. Kennedy in the White House* (Boston: 1965) and Theodore C. Sorensen, *Kennedy* (New York: 1965). A revisionist interpretation of the Kennedy presidency is provided by Bruce Miroff, *Pragmatic Illusions: The Presidential Politics of John F. Kennedy* (New York: 1976).

Works dealing with Kennedy's foreign policy are Alexander L. George and Richard Smoke, *Deterrence and American Foreign Policy: Theory and Practice* (New York: 1974) and Roger Hilsman, *To Move A Nation: The Politics of Foreign Policy in the Administration of John F. Kennedy* (Garden City, N.Y.: 1967). Revisionist interpretations are provided by Louise FitzSimons, *The Kennedy Doctrine* (New York: 1972) and Richard J. Walton, *Cold War and Counterrevolution: The Foreign Policy of John F. Kennedy* (New York: 1972).

For works dealing with the Cuban missile crisis, see Elie Abel, *The Missile Crisis* (New York: 1966); Herbert S. Dinerstein, *The Making of a Missile Crisis: October, 1962* (Baltimore: 1976); Robert F. Kennedy, *Thirteen Days: A Memoir of the Cuban Missile Crisis* (New York: 1969); Robert A. Divine, ed., *The Cuban Missile Crisis* (Chicago: 1971); and Graham T. Allison, *Essence of Decision: Explaining the Cuban Missile Crisis* (Boston: 1971).

An excellent overall view of the Johnson presidency is provided by Vaughn Davis Bornet, *The Presidency of Lyndon B. Johnson* (Lawrence, Kans.: 1983). For the President's memoir, see Lyndon Baines Johnson, *The Vantage Point: Perspectives of the Presidency, 1963–1969* (New York: 1971). Useful as an introduction to the foreign policy of Lyndon Johnson is Philip L. Geyelin, *Lyndon B. Johnson and the World* (New York: 1966).

The defense policies of Kennedy and Johnson are discussed in Robert S. McNamara, *The Essence of Security: Reflections in Office* (New York: 1968);

William W. Kaufmann, *The McNamara Strategy* (New York: 1964); Desmond
Ball, *Politics and Force Levels: The Strategic Missile Program of the Kennedy
Administration* (Berkeley, Calif.: 1980); Harland B. Moulton, *From Superiority
to Parity: The United States and the Strategic Arms Race, 1961–1971* (Westport,
Conn.: 1973); Edgar M. Bottome, *The Balance of Terror: A Guide to the Arms
Race* (Boston: 1971); Henry L. Trewhitt, *McNamara, His Ordeal in the Pentagon*
(New York: 1971); John Lewis Gaddis, *Strategies of Containment: A Critical
Appraisal of Postwar American National Security Policy* (New York: 1982); Alain
C. Enthoven and K. Wayne Smith, *How Much Is Enough? Shaping the Defense
Program, 1961–1969* (New York: 1971); Jerome H. Kahan, *Security in the Nu-
clear Age: Developing U.S. Strategic Arms Policy* (Washington, D.C.: 1975); Her-
bert York, *Race to Oblivion: A Participant's View of the Arms Race* (New York:
1970); Ralph E. Lapp, *Arms Beyond Doubt: The Tyranny of Weapons Technology* (New York: 1970).

The MIRV decision is discussed in Ted Greenwood, *Making the MIRV: A
Study in Decision Making* (Cambridge, Mass.: 1975) and Ronald L. Tammen,
MIRV and the Arms Race: An Interpretation of Defense Strategy (New York:
1973).

The ABM decision is discussed in Morton H. Halperin, "The Decision to
Deploy the ABM: Bureaucratic and Domestic Politics in the Johnson Administration," *World Politics* 25 (October 1972), 62–95; and Abram Chayes and Jerome B. Wiesner, *ABM: An Evaluation of the Decision to Deploy an Antiballistic
Missile System* (New York: 1969).

The arms control policies of the Kennedy and Johnson administrations are
discussed in John H. Barton and Lawrence D. Weiler, eds., *International Arms
Control: Issues and Agreements* (Stanford, Calif.: 1976). The major published
source of government documents dealing with the arms control policies of Kennedy and Johnson is U.S. Arms Control and Disarmament Agency, *Documents
on Disarmament, 1961–1969* (Washington, D.C.: 1962–1970).

The test ban negotiations are covered in Harold Karan Jacobsen and Eric
Stein, *Diplomats, Scientists and Politicians: The United States and the Nuclear
Test Ban Negotiations* (Ann Arbor, Mich.: 1966); Glenn T. Seaborg, *Kennedy,
Khrushchev and the Test Ban* (Berkeley, Calif.: 1981); Bernard J. Firestone, *The
Quest for Nuclear Stability: John F. Kennedy and the Soviet Union* (Westport,
Conn.: 1982).

Soviet military and diplomatic policies during the Kennedy-Johnson years are
discussed in Lincoln P. Bloomfield, Walter C. Clemens, Jr., and Franklyn Griffiths, *Khrushchev and the Arms Race: Soviet Interests in Arms Control and Disarmament, 1954–1964* (Cambridge, Mass.: 1966); Arnold L. Horelick and Myron
Rush, *Strategic Power and Soviet Foreign Policy* (Chicago: 1966); Thomas W.
Wolfe, *Soviet Power and Europe: 1945–1970* (Baltimore: 1970); Thomas B. Larson, *Disarmament and Soviet Policy, 1964–1968* (Englewood Cliffs, N.J.: 1969);
and Raymond L. Garthoff, *Soviet Military Policy: A Historical Analysis* (New
York: 1966).

The Nixon-Ford Years, 1969–1977 (Chapters 9–10)

Henry Kissinger's two-volume memoir, *White House Years* (Boston: 1979) and
Years of Upheaval (Boston: 1982) are indispensable to any study of the diplo-

macy and military strategy of the Nixon-Ford years. Less helpful are the presidential memoirs: Richard Nixon's *RN: The Memoirs of Richard Nixon* (New York: 1978) and Gerald Ford's *A Time to Heal: The Autobiography of Gerald R. Ford* (New York: 1979).

Among the secondary works on the Nixon-Ford years are Raymond L. Garthoff, *Détente and Confrontation: American-Soviet Relations from Nixon to Reagan* (Washington, D.C.: 1985); Robert S. Litwak, *Détente and the Nixon Doctrine: American Foreign Policy and the Pursuit of Stability, 1969–1976* (Cambridge, Eng.: 1984); and Seymour M. Hersh, *The Price of Power: Kissinger in the White House* (New York: 1983). Gerard Smith's *Doubletalk: The Study of the First Strategic Arms Limitation Talks* (New York: 1980) is an important firsthand account of SALT I. An annotated source of documents on SALT is Roger P. Labrie, ed., *SALT Hand Book: Key Documents and Issues, 1972–1979* (Washington, D.C.: 1979). Very helpful secondary sources were Thomas W. Wolfe, *The SALT Experience* (Cambridge, Mass.: 1979); Alan Platt, *The U.S. Senate and Strategic Arms Policy, 1969–1977* (Boulder, Colo.: 1978); Jerome H. Kahan, *Security in the Nuclear Age: Developing U.S. Strategic Arms Policy* (Washington, D.C.: 1975); John H. Barton and Lawrence D. Weiler, eds., *International Arms Control: Issues and Agreements* (Stanford, Calif.: 1976); John Newhouse, *Cold Dawn: The Story of SALT* (New York: 1973); Duncan L. Clarke, *Politics of Arms Control: The Role and Effectiveness of the U.S. Arms Control and Disarmament Agency* (New York: 1979); W. H. Kincade and J. D. Porro, *Negotiating Security: An Arms Control Reader* (Washington, D.C.: 1979); and Mason Willrich and John B. Rhinelander, eds., *SALT: The Moscow Agreements and Beyond* (New York: 1974).

U.S. Arms Control and Disarmament Agency, *Documents on Disarmament, 1969–1976* (Washington, D.C.: 1970–1977) is a collection of the most important primary sources for this period. Nuclear proliferation during the Nixon-Ford years is discussed in William Epstein, *The Last Chance: Nuclear Proliferation and Arms Control* (New York: 1976); Michael J. Brenner, *Nuclear Power and Non-Proliferation* (Cambridge, Eng.: 1981); Ashok Kapur, *International Nuclear Proliferation: Multilateral Diplomacy and Regional Aspects* (New York: 1979).

Soviet military and diplomatic policies during this period are discussed in Stephen S. Kaplan, ed., *Diplomacy of Power: Soviet Armed Forces as a Political Instrument* (Washington, D.C.: 1981); Robin Edmonds, *Soviet Foreign Policy— The Brezhnev Years* (New York: 1983); William E. Griffith, ed., *The Soviet Empire: Expansion and Détente: Critical Choices for Americans* (Lexington, Mass.: 1976); Derek Leebaert, ed., *Soviet Military Thinking* (London: 1981).

The Carter Years, 1977–1981 (Chapter 11)

Primary source materials dealing with the foreign, defense, and arms control policies of the Carter administration can be found in *Public Papers of the Presidents of the United States: Jimmy Carter, 1977–1981* (Washington, D.C.: 1977–1982) and U.S. Arms Control and Disarmament Agency, *Documents on Disarmament, 1977–1980* (Washington, D.C.: 1979–1983).

Indispensable are three memoirs: Zbigniew Brzezinski, *Power and Principle: Memoirs of the National Security Adviser, 1977–1981* (New York: 1983); Cyrus Vance, *Hard Choices: Critical Years in America's Foreign Policy* (New York:

1983); and Jimmy Carter, *Keeping Faith: Memoirs of a President* (New York: 1982).

Useful secondary works dealing with the arms control policies of the Carter administration are Strobe Talbott, *Endgame: The Inside Story of SALT II* (New York: 1979) and Roger P. Labrie, ed., *SALT Hand Book: Key Documents and Issues, 1972–1979* (Washington, D.C.: 1979).

The Reagan Years, 1981 to the Present (Chapters 12–13)

A number of studies have appeared on the early years of Reagan's presidency. Among them are Lou Cannon, *Reagan* (New York: 1982); Robert Dallek, *Ronald Reagan: The Politics of Symbolism* (Cambridge, Mass.: 1983); Ronnie Dugger, *On Reagan: The Man and His Presidency* (New York: 1983); Laurence I. Barrett, *Gambling with History: Reagan in the White House* (New York: 1983); and an early memoir, Alexander M. Haig, Jr., *Caveat: Realism, Reagan, and Foreign Policy* (New York: 1984). See also *Public Papers of the Presidents of the United States: Ronald W. Reagan, 1981–1983* (Washington, D.C.: 1982–1984).

The early diplomatic and military policies of the Reagan administration are discussed in Robert Sheer, *With Enough Shovels: Reagan, Bush, and Nuclear War* (New York: 1982); Louis René Beres, *Mimicking Sisyphus: America's Countervailing Nuclear Strategy* (Lexington, Mass.: 1983). For an important, albeit hostile, account of the activities of the Committee on the Present Danger, see Jerry W. Sanders, *Peddlers of Crisis: The Committee on the Present Danger and the Politics of Containment* (Boston: 1983). The papers of the committee appear in Charles Tyroler II, ed., *Alerting America: The Papers of the Committee on the Present Danger* (Washington, D.C.: 1984).

A useful survey of the Reagan defense and arms control policies is provided by Michael D. Wormser, ed., *U.S. Defense Policy*, 3rd ed. (Washington, D.C.: 1983). A work that relied extensively on interviews with administration officials is Strobe Talbott's *Deadly Gambits: The Reagan Administration and the Stalemate in Nuclear Arms Control* (New York: 1984). See also Talbott's *The Russians and Reagan* (New York: Vintage, 1984).

Books on the nuclear arms race in space have proliferated. Among them are Lt. General Daniel O. Graham, *High Frontier: A New National Strategy* (Washington, D.C.: 1982); Jeff Hecht, *Beam Weapons: The Next Arms Race* (New York: 1984); Ashton B. Carter and David N. Schwartz, eds., *Ballistic Missile Defense* (Washington: 1984); Jack Manno, *Arming the Heavens: The Hidden Military Agenda for Space, 1945–1995* (New York: 1984); John Tirman, ed., *The Fallacy of Star Wars* (New York: 1984); Sidney D. Drell, Phillip J. Farley, and David Holloway, *The Reagan Strategic Defense Initiative: A Technical, Political and Arms Control Assessment* (Stanford, Calif.: 1984); and William J. Durch, ed., *National Interests and the Military Use of Space* (Cambridge, Mass., 1984).

Books dealing with the possible effects of a nuclear war are Michael Riordan, ed., *The Day After Midnight: The Effects of Nuclear War* (Palo Alto, Calif.: 1982); Jeannie Petersen, *The Aftermath: The Human and Ecological Consequences of Nuclear War* (New York: 1983); and Carl Sagan, Paul R. Ehrlich, Donald Kennedy, and Walter Orr Roberts, *The Cold and the Dark: The World After Nuclear War* (New York: 1984).

INDEX

A

ABMs. *See* Anti-ballistic missiles
ABM Treaty, 1972, 142, 211–12
  and Iceland summit conference, 1986, 220
  and nuclear arms talks, 1986, 217
Acheson, Dean
  and Acheson-Lilienthal plan, 1946, 40–42
  and hydrogen bomb decision, 55–56
Action-reaction cycle, 114–15
Afghanistan invasion, and demise of arms
  control efforts, 180–81
Agreement on Preventing Nuclear War,
  1973, 148
Aide-mémoire of Roosevelt and Churchill,
  10, 11, 13
Air-launched cruise missiles (ALCMs), 168,
  188, 218, 219
Allen, Richard V., 185
ALSOS, 12
American Civil Liberties Union, and nuclear
  freeze movement, 195
American Friends Service Committee, and
  Pugwash conference on nuclear arms
  race, 84
American Security Council, 178
Andropov, Yuri, 209
  and intermediate-range nuclear forces
  (INF), 199–201
Antarctic Treaty, 1959, 90
Anti-ballistic missiles (ABMs), 110–11, 116,
  118–21, 125, 135, 138–39

  and resumption of testing, 1961, 102
  see also ABM Treaty
Anti-missile research, 217
Anti-nuclear movement, and counterforce
  strategy, 192
Anti-satellite weapons, 188
Anti-Sovietism, of Reagan, 185–87
Arab-Israeli War, 1973, 154, 224
Argentina, Reagan and sales of nuclear
  weapons to, 205
Argus test series, 88
Arisue, Seijo, on bombing of Hiroshima, 26
Armed forces
  509th Composite Group, Air Force, and
  atomic bomb, 13, 25
  intermediate defense condition (DEFCON)
  level alert, 149
  and military-industrial complex, 71–72
  New Look, 63–65
  and nuclear deterrence policy, 50–51
Arms control
  actions, 1982, 204–5
  Baruch plan, 42–45
  Bohr and, 7–8
  Bush plan, 33
  domestic control of, 38–40
  failure of, 181, 227–30
  and Moscow Conference of Foreign
  Ministers, 1945, 34–36
  and nuclear deterrence policy, 50–51
  opportunity, 1961, 93
  Reagan's deemphasis of, 184

Arms control (*continued*)
 Stimson plan, 30–31
 umbrella talks proposal, 1984, 213–14
 see also Atomic bomb; Disarmament;
   Nuclear weapons; Test ban
Arms Control and Disarmament Agency, 185
 and nuclear arms talks, 1986, 217
 purge, 1972–1973, 146–47
ASAT (anti-satellite) program, 212–13
Aspin, Les, and Iceland summit conference,
   1986, 220
Atlantic Conference (1941), 10
Atlas missiles, 63–64, 69, 71, 73
Atomic bomb
 China, 75
 dropped on Hiroshima, 26–27
 dropped on Nagasaki, 27–28
 France, 73
 India, 75, 159–60
 Interim Committee recommends use against
   Japan, 15–17
 Japan as target of first, 12–13
 and Korean War, 58–59
 preparation for use on Japan, 12–13,
   25–26
 Roosevelt, Soviet Union, and, 8–11
 SANDSTONE test series, 1946, 51–52
 scientific opposition to use of, 17–18
 Soviet, 53
 Stimson reconsiders, 29–30
 TRINITY exploded, 1945, 22–23
 Truman and international control of, 32–33
 see also Arms control; Disarmament;
   Nuclear weapons; Test ban
Atomic Development Authority (ADA), and
   Acheson-Lilienthal plan, 41–42
Atomic diplomacy, at London and Moscow
   Conferences of Foreign Ministers, 1945,
   31, 34–36
Atomic energy. *See* Atomic bomb; Arms
   control; Nuclear weapons; Test ban
Atomic Energy Act, 1954, 74–75
Atomic Energy Commission (AEC), 6
 and hydrogen bomb decision, 55–58
 and Oppenheimer's security clearance,
   60–61
 and test ban issue, 1958, 85–86
 and threat of radioactive fallout, 75–76
Atomic Scientists of Chicago, 32
Atoms-for-Peace Plan, 74–75
Attlee, Clement, and international control of
   atom, 32, 33
Augusta, U.S. cruiser, 26
Australia, and horizontal nuclear weapons
   proliferation, 160

B

Backfire bomber, 157–59, 164, 178, 179
Ballistic missiles
 defense system (BMD), 210, 215, 216,
   217, 219, 221
 see also names of missiles (Atlas;
   Minuteman) *and types of missiles*
   (Intercontinental; Sea-launched)
Bard, Ralph A., and Interim Committee, 15
Barnard, Charles, and Acheson-Lilienthal
   plan, 1946, 40
Barnet, Richard, and nuclear freeze
   movement, 193
Baruch, Bernard, and Baruch plan, 1946,
   42–45
Berkner, Lloyd, and Berkner Panel, test ban
   conference, Geneva, 1958–1959, 89–90
Berlin
 blockade, 1948, 49, 224
 crisis, 1961, 100–101, 224
 and division of Germany, 48–50
 and Sputnik diplomacy, 70
 wall, 101
Bethe, Hans
 and Safeguard ABM, 132
 and Teller's campaign for hydrogen bomb,
   54
 and test ban issue, 1958, 85
Bevin, Ernest, and division of Germany, 48
Bikini Atoll, testing at, 44, 75
Bison bomber, 177
Bissel, Raymond Jr., and U-2 photo-
   reconnaisance program, 67
BMD. *See* Ballistic missiles, defense system
Boeing Corp., and military-industrial
   complex, 71
Bohr, Niels, 3–4, 10, 226
 and international control of atomic energy,
   7–8
Bombers
 Bison, 177
 B-1, 132, 162, 166–68, 188, 203
 bonfire, 124
 and decline of SALT II, 157–58
 gap, 65–67
 and multiple warheads, 117
Borden, William, and Teller's campaign for
   hydrogen bomb, 54
Bradbury, Norris
 and Limited Test Ban Treaty, 1963, 111
 and Teller's campaign for hydrogen
   bomb, 54
Bradley, Omar
 and Korean War, 59

and Teller's campaign for hydrogen
bomb, 54
BRAVO nuclear test, 75, 76
Brazil, and horizontal nuclear weapons
proliferation, 160
Breslauer, George W., and debate over
détente, 154
Brezhnev, Leonid, 199
and Cuban missile crisis, 1962, 106
and SALT I, 141
and SALT II, 148, 157–159, 166, 173,
174
and START, 203–4
and Vienna summit conference, June 1979,
174, 176–78
and Vladivostok Accord, 1974, 151–52
Briggs, Lyman C., 5
Brinksmanship, Khrushchev on, 65–66
Broomfield, William S., and nuclear freeze
movement, 195
Brown, Harold
and ABM system, 121
and crisis in Horn of Africa, 170–71
and Limited Test Ban Treaty, 1963,
110–11
and MX ICBMs, 175
and SALT II, 173, 174, 177
Brzezinski, Zbigniew
in Carter administration, 162–64
and crisis in Horn of Africa, 170–71
and MX ICBMs, 175–76
and SALT II, 173, 177, 180
Buchanan, Patrick, and Iceland summit
conference, 1986, 220
Build-down proposal, and START, 1983,
208–9
Bulganin, Nicolai
and Open Skies Plan, 1955, 78–79
and test ban issue, 79–80, 84–85
Bulgaria, and atomic diplomacy, 1945, 31
Bundy, Harvey, 11
Bush, Vannevar
and Acheson-Lilienthal plan, 40–41
and Bush plan, 33
and Interim Committee, 15
and Manhattan Project, 5–10
and Stimson plan, 30
Byrd, Robert, and demise of SALT II, 180
Byrnes, James, 24, 29, 47
and division of Germany, 48
and Interim Committee, 15–16
at London Conference of Foreign
Ministers, 1945, 31
and Moscow Conference of Foreign
Ministers, 1945, 34–37

and scientists' opposition to using atomic
bomb, 17

C

Cadell, Pat, and demise of SALT II, 180
Canada
and Bush plan, 33–34
and horizontal nuclear weapons
proliferation, 159–61
Soviet spy ring, and U.S. domestic control
of atomic energy, 39–40
and Three Nation Declaration, 1945, 34
Caron, George, and bombing of Hiroshima,
26
Carter, Jimmy, 230
and arms control efforts, 181
and crisis in Horn of Africa, 170–71
decision to cancel B-1 bomber, 1977,
166–68
and decline of nuclear nonproliferation
regime, 181–83
and MX ICBMs, 174–76
and SALT II, 164–69, 172–73
and SALT II Treaty, 229; ratification
debate, 179
transformation of idealist, 162
Vance and Brzezinski in administration of,
162–64
and Vienna summit conference, June 1979,
174, 176–78
CASTLE nuclear test series, 75, 76
Castro, Fidel, 92
and Bay of Pigs invasion, 100
Central Intelligence Agency, and Bay of Pigs
invasion, 100
Chain reaction, 4
Chamber of Commerce, U.S., and military-
industrial complex, 72
Chambers, Whittaker, 57
China
and ABM system, 121
and disarmament talks, 1954–1955, 77
and horizontal nuclear weapons
proliferation, 161
and Japanese surrender, 1945, 28
and Korean War, 58–59
nuclear device, 75
and nuclear nonproliferation treaty, 123
offshore islands crisis, 1950s, 63, 70, 224
Reagan and sales of nuclear weapons to,
205
and SALT II, 173
and Sputnik diplomacy, 70

Church, Frank, and SALT II, 179
Churchill, Winston, 6, 8, 10–11
    and first atomic bomb, TRINITY, 13,
        22–23
    Iron Curtain speech, 38
    and Potsdam Declaration, 24–25
Citizen's Committee for a Nuclear Test Ban,
    111
Clark, Dick, and debate over détente, 155
Clark, William P., on goal of Reagan
        strategy, 186
Clayton, William L., and Interim Committee,
    15
Clinch River breeder reactor, 182
Coalition for Peace Through Strength, 178
Coexistence, peaceful, 129
Cohen, Benjamin V., 35
Cohen, William, build-down proposal, 1983,
    208
Cohen-Pasvolsky plan, and Moscow
        Conference of Foreign Ministers, 1945,
        35, 37
Cold War, 223–24
    and Berlin blockade, 49
    Kennedy and, 98–100
    Nixon and, 127–28
    and U.S. containment policy, 38
Collins, John M., and Iceland summit
        conference, 1986, 220
Cominform, and Molotov Plan, 48
Committee on the Present Danger, 178, 185,
    187
Compartmentalization, 6
Comprehensive test ban treaty (CTB), 205
    demise, 1978, 181
Compton, Arthur H., and Interim Committee,
    15–16
Compton, Karl T., and Interim Committee,
    15–16
Conant, James B.
    and Acheson-Lilienthal plan, 1946, 40
    and Interim Committee, 15
    and Manhattan Project, 5–11
Conference of Foreign Ministers. *See* Foreign
        ministers conference
Conference on the Discontinuance of Nuclear
        Weapons Tests. *See* Test ban
Congressional Office of Technology
        Assessment, 189
Connally, Tom, and Moscow Conference of
        Foreign Ministers, 1945, 35
Containment policy, 37–38
    and Truman Doctrine, 47
Council of Foreign Ministers. *See* Foreign
        ministers conference

Counterforce option, 131
    counterattack on, 189–91
    and nuclear freeze movement, 193–96
    reaction of Europe to, 191–93
Cousins, Norman, and test ban talks, 1962,
    108
Critical mass, 4
Cruise missiles, 149, 157–59, 162, 168, 203,
    219
Cuba
    and Bay of Pigs invasion, 100
    and crisis in Horn of Africa, 170–71
    and demise of SALT II, 179–80
    missile crisis, 1962, 103–6
    and Sputnik diplomacy, 71
Czechoslovakia, communist coup, 1948, 48,
    126

D

Dallek, Robert, 10, 11, 185
    on nuclear freeze movement, 196
Damage limitation strategy, 95–97
Daniloff, Nicholas S., 218
Davies, Joseph E., and Truman's policy
        toward Soviets, 14
"The Day After," film, 192
Dean, Arthur, and resumption of test ban
        talks, 1962, 108
De Gaulle, Charles, 73
    and test ban efforts, 1960, 92
Deng Xiaoping, and SALT II, 173
Dense-pack deployment plan for MX, 205–6
Détente
    and invasion of Afghanistan, 181
    and Reagan, 186
    U.S. debate over, 153–57
Dien Bien Phu, and massive retaliation, 62
Diplomacy
    atomic, 31, 34–36
    back channel, 137–38, 141
    Sputnik, 70–71
Disarmament
    and Khrushchev, 1955–1956, 77–78
    talks, 1954–1955, 76–77
    see also Arms control; Atomic bomb;
        Nuclear weapons; Test ban
Disarmament Subcommittee, London, 1957,
    83
Dobrynin, Anatoly
    and back channel, 137
    and SALT I, 126
    and SALT II, 174
Double build-down, 208

Draft, peacetime, and U.S. nuclear
  deterrence policy, 50
Dubcek, Alexander, and SALT, 126
Dulles, John Foster
  and ballistic missiles, 64
  and massive retaliation, 61–63
  as secretary of state, 61
  and test ban issue, 80–86

E

Eaton, Cyrus, and Pugwash conference on
  nuclear arms race, 84
Eden, Anthony, and Open Skies Plan, 1955,
  78–79
Egypt
  and bomber gap, 66–67
  and Sputnik diplomacy, 70
Einstein, Albert, 3, 4
  calls for end to nuclear testing, 75
Eisenhower, Dwight D., 225, 227
  and Atoms-for-Peace plan, 74–75
  and bomber gap, 65–67
  and bombing of Japan, 25–26
  and disarmament talks, 1954–1955, 76–77
  and massive retaliation, 61–63
  and military-industrial complex, 71–72
  and missile gap, 69
  and New Look for armed forces, 63–65
  and nuclear arms race, 60
  and nuclear peril, 74
  and Open Skies Plan, 1955, 78–79
  and Sputnik diplomacy, 70–71
  and temper of times, 60–61
  and test ban conference, Geneva, 1958–
    1959, 87–90
  test ban efforts, 1960, 90–92
  and test ban issue, 1956–1958, 79–87
  and threat of radioactive fallout, 75
Election
  of 1960, 92
  of 1964, 115
  of 1984, and umbrella talks proposal,
    213–14
Elugelab, and MIKE test, 59
Encryption, 172, 173
Enhanced radiation (neutron) weapons, 162,
  169
Enola Gay, B-29, and bombing of
  Hiroshima, 26
Ethiopia, and crisis in Horn of Africa,
  170–71
Ethridge, Mark, and Truman Doctrine, 47
EURATOM (European Atomic Energy
  Agency), 122–23

Europe
  and Marshall Plan, 47–48
  reaction to Reagan's counterforce strategy,
    191–93
  and zero option plan for intermediate-range
    nuclear forces (INF), 197–99
Explorer satellite, 68
Export-Import Bank, 152

F

Fallout shelters, and ABM system, 119
"Fat Man" plutonium bomb, 24, 28
Faure, Edgar, and Open Skies Plan, 1955, 79
Federation of Atomic Scientists, 32
F-18 Hornets, 188
Fermi, Enrico, 4, 6
  and hydrogen bomb, 53, 55
  and Interim Committee, 15
F-4 Phantom attack planes, 188
Finletter, Thomas, and U.S. nuclear
  deterrence policy, 50–51
Firestone, Bernard, on Limited Test Ban
  Treaty, 1963, 112
First-strike capability, 114
"Fiscal Year 1984–1988 Defense Guidance,"
  187–88
FitzSimons, Louise
  on Bay of Pigs invasion, 100
  on Cuban missile crisis, 1962, 105
Flexible response strategy, 95, 131
Ford, Gerald, 150
  and debate over détente, 154
  and SALT II, 158
  and Vladivostok Accord, 1974, 151–52
Foreign ministers conference
  London, 1945, 29, 31
  Moscow, 1945, 34–37
Forrestal, James V.
  and Japanese surrender, 28
  and Moscow Conference of Foreign
    Ministers, 1945, 35
  and Stimson plan, 30
  and U.S. option to negotiate end of war
    with Japan, 21
  view of Soviet Union, 14
Forsberg, Randall, and nuclear freeze
  movement, 193
Forward-based systems (FBS), 133
Foster, John S. Jr.
  and ABM system, 119
  and multiple warheads, 116
Foster, William, and test ban talks, 1962,
  107

Fractionation freeze, 171
France
  and Antarctic Treaty, 1959, 90
  and division of Germany, 48
  and horizontal nuclear weapons
    proliferation, 160–61
  and intermediate-range nuclear forces
    (INF), 200–201
  nuclear device, 73, 75
  and nuclear nonproliferation treaty, 123
  nuclear weapon test, 1962, 107
  and Open Skies Plan, 1955, 79
  and test ban issue, 1957, 82
Franck, James, 226
  and international control of atom, 32
  and scientists' opposition to using atomic
    bomb, 17–18
Frankfurter, Felix, 8
Frisch, Otto R., 3
FRODs (functionally related observable
  differences), 177
Fuchs, Klaus, 57
  and first atomic bomb, TRINITY, 23
Fulbright, William, and SALT, 126

G

Gaither, Rowan, and report on missile gap,
  68–69
Gallery, Daniel, and U.S. nuclear weapons
  arsenal, 1947–1949, 52
Galosh missiles, 116, 121
Garcia Robles, Alfonso, 161
  and horizontal nuclear weapons
    proliferation, 161
Gavin, James M., and armed forces New
  Look, 65
General Advisory Committee (GAC), Atomic
  Energy Commission and hydrogen bomb
  decision, 55
General Dynamics Corp., and military-
  industrial complex, 71
Geneva
  and SALT II, 1973–1974, 148
  START talks, 1985, 214–16
  summit conference, 1955, and Open Skies
    Plan, 78–79
  summit conference, 1985, 216
  test ban conference, 1958–1959, 86,
    87–90
George, Alexander, and debate over détente,
  154
George Washington, U.S.S., 64
Germany

division into Democratic Republic (East)
  and Federal Republic (West), 48–50
and European reaction to counterforce
  strategy, 192
see also Berlin; West Germany
Gilpatric, Roswell, and Cuban missile crisis,
  1962, 103
Ginsburg, Alexander, 172
GLCMs. *See* Ground-launched cruise missiles
Goldwater, Barry
  and multiple warheads, 118
  and Vladivostok Accord, 1974, 152
Gorbachev, Mikhail S.
  and Geneva summit conference, 1985, 216
  and Iceland summit conference, 1986,
    218–21
  and nuclear arms talks, 1986, 216–18
  and START talks, 1985, 215–16
Gore, Albert, and test ban conference,
  Geneva, 1958–1959, 88–89
Graham, Billy, and nuclear freeze movement,
  194
Graham, Daniel O., and Star Wars, 210
Gray, Colin, and preparation for protracted
  nuclear warfare, 187
Great Britain
  and Antarctic Treaty, 1959, 90
  atomic research, 5
  and Bush plan, 33–34
  and division of Germany, 48–49
  and European reaction to counterforce
    strategy, 192
  and horizontal nuclear weapons
    proliferation, 160
  and intermediate-range nuclear forces
    (INF), 200–201
  and Japanese surrender, 1945, 28
  and Manhattan Project, 5–11
  missile launch sites, 73
  and nuclear nonproliferation treaty, 123
  nuclear tests, 1958, 87
  Polaris missiles sold to, 96
  and test ban issue, 1957, 82
  and Three Nation Declaration, 1945, 34
Greece, and Truman Doctrine, 46–47
Green Party, West Germany, and European
  reaction to counterforce strategy, 192
Grew, Joseph
  and Potsdam Declaration, 24
  and U.S. strategy toward Japan, 1945,
    20–22
Gromyko, Andrei
  and SALT I, 129
  and SALT II, 165, 168–69, 171–72
  and START, 1985, 214

and test ban issue, 1958, 86
and umbrella talks proposal, 1984, 214
Gromyko plan, and Baruch plan, 1946,
    44–45
Ground-launched cruise missiles (GLCMs),
    162, 168, 219
Ground Zero education week, and nuclear
    freeze movement, 194
Groves, Leslie
    and Acheson-Lilienthal plan, 1946, 40
    and ALSOS, 12
    and Baruch plan, 1946, 43
    and bombing of Japan, 25, 26
    and domestic control of atomic energy, 39
    and first atomic bomb, TRINITY, 22
    and Manhattan Project, 6–7, 12–13
    and Stimson plan, 30–31
Guhin, Michael A., on Dulles, 61

H

Hahn, Otto, 3
    and test ban issue, 1957, 81
Haig, Alexander, 185
    and START, 202
    and zero option plan for intermediate-range
        nuclear freeze forces (INF), 197–98
Halperin, Morton, and nuclear freeze
    movement, 195
HARDTACK nuclear test series, 84, 85
    and test ban conference, Geneva, 1958–
        1959, 87–88, 90
Harmon, Hubert R., and U.S. nuclear
    arsenal, 1947–1949, 52
Harriman, W. Averell
    and limited test ban agreement, 1963, 109
    and Moscow Conference of Foreign
        Ministers, 1945, 34
    view of Soviet Union, 14
Harrison, George
    and Interim Committee, 15
    and scientists' opposition to using atomic
        bomb, 18
Hatfield, Mark, and nuclear freeze
    movement, 194
Helsinki, and SALT I, 133, 136–38
Heritage Foundation, 178
    and Star Wars, 210
Hirohito, Emperor
    and Japanese surrender, 28
    and U.S. option to negotiate end of war
        with Japan, 20–21
Hiroshima, bombed, 26–27
Hiss, Alger, 57

Hooper, S. C., 4
Hopkins, Harry, 15
    and U.S. option to negotiate end of war
        with Japan, 21
Horizontal nuclear proliferation, 231
Hostage crisis, Iran, and SALT II, 180
Huffman, James, and atomic bomb tests at
    Bikini Atoll, 1946, 44
Hull, Cordell, and Potsdam Declaration,
    1945, 24
Humphrey, Hubert, and test ban, 85, 88, 91
Hungary
    and bomber gap, 66–67
    and Sputnik diplomacy, 70
    and test ban issue, 1956, 80–81
Hunthausen, Raymond, and nuclear freeze
    movement, 194
Hydrogen bomb
    decision on, 55–58
    Soviet, 64
    Teller's campaign for, 53–55

I

ICBMs. *See* Intercontinental ballistic missiles
Iceland, summit conference, 1986, 218–21
Iklé, Fred, and ACDA purge, 1972–1973,
    146
India
    nuclear device, 75, 159–60
    and nuclear nonproliferation treaty, 123–24
    Reagan and sales of nuclear weapons to,
        205
    war with Pakistan, 1971, 141
Indianapolis, U.S. cruiser, 25
INF. *See* Intermediate-range nuclear forces
Intercontinental ballistic missiles (ICBMs),
    63–64, 69, 73, 104, 114, 116, 125,
    134, 135, 137, 138, 139, 149, 157, 168,
    176, 177, 189, 203, 215
    Atlas, 63–64, 69, 71, 73
    and damage limitation strategy, 96–97
    and Interim Offensive Arms Agreement,
        1972, 143
Interim Committee, 15–17
Interim Offensive Arms Agreement, 1972,
    143, 168
Intermediate-range ballistic missiles (IRBMs),
    64, 70, 73, 103, 162, 198
Intermediate-range nuclear forces (INF), 197
    stalemate in negotiations over, 201
    Walk-in-the-Woods formula for, 199
    zero option plan for, 198
International Atomic Energy Agency (IAEA),
    75, 182

Int'l/Atomic Energy Agency (*continued*)
  and horizontal nuclear weapons
    proliferation, 159–60
  and nuclear nonproliferation treaty, 122–23
Iran
  hostage crisis, and demise of SALT II, 180
  and last phase of SALT II, 1978–1979,
    173
IRBMs. *See* Intermediate-range ballistic
    missiles
Iron Curtain speech of Churchill, and U.S.
    containment policy, 38
Italy, missile launch sites, 73

J

Jackson, Henry
  and ABM system, 120
  and ACDA purge, 1972–1973, 146–47
  and Carter's decision to cancel B-1
    bomber, 167
  and Limited Test Ban Treaty, 1963, 111
  and military-industrial complex, 71
  and Moscow summit, 1972, 143
  and nuclear freeze movement, 195
  and SALT I, 145
  and SALT II, 158, 168–69, 173
  and U.S.-Soviet trade, 1972–1975, 153
  and Vladivostok Accord, 164
Jackson-Vanik Amendment
  and debate over détente, 154
  and Soviet-American trade, 153
Jackson-Warner Amendment, and nuclear
    freeze movement, 195
Japan
  Hiroshima bombed, 26–27
  Interim Committee recommends atomic
    bombing of, 15–17
  Nagasaki bombed, 27–28
  option to negotiate end of war with, 20–22
  and Potsdam Declaration, 24–25
  preparation for bombing, 1945, 25–26
  Soviet participation in war against, 14–15
  surrender, 28
  U.S. strategy toward, 18–20
Jerusalem Post, 193
Johnson, Edwin C., 38–39
Johnson, Louis, and hydrogen bomb
    decision, 55–56
Johnson, Lyndon Baines
  and ABM system, 119–20
  and military-industrial complex, 72
  and multiple warheads, 115–18
  nuclear doctrine, 113–15
  and nuclear nonproliferation treaty, 124

  and SALT I, 1964–1968, 124–26
  world view, 113
Joint Chiefs of Staff
  and ABM system, 120
  and arms control, 204, 230
  and Baruch Plan, 43
  and nuclear arsenal, 52
  and resumption of nuclear testing, 1961,
    102
  and SALT I, 140, 145
  and SALT II, 147–48
  and Teller's campaign for hydrogen bomb,
    54–55
  and Vladivostok Accord, 1974, 151–52
  and Walk-in-the-Woods formula for
    intermediate-range nuclear forces (INF),
    199
Joint Statement of Principles and Basic
    Guidelines for Subsequent Negotiations
    on the Limitations of Strategic Arms,
    1979, 176
Jones, David
  and counterattack on Reagan's counterforce
    strategy, 189–90
  and SALT II agreements, June 1979, 177
Jupiter missiles, 64, 73

K

Kampelman, Max M., and START talks,
    1985, 214
Katzenbach, Nicholas, and SALT I, 126
Kendall, Henry, and counterattack on
    Reagan's counterforce strategy, 189–90
Kennan, George
  on threat of nuclear war, 193
  and Truman Doctrine, 47
  and U.S. containment policy, 38
  and U.S. nuclear weapons arsenal, 1947–
    1949, 52
Kennedy, Edward, and nuclear freeze
    movement, 194
Kennedy, John F., 230
  and arms control, 1961, 93
  and Bay of Pigs invasion, 100
  and Berlin crisis, 1961, 100–101
  and Cold War, 98–100
  and Cuban missile crisis, 1962, 103–6
  damage limitation strategy, 95–97
  and flexible response strategy, 95
  and limited test ban agreement, 1963,
    109–12
  and military-industrial complex, 72
  and missile gap, 1961, 94

and nuclear superiority, 97–98
and nuclear testing, 1961, 101–2
and Oppenheimer's security clearance, 61
and test ban talks, 1962, 106–9
Kennedy-Hatfield Amendment, 194
and nuclear freeze movement, 195
Kent, Glenn, and START build-down
proposal, 1983, 208
Khomeini, Ayatollah, and last phase of
SALT II, 1978–1979, 173
Khrushchev, Nikita
and arms control, 1961, 93
and Berlin crisis, 1961, 101
and bomber gap, 65–67
and brinksmanship, 65–66
and Cuban missile crisis, 1962, 103–6
and disarmament, 1955–1956, 77–78
and Eisenhower test ban efforts, 1960, 92
and limited test ban agreement, 1963,
109–10
and nuclear testing, 1961, 101–2
and Open Skies Plan, 1955, 79–80
and peaceful coexistence, 129
and SALT I, 124, 128–29
and Sputnik diplomacy, 70–71
and test ban, 80, 85–86, 89, 108–9
Kido Koichi, and U.S. option to negotiate
end of war with Japan, 21
Killian, James
and Safeguard ABM, 132
and test ban, 83, 85, 89
King, Ernest J., and U.S. strategy toward
Japan, 20
Kissinger, Henry
and back channel to Soviets, 137–38
and détente, 153–57
and horizontal nuclear weapons
proliferation, 159–61
and massive retaliation, 64
and Moscow summit, 1974, 150, 150
and pressure for MIRV ban, 134
and SALT I, 128–45
and SALT II, 147–50, 157–59
and summit conference, Moscow, 1972,
142–44
and trade with Soviets, 1972–1975,
152–53
and Vladivostok Accord, 1974, 151–52
Kistiakowsky, George
and first atomic bomb, TRINITY, 22
and Safeguard ABM, 132
Korean airliner, shot down by Soviet air
force, 1983, 209
Korean War
and massive retaliation, 62

and NSC-68, 58–59
Kornienko, Georgy, and SALT II proposals,
March 1977, 166
Kosygin, Alexei
and ABM system, 120
and SALT I, 125
Kuboyama, Aikichi, 75
Kurchatov, Igor, and Soviet atomic bomb, 23
Kuznetsov, Vasily
and arms control, 1961, 93
and test ban talks, 1962, 108
Kvitsinsky, Yuli
and stalemate in negotiations over INF,
201
and Walk-in-the-Woods formula for INF,
199
and zero option plan for INF, 198

L

Laird, Melvin
and pressure for MIRV ban, 134
and SALT I, 140
Lapp, Ralph, and threat of radioactive
fallout, 76
Lasers, 210, 211
Latin America, and Treaty of Tlatelolco,
1967, 121
Latter, Albert, and test ban conference,
Geneva, 1958–1959, 89
Lawrence, Ernest O.
and Interim Committee, 15
and test ban issue, 1957, 82
Leahy, William D., and U.S. strategy toward
Japan, 19
Lehman Brothers Investment Corp., 4
Lend-Lease Agreement, 1972, 152–53
Leningrad ABM, 121
Lilienthal, David
and Acheson-Lilienthal plan, 1946, 40–41
and Baruch plan, 1946, 43
and hydrogen bomb decision, 56
and U.S. nuclear weapons arsenal, 1947–
1949, 51–52
Limited Test Ban Treaty, 1963, 110–12, 211
Lippmann, Walter, and Cuban missile crisis,
1962, 105
"Little Boy" uranium bomb, 13, 24, 26
Lockheed Corp., and military-industrial
complex, 71
Long Beach, nuclear-powered cruiser, 73
Long Telegram, of George Kennan, 38
Lucas, Scott, and atomic bomb tests at Bikini
Atoll, 1946, 44

Lucky Dragon, tuna trawler, 75
Lutheran World Federation, and Pugwash
    conference on nuclear arms race, 84

M

MacArthur, Douglas
    and Korean War, 59
    and U.S. strategy toward Japan, 19
Mackenzie King, William Lyon, 33
Macmillan, Harold
    and Geneva test ban conference,
        1958–1959, 89
    and nuclear testing, 1961, 102
    and test ban, 82, 91–92
Manhatten Project, 5–7
Marshall, George C., 5, 6, 14, 47
    and bombing of Japan, 25
    and first atomic bomb, TRINITY, 23
    and Marshall Plan, 47–48, 224
    and U.S. strategy toward Japan, 19–21
Martin, Joseph W., 7
MaRVs (maneuverable reentry vehicles), 149
Massive retaliation, Eisenhower policy,
    61–63
Maud Committee, Great Britain, 5
May, Alan Nunn, and U.S. domestic control
    of atomic energy, 39
May, Andrew, 38–39
May-Johnson bill, on domestic control of
    atomic energy, 38–39
McCarthy, Joseph
    and secrecy about hydrogen bomb
        decision, 57–58
    witch hunt, 60–61
McCloy, John J., 14
    and Acheson-Lilienthal plan, 40
McCone, John, and test ban, 88, 91
McGovern, George, and nuclear
    superiority, 98
McMahon, Brien, 39
    and Teller's campaign for hydrogen
        bomb, 54
McMahon bill on domestic control of atomic
    energy, 39
McNamara, Robert S., 131–32
    and ABM system, 118–21
    and Cuban missile crisis, 1962, 103
    and damage limitation strategy, 1962,
        95–97
    and Johnson's nuclear doctrine, 114–115
    and missile gap, 1961, 94
    and multiple warheads, 115–17
    and nuclear superiority, 98
    and nuclear testing, 1961, 102

and test ban talks, 1962, 107
NcNaughton, Frank, and domestic control of
    atomic energy, 39
Meitner, Lise, 3
Mexico
    and horizontal nuclear weapons
        proliferation, 161
    and Tlatelolco Treaty, 1967, 121
Middle East war, and SALT II stalemate,
    1973–1974, 148–49
Midgetman missiles, 188, 215–16
MIKE, 59
Military-industrial complex, 71–72, 225–27
Mills, Mark, and test ban issue, 82
Minuteman missiles, 64, 73, 115, 116, 149,
    165
Miroff, Bruce, on Kennedy and Cold War,
    99
MIRVs. *See* Multiple independently
    targetable reentry vehicles
Missiles
    and decline of SALT II, 157–58
    gap, 68–73, 94
    see also names of missiles (Atlas;
        Minuteman) *and types of missiles*
        (Intercontinental; Sea-launched)
Missouri, U.S.S., 28
Molotov, Vyacheslav, 14
    and Bush plan, 34
    and first atomic bomb, TRINITY, 23
    and London Conference of Foreign
        Ministers, 1945, 31
    and Molotov plan, 48, 224
    and Moscow Conference of Foreign
        Ministers, 1945, 36
Mondale, Walter, 213
Moscow
    foreign ministers conference, 1945, 34–37
    summit conference, 1972, 142–44
    summit conference, 1974, 150
Multilateral nuclear force (MLF), NATO,
    122
Multiple independently targetable reentry
    vehicles (MIRVs), 115–18, 149–50,
    157, 165, 167, 168, 171, 176, 203, 207,
    209
    and Interim Offensive Arms Agreement,
        1972, 143
    pressure for ban on, 134
Multiple reentry vehicles (MRVs), 115
Multiple warheads, 115–18
Murrow, Edward R., on U.S. public's
    response to Sputnik, 68
Mutual assured destruction (MAD) doctrine,
    114, 187

MX missiles, 149, 155, 162, 174–76, 188, 165, 203, 215, 230
  and Scowcroft Commission, 205–7

N

Nagasaki bombed, 27–28
The Nation, 72
National Defense Research Committee (NDRC), 5
National Intelligence Estimate, 1961, 69
National Security Council, 49, 50, 51, 52
  and hydrogen bomb decision, 55–56
  and massive retaliation, 62–63
  and MX ICBMs, 175–76
  Verification Panel, 129
NATO. *See* North Atlantic Treaty Organization
Navstar navigation satellites, 212
Nehru, Jawaharlal, calls for end to nuclear testing, 75
Neutron weapons, 162, 169
  and European reaction to counterforce strategy, 191
New Republic, 83
New York Times, 72, 83, 187
Nigeria, and horizontal nuclear weapons proliferation, 161
Nimitz, Chester, and Baruch plan, 43
Nitze, Paul, 185
  and intermediate-range nuclear forces (INF), 198, 199, 201
Nixon, Richard
  and ABM system, 120
  and ACDA purge, 1972–1973, 146–47
  and Cold War, 127–28
  and Moscow summit, 1972, 142–44
  and Moscow summit, 1974, 150
  and Nixon doctrine, 130–32
  and pressure for MIRV ban, 134
  resigns, 150
  and SALT I, 127–30, 133–37, 139–42, 145
  and SALT II, 147–49
Nonproliferation Act (1978), 182
North American Defense Command (NORAD), and counterattack on Reagan's counterforce strategy, 190
North Atlantic Treaty Organization (NATO), 49, 73, 224
  and Berlin blockade, 49
  and damage limitation strategy, 1962, 96–97
  multilateral nuclear force (MLF), 122

  and nuclear weapons, 64
  and Open Skies Plan, 1955, 78–79
  and SALT II, 169–70, 178
  and zero option plan for intermediate-range nuclear forces (INF), 197–99
NSC-20, 51
NSC-30, 51
NSC-57, 52
NSC-68, and Korean War, 58–59
Nuclear fission, 3
Nuclear freeze movement, 193–96
Nuclear Nonproliferation Treaty (NPT), 1967, 122–23, 205
  and horizontal proliferation, 159–61
  and sales of nuclear weapons, 205
Nuclear Regulatory Commission, U.S., and horizontal nuclear weapons proliferation, 160
Nuclear Threshold Test Ban Treaty, 204
Nuclear weapons
  "clean" bomb, 82–83
  curbing worldwide dissemination of, 121–24
  decline of nonproliferation regime, 181–83
  deterrence, 50–51, 224–25
  Eisenhower and arms race, 60
  embryonic arsenal, 1948–1950, 51–52
  flexible response strategy, 95
  HARDTACK tests, 84, 85
  horizontal proliferation, 159–61, 231
  Johnson's doctrine, 113–15
  and massive retaliation, 61–63
  and power politics, 222–23
  Reagan's buildup of, 184, 187–89
  superiority, 97–98
  testing, 87, 101–2, 219
  see also Arms control; Atomic bomb; Disarmament; Test ban
Nunn, Sam
  build-down proposal, 1983, 20
  and Iceland summit conference, 1986, 220

O

Office of Scientific Research and Development (OSRD), 5
Ohio, Trident submarine, 140
Olympic Games
  and demise of SALT II, 180
  Soviet refusal to participate, 1984, 209
On the Beach (Shute), 84
Open Skies Plan, 1955, 78–79
Oppenheimer, J. Robert, 6, 226
  and Acheson-Lilienthal plan, 40–41

Oppenheimer, J. Robert (*continued*)
  dismissal, 60–61
  and first atomic bomb, TRINITY, 22
  and hydrogen bomb decision, 55, 55–57
  and Interim Committee, 15
  on Manhattan Project after German
    surrender, 12
  and McCarthy red scare, 60–61
  and Teller's campaign for hydrogen bomb,
    54
Outer Space Treaty, 1967, 122, 211

P

Pachter, Henry, on Cuban missile crisis,
    1962, 105
Pakistan, war with India, 1971, 141
Panofsky, Wolfgang
  and demise of arms control efforts, 181
  and test ban conference, Geneva,
    1958–1959, 88
Paris Peace Agreement, and Vietnam War,
    154
Pasvolsky, Leo, and Moscow Conference of
    Foreign Ministers, 1945, 35
Pauling, Linus, and test ban issue, 1957, 81
Peaceful coexistence, 129
Peaceful Nuclear Explosions (PNE) Treaty,
    150, 204
  and Iceland summit conference, 1986, 218
Pearson, Drew, and domestic control of
    atomic energy, 39
Pegram, George B., 4
Pentagon
  and failure of arms control, 230
  and multiple warheads, 116
  and nuclear weapons arsenal,
    1947–1949, 52
  and SALT I, 136
  and SALT II, 158
  and test ban issue, 81–83, 85–86
Perle, Richard
  and SALT II Washington agreements,
    1977, 169
  and Vladivostok Accord, 164
  and Walk-in-the-Woods formula for INF,
    199
  and zero option plan for INF, 197
Pershing missiles, 162, 169, 191, 197, 198,
    199, 200, 201, 203, 204, 207, 215, 219,
    230
Pius XII, Pope
  calls for end to nuclear testing, 75
  and test ban issue, 1956, 80
Poland, Stalin's policy toward, 14–15

Polaris missiles, 64, 85, 96–97, 115
Polaris submarines, 140
Political system, and nuclear weapons, 227
Poseidon missiles, 115, 116, 130
Potsdam Conference, and first atomic bomb,
    TRINITY, 22–23
Potsdam Declaration, 24–25
  and Japanese surrender, 28
Power, Thomas, and Limited Test Ban
    Treaty, 1963, 111
Power politics
  and arms race, 222–23
  and nuclear deterrence, 225
Powers, Francis Gary, and U-2 incident, 92
Pravda, and Baruch plan, 44
Presidential Directive 59, 180
President's Air Policy Commission, 50
Profiles in Courage (Kennedy), 99
Project Argus nuclear tests, and Geneva
    conference, 1958–1959, 88
Pugwash conference on nuclear arms race, 84

Q

Quebec Agreement, 34
Quebec Conference, 1943, 8
Quemoy and Matsu, and massive retaliation,
    63

R

Rabi, I. I., and hydrogen bomb decision, 55
Radford, Arthur, and test ban issue, 1957, 82
Radioactive fallout, threat of, 75–76
Raimond, Jean-Bernard, and Iceland summit
    conference, 1986, 220
Rand Corporation, and test ban conference,
    Geneva, 1958–1959, 89
RANIER underground test, 85
Reagan, Ronald
  and ASAT, 212–13
  and counterattack on counterforce strategy,
    189–91
  and Geneva summit conference, November
    1985, 216
  and Iceland summit conference, 1986,
    218–21
  idealogue as president, 185–87
  and intermediate-range nuclear forces
    (INF), 198–201
  leadership team, 184–85
  and MX, 205–7
  and nuclear arms control, 204–5, 230

and nuclear arms talks, 1986, 216–18
nuclear buildup, 184
and nuclear freeze movement, 195–96
and preparation for protracted nuclear
  warfare, 187–89
and START, 201–4, 207–9
and Star Wars, 1983–1984, 210–12
and umbrella talks proposal, 1984, 213–14
and U.S. debate over détente, 153–54
Religions, and nuclear freeze movement, 194
Republican Conference Committee on
  Nuclear Testing, and test ban talks,
  1962, 108
Research Institute for Nuclear Medicine, 27
Reston, James, and test ban issue, 1957, 83
Reuther, Walter, and military-industrial
  complex, 72
Review Conference, NPT, and horizontal
  nuclear weapons proliferation, 161
Rickover, Hyman, and threat of nuclear war,
  193
Ridgeway, Matthew, and armed forces' New
  Look, 65
Rogers, William, and SALT I, 141–42
Roosevelt, Franklin D., 4, 5, 6
accommodationist approach to Soviet
  Union, 223
and plan to use first atomic bomb, 8–13
and Soviet influence in Eastern Europe,
  223
Rostow, Eugene, 185
and START, 202
Rowny, Edwin, 185
Rumania
and atomic diplomacy, 1945, 31
and horizontal nuclear weapons
  proliferation, 161
Rusk, Dean, and test ban talks, 1962, 107
Russell, Bertrand, and Pugwash conference
  on nuclear arms race, 84
Russell, Richard, and ABM system, 120

S

Sachs, Alexander, 4
and Manhattan Project, 13
Safeguard ABM, 132–33, 134
SALT I
and ABM Treaty, 142
and back channel to Soviets, 137–38
and Interim Offensive Arms Agreement,
  143, 168
and Moscow summit, 1972, 142–44
Nixon-Kissinger approach to, 129–30
reaction to, 144–45

round one, 1969, 133
round two, 1970, 134–36
round three, 1970, 136–37
round four, 1971, 138
round five, 1971, 138–40
round six, 1971–1972, 140–41
round seven, 1972, 141–42
Soviet approach to, 128–29
SALT II
beginning, 1972–1973, 147–48
Carter and, 162
and crisis in Horn of Africa, 170–71
decline, 1975–1976, 157–59
demise of, 179–81
Geneva breakthrough and Carter's B-1
  decision, May–June 1977, 166–68
last phase, 1978–1979, 173–74
March 1977 proposals, 164–66
meetings, late 1978, 172–73
meetings, spring–fall 1978, 171–72
and Moscow summit, 1974, 150
stalemate, 1973–1974, 148–50
and trouble with NATO, 169–70
and Vienna summit conference, June 1979,
  174, 176–78
Washington agreements, September 1977,
  168–69
SALT II Treaty, 204, 218, 229, 230
and Iceland summit conference, 1986, 218
ratification debate, 178–79
and Reagan nuclear arms control actions,
  1982, 204–5
SANDSTONE atomic weapons test series,
  51–52
Satellites, 67–68
Sato Naotake, and U.S. option to negotiate
  end of war with Japan, 21
Schindler, Alexander, and nuclear freeze
  movement, 194
Schlesinger, Arthur M. Jr.
and Berlin crisis, 1961, 101
on Kennedy and Cold War, 99
Schlesinger, James
and Cuban missile crisis, 1962, 104–5
and SALT II, 149, 151, 158
and Vladivostok Accord, 1974, 151–52
Schmidt, Helmut, and European reaction to
  counterforce strategy, 192
Schweitzer, Albert, calls for end to nuclear
  testing, 75, 81
Scoville, Herbert, and MX program, 206
Scowcroft, Brent, and Scowcroft
  Commission, 205–7
Seabed Treaty, 1972, 138
Seaborg, Glenn, and test ban, 107–10

Sea-launched cruise missiles (SLCMs), 168, 188
Second-strike capability, 114
Secrecy, and hydrogen bomb decision, 56–58
Seignious, George, and MX ICBMs, 176
Seijo Arisue, and bombing of Hiroshima, 26
Seismographs, and test ban, 85, 107
Seismometers, "deep hole," and test ban, 89–90
Semenov, Vladimir, and SALT I, 137, 139
Senate Disarmament Subcommittee, and test ban issue, 1958, 85
Senate Subcommittee on Preparedness, and military-industrial complex, 72
Sentinel ABM, 132
Shcharansky, Anatoly, 172
Sheer, Robert, and preparation for protracted nuclear warfare, 188
Sherwin, Martin, 10
Shultz, George, 185
    and threat of nuclear war, 193
    and umbrella talks proposal, 1984, 214
    and Walk-in-the-Woods formula for intermediate-range nuclear forces (INF), 199
Shute, Nevil, 84
Sikes, Robert L. F., and ABM system, 119
Silo-killer, 149
Single Integrated Operational Plan (SIOP), 95–96
Skybolt air-to-ground missile, 96
SLBMs. *See* Submarine-launched ballistic missiles
Smith, Gerard
    and ACDA purge, 1972–1973, 146
    and back channel to Soviets, 137–38
    and Moscow summit, 1972, 142
    on Nixon, 127
    and pressure for MIRV ban, 134
    and SALT I, 129–30, 136, 138–42
Smith, Harold, 31
Solzhenitsyn, Alexander, and debate over détente, 154
Somalia, and crisis in Horn of Africa, 170–71
S-1 (Uranium Section of NDRC), 5–6
Sorensen, Theodore
    and arms control, 1961, 93
    and Limited Test Ban Treaty, 1963, 111–12
South Africa, Reagan and sales of nuclear weapons to, 205
Soviet missiles, 135, 149, 165, 169, 171, 178, 197–201, 203, 215, 216
Soviet Union
    and ABM systems, 120–21

    and Acheson-Lilienthal plan, 1946, 40–42
    and Antarctic Treaty, 1959, 90
    and ASAT, 212–13
    atomic bomb, 53
    and atomic diplomacy, 1945, 31
    and Atoms-for-Peace plan, 74–75
    attack on Japanese in Manchuria, 27
    and Baruch plan, 1946, 42–45
    and bomber gap, 65–67
    and Bush plan, 33–34
    and crisis in Horn of Africa, 170–71
    and damage limitation strategy, 1962, 95–97
    and détente, 153–57
    and disarmament talks, 1954–1955, 76–77
    and division of Germany, 48–50
    and failure of nuclear arms control, 227–30
    and horizontal nuclear weapons proliferation, 159–61
    hydrogen bomb, 64, 80
    invasion of Afghanistan, 180
    invasion of Czechoslovakia, 126
    invasion of Hungary, 80–81
    and Japanese surrender, 1945, 28
    and Marshall Plan, 47–48
    and massive retaliation, 61–65
    and missile gap, 68–69
    and Molotov Plan, 48
    and Moscow Conference of Foreign Ministers, 1945, 34–36
    and multiple warheads, 115–17
    and nuclear arms race, 230
    and nuclear deterrence, 224–25
    and nuclear nonproliferation treaty, 123
    nuclear tests, 1958, 87
    Reagan strategy toward, 185–87
    Roosevelt, atomic bomb, and, 8–11
    and SALT I, 128–29, 135–40
    and SALT II, 165–66, 180
    and Sputnik, 67–68
    and Sputnik diplomacy, 70–71
    and START, 202–4, 207–9
    and Star Wars, 210–12
    and test ban, 81–90, 106–9
    and Three Nation Declaration, 1945, 34
    trouble over trade with U.S., 1972–1975, 152–53
    Truman and, 13–15
    and U.S. containment policy, 37–38
    and U.S. nuclear deterrence policy, 50–51
    and U.S. option to negotiate end of war with Japan, 21–22
    and U.S. preparation for protracted nuclear warfare, 187–89
    and U.S. strategy toward Japan, 19–20

and Vladivostok Accord, 1974, 151–52
Sputnik, 67–68
Sputnik diplomacy, 70–71
Stalin, Josef, 9, 14
    and Cold War, 223
    and first atomic bomb, TRINITY, 23
    and Marshall Plan, 47
    and U.S. option to negotiate end of war
        with Japan, 21
    and Yalta agreement, 14
Standard Oil Co., 5
Standing Consultative Commission, 142
    and SALT I, 139
START (Strategic Arms Reduction Talks)
    and build-down proposal, 1983, 208–9
    Eureka proposal for, 202–4
    meetings, 1983, 207
    meetings, 1985, 214–16
    meetings, 1986, 216–18
    proposed new name for SALT, 186
    slow beginnings for, 201–2
Star Wars, 210–12, 217
Stassen, Harold
    and Open Skies Plan, 1955, 79
    and test ban issue, 1957–1958, 82–85
Stealth bomber, 188
Steel, Ronald, and Cuban missile crisis,
    1962, 105–6
Stennis, John, and ABM system, 120
Stevenson, Adlai
    and Cuban missile crisis, 1962, 104
    and nuclear testing, 1961, 102
    and test ban issue, 1956, 80
Stimson, Henry L.
    and bombing of Japan, 25
    and first atomic bomb, TRINITY, 22–23
    and Interim Committee, 15–17
    and Manhattan Project, 5–11
    and plan to use first atomic bomb, 12–13
    reconsideration of atomic bomb, 29–30
    and Stimson plan, 30–31
    and Truman's dealings with Soviets, 14–15
    and U.S. strategy toward Japan, 19–21
Stone, I. F.
    and Cuban missile crisis, 1962, 105
    and nuclear superiority, 98
Stone, Jeremy, and MX program, 206
Strassmann, Fritz, 3
Strategic Air Command (SAC) bombers, 69
Strategic arms limitation talks (SALT). *See*
    SALT I; SALT II
Strategic Defense Initiative (SDI), or Star
    Wars, 210–12, 217
Strauss, Lewis
    and Oppenheimer's dismissal, 61
    and test ban issue, 81, 83, 85–86

and threat of radioactive fallout, 75–76
Subcommittee on Disarmament, U.N., and
    disarmament talks, 1954–1955, 76–77
Submarine-launched ballistic missiles
    (SLBMs), 64, 116, 125, 134, 135, 138,
    140, 165, 176, 177, 189, 203, 215
    and Interim Offensive Arms Agreement,
        1972, 143
Submarines
    and Interim Offensive Arms agreement,
        1972, 143
    missile-launching, 130
Suez War
    and bomber gap, 66–67
    and Sputnik diplomacy, 70
Summit conference
    Geneva, 1955, 78
    Geneva, 1985, 216
    Glassboro, New Jersey, 120
    Iceland, 1986, 218–21
    Moscow, 1972, 142–44
    Moscow, 1974, 150
    Paris, 1960, 92
    Vienna, 1961, 100–101
    Vienna, 1979, 174, 176–78
    Washington, 1973, 148
Supreme War Council, Japan, 27, 28
Suzuki Kantaro, and U.S. option to negotiate
    end of war with Japan, 21
Szilard, Leo, 4
    and scientists' opposition to using atomic
        bomb, 17–18, 226

T

Talbott, Strobe
    and START, 202
    and Walk-in-the-Woods formula for
        intermediate-range nuclear forces (INF),
        199
Tallinn Line, 116, 121
Tass, 207
Taylor, Maxwell
    and armed forces' New Look, 65
    and flexible response strategy, 95
    and nuclear superiority, 98
Technical conference, Geneva, and test ban,
    86–90
Telemetry encryption, 172
Teller, Edward
    campaign for hydrogen bomb, 53–55
    and Limited Test Ban Treaty, 1963, 110
    and MIKE test, 59
    and Oppenheimer's dismissal, 61

Teller, Edward (*continued*)
  and resumption of nuclear testing, 1961,
    102
  and Star Wars, 210
  and test ban issue, 82, 85, 88–90
Test ban
  and Dulles, 83–84
  Eisenhower efforts, 1960, 90–92
  Geneva conference, 1958–1959, 86, 87–90
  issue, 1956, 79–81
  issue, 1957, 81–83
  issue, 1958, 84–87
  limited agreement negotiated, 1963,
    109–10
  talks resume, 1962, 106–9
  treaty ratified, 110–12
  see also Arms control; Atomic bomb;
    Disarmament; Nuclear weapons
Theoharis, Athan, and secrecy about
  hydrogen bomb decision, 58
Thermonuclear weapons, 55–57
"Thin Man" uranium bomb, 13
Third World
  and Cold War, 224
  and horizontal nuclear weapons
    proliferation, 161
Thomas, Charles, and Acheson-Lilienthal
  plan, 1946, 40
Thompson, E. P., and European reaction to
  counterforce strategy, 192
Thor missiles, 64, 73
Three Nation Declaration, 1945, 34, 35
Threshold Test Ban Treaty (TTBT), 1974,
  150
  and Iceland summit conference, 1986, 218
Throw-weight, 202, 203, 208
Titan missiles, 71, 73
Tlatelolco Treaty, 1967, 121
Togo Shigenori
  and bombing of Hiroshima, 26
  and bombing of Nagasaki, 27, 28
  and U.S. option to negotiate end of war
    with Japan, 21
Tomahawk missiles, 162, 169, 191, 197,
  198, 200, 201, 204, 207, 215, 219, 230
Top Policy Group, 5–6
Trade, 1972–1975, trouble over, 152–53
Trewitt, Henry, and missile gap, 1961, 94
Trident missiles, 134, 162, 188, 203
Trident submarine, 132, 140, 188
TRINITY, first atomic bomb, 22–23
Truman, Harry, 7
  accommodationist approach to Soviet
    Union, 223
  and Baruch plan, 1946, 43

and bombing of Japan, 26–28
and containment policy, 1946, 37–38
and domestic control of atomic energy, 39
and first atomic bomb, TRINITY, 22–23
and hydrogen bomb decision, 55–58
and Interim Committee, 15–17
and international control of atom, 32–33
and Japanese surrender, 28
and London Conference of Foreign
  Ministers, 1945, 31
and Marshall Plan, 47–48
and Moscow Conference of Foreign
  Ministers, 1945, 35–37
and nuclear arsenal, 1947–1949, 51–52
and option to negotiate end of war with
  Japan, 21–22
and Potsdam Declaration, 24–25
and Soviet Union, 13–15
and Stimson plan, 29–30
and Truman Doctrine, 46–47, 224
and Yalta agreement, 14–15
Tsaldares, Constantine, and Truman
  Doctrine, 46
Tsarapkin Semyen, and test ban, 88–89, 92
Tully, Grace, 13
Turkey
  missile launch sites, 73
  and Truman Doctrine, 47
Turner, Stansfield, and SALT II, 173
Twelve Basic Principles, 1972, 154
Twining, Nathan, and massive retaliation, 62

U

Ulam, Stanislaw, and MIKE test, 59
Umbrella talks proposal, 1984, 213–14
Underground testing
  ban, 125
  moratorium, 91, 161
Union of Concerned Scientists, and nuclear
  freeze movement, 194
United Nations
  and ASAT, 213
  Atomic Energy Commission, 40
  and Baruch plan, 1946, 42–45
  and disarmament talks, 1954–1955, 76–77
  and nuclear nonproliferation treaty, 123
United States
  and ABM system, 118–21
  and Acheson-Lilienthal plan, 40–42
  and Antarctic Treaty, 1959, 90
  and atomic bombing of Japan, 18–20,
    25–28

atomic bomb tests at Bikini Atoll, 1946, 44
and Atoms-for-Peace plan, 74–75
and Baruch plan, 42–45
and bomber gap, 65–67
containment policy, 37–38
and damage limitation strategy, 1962, 95–97
and détente, 153–57
and disarmament talks, 1954–1955, 76–77
and division of Germany, 48–50
domestic control of atomic energy, 38–40
and failure of nuclear arms control, 227–30
and horizontal nuclear weapons proliferation, 159–61
and Marshall Plan, 47–48
and missile gap, 68–69
and Moscow Conference of Foreign Ministers, 1945, 34–36
and multiple warheads, 115–18
nuclear arsenal, 1947–1949, 51–52
and nuclear deterrence, 50–51, 224–25
and nuclear nonproliferation treaty, 123
nuclear tests, 1958, 87
option to negotiate end of war with Japan, 20–22
and SALT. *See* SALT I; SALT II
and Soviet atomic bomb, 53
and test ban, 81–90, 106–9
and trade with Soviet Union, 1972–1975, 152–53
Truman Doctrine, 46–47
and Vladivostok Accord, 1974, 151–52
Universal Military Training (peacetime draft), and U.S. nuclear deterrence policy, 50
Uranium Committee, 5
U-2 photo-reconnaissance missions, 67
aircraft shot down over Soviet Union, 1960, 92
and Eisenhower test ban efforts, 1960, 92
and Open Skies Plan, 1955, 79

V

Vance, Cyrus
in Carter administration, 162–64
and crisis in Horn of Africa, 170–71
and MX ICBMs, 175–76
and SALT II, 165–66, 169, 171–78
Vandenberg, Arthur
and domestic control of atomic energy, 39
and Moscow Conference of Foreign Ministers, 1945, 35–36

and Truman Doctrine, 47
Vanguard satellite, 67, 68
Vanik, Charles, A., and U.S.-Soviet trade, 1972–1975, 153
Verification Panel, 137, 147
Vessey, John, and Walk-in-the-Woods formula for intermediate-range nuclear forces (INF), 199
Vienna
and SALT I, round four, 1971, 138
summit conference, 1961, 100–101
summit conference, 1979, 174, 176–78
Vietnam War, and Paris Peace Agreement, 154
Vladivostok Accord, 1974, 151–52
and SALT II, 1975–1976, 157–59
and horizontal nuclear weapons proliferation, 161

W

Wadsworth, James J., and test ban conference, Geneva, 1958–1959, 88
Walk-in-the-Woods formula, intermediate-range nuclear forces (INF), 199
Wallace, Henry, 6
and domestic control of atomic energy, 39
and secrecy about hydrogen bomb decision, 58
and Stimson plan, 30
Walton, Richard, on Limited Test Ban Treaty, 1963, 112
Warner, John, and nuclear freeze movement, 195
Warnke, Paul
and MX ICBMs, 175–76
and SALT II, 164, 166, 173
Warsaw Pact, 224
Watergate
and Moscow summit, 1974, 150
and SALT II, 1972–1973, 147–48
Weinberger, Caspar, 185
and European reaction to counterforce strategy, 191
and nuclear arms talks, 1986, 217
and preparation for protracted nuclear warfare, 188
and Star Wars, 210
and Walk-in-the-Woods formula for intermediate-range nuclear forces (INF), 199
West Germany
and European reaction to counterforce strategy, 192

and horizontal nuclear weapons
    proliferation, 160
and nuclear nonproliferation treaty, 122
and negotiations over intermediate-range
    nuclear forces (INF), 201
    see also Berlin; Germany
White, Thomas K., and preparation for
    protracted nuclear warfare, 188–89
Wiesner, Jerome
    and ABM system, 121
    and arms control, 1961, 93
    and nuclear superiority, 98
Wilson, Charles, and massive retaliation, 62
Winne, Harry A., and Acheson-Lilienthal
    plan, 1946, 40
Women's International League for Peace and
    Freedom, and Pugwash conference on
    nuclear arms race, 84
World Council of Churches, and Pugwash
    conference on nuclear arms race, 84

Y

Yalta agreement, 13–14
York, Herbert
    and nuclear superiority, 97
    and Safeguard ABM, 132
Yucca Flats, HARDTACK II nuclear tests,
    1958, 87
Yugoslavia, and horizontal nuclear weapons
    proliferation, 161

Z

Zacharias, Ellis M., and U.S. option to
    negotiate end of war with Japan, 21
Zakharov, Gennadi F., 218
Zhukov, Georgi, and first atomic bomb,
    TRINITY, 23
Zorin, Valerian, and test ban issue, 1957, 82
Zumwalt, Elmo R., and SALT II, 158